AF702661

Medicinal Plants of Himalayas

The healing properties of medicinal plants have been known to mankind since time immemorial. Although the reasons contributing healing power to medicinal plants were known to people of that time purely based on experience and not because of any scientific inference. The gradual trend of decreasing drug efficacy of synthetic drugs along with the associated contraindications due to their excessive usage, has opened up the gates of knowledge for preferring the use of natural drugs once again in pharmacology.

Medicinal Plants of Himalayas: Nutrition, Active Ingredients, and Pharmacological Applications presents glimpses of the present status of metabolite diversity within different medicinal plants from the Himalayan regions. Apart from the metabolite diversity, the book provides glimpses of current isolation techniques and pharmacological activity of the bioactive components. The book tries to plug the gap by focusing on compiling different extraction and isolation procedures, pharmaceutical applications of different active ingredients along with providing the detailed introduction of the plants, altogether in a single book.

Key Features:

- Provides the bioactive profile of listed medicinal plants available in the literature until now necessary for drug development.
- Therapeutic value of the medicinal plants further explores the possible disease mitigation potential of these plants.
- Nutritional and anti-nutritional factors are of important consideration when choosing a particular plant as a source or remedy for disease.

The current book is ideal for students, academicians as well as researchers to deepen their knowledge about medicinal plants present around the globe, and particularly the ones distributed in Himalayan region.

Medicinal Plants of Himalayas

Nutrition, Active Ingredients, and Pharmacological Applications

Edited by
Raees ul Haq
Sajad Ahmad Wani
Jasmeet Kour
Shiv Kumar

CRC Press
Taylor & Francis Group
Boca Raton London New York

CRC Press is an imprint of the
Taylor & Francis Group, an **informa** business

Designed cover image: Getty Images [Image ID: 876868296]

First edition published 2026
by CRC Press
2385 NW Executive Center Drive, Suite 320, Boca Raton FL 33431

and by CRC Press
4 Park Square, Milton Park, Abingdon, Oxon, OX14 4RN

CRC Press is an imprint of Taylor & Francis Group, LLC

© 2026 selection and editorial matter, Raees ul Haq, Sajad Ahmad Wani, Jasmeet Kour, and Shiv Kumar; individual chapters, the contributors

Reasonable efforts have been made to publish reliable data and information, but the author and publisher cannot assume responsibility for the validity of all materials or the consequences of their use. The authors and publishers have attempted to trace the copyright holders of all material reproduced in this publication and apologize to copyright holders if permission to publish in this form has not been obtained. If any copyright material has not been acknowledged please write and let us know so we may rectify in any future reprint.

Except as permitted under U.S. Copyright Law, no part of this book may be reprinted, reproduced, transmitted, or utilized in any form by any electronic, mechanical, or other means, now known or hereafter invented, including photocopying, microfilming, and recording, or in any information storage or retrieval system, without written permission from the publishers.

For permission to photocopy or use material electronically from this work, access www.copyright.com or contact the Copyright Clearance Center, Inc. (CCC), 222 Rosewood Drive, Danvers, MA 01923, 978-750-8400. For works that are not available on CCC please contact mpkbookspermissions@tandf.co.uk

Trademark notice: Product or corporate names may be trademarks or registered trademarks and are used only for identification and explanation without intent to infringe.

ISBN: 978-1-032-27341-9 (hbk)
ISBN: 978-1-032-27348-8 (pbk)
ISBN: 978-1-003-29234-0 (ebk)

DOI: 10.1201/9781003292340

Typeset in Times
by SPi Technologies India Pvt Ltd (Straive)

Contents

About the Editors vii
List of Contributors ix

Chapter 1 Overall Overview of the Medicinal Plants Found in Himalayan Region 1

Raees ul Haq and Shiv Kumar

Chapter 2 Wormwood (*Artemisia absinthium*) 9

Bharti Mittu, Mohammed Shafiq Alam, Maninder Kaur, Zerina Begum, and Jasmeet Kour

Chapter 3 Atis (*Annona squamosa*) 34

Bharti Mittu, Zerina Begum, Sandaldeep Kaur, Mahaldeep Kaur, Abida Bhat, Neha Sharma, and Raees ul Haq

Chapter 4 Indian Belladonna (*Atropa belladonna*) 50

Ramandeep Kaur, Sukhcharn Singh, and Raees ul Haq

Chapter 5 Redwood Pigweed (*Amaranthus retroflexus*) 60

Divyang Solanki, Qiu Yi Tan, Anima Manhswita, Srishti, Shreedevi Indapwar, Ritika Rathi, Gareema Pandey, Sangeeta Prakash, and Raees ul Haq

Chapter 6 Corn/Wild Mint (*Mentha arvensis*) 78

Kanchan Sharma and Raees ul Haq

Chapter 7 Basil (*Ocimum basilicum*) 93

Gareema Pandey, Divyang Solanki, Anima Manhswita, Qiu Yi Tan, and Sangeeta Prakash

Chapter 8 Wound Wort/Common Self-Heal (*Prunella vulgaris*) 115

Santwana Palai and Raees ul Haq

Chapter 9 Himalayan Rhubarb (*Rheum emodi*) 128

Pratiksha, Elizabeth Thomas, and C. G. Harshitha

Chapter 10 Costus (*Saussurea costus*) 151

Kawaljeet Kour, Ilknur Ucak, and Raees ul Haq

Chapter 11 Dandelion (*Taraxacum officinale*) 175
Santwana Palai and Raees ul Haq

Chapter 12 Sweet Violet (*Viola odorata*) 190
Kanchan Sharma and Raees ul Haq

Index 199

About the Editors

Raees ul Haq is currently engaged as a Lecturer at Government Degree College Shopian, J&K India. He completed his PhD at the Department of Food Engineering and Technology at Sant Longowal Institute of Engineering and Technology, Longowal, Sangrur, Punjab, India. He earned his Master's degree in Food Technology from the Islamic University of Science and Technology, Awantipora, Jammu and Kashmir, India. He has authored and co-authored different books as well as book chapters published by prominent international publishers. He has presented his research work at various national and international conferences, seminars, and workshops. He is actively engaged with the scientific community by publishing his work as well as working as an editorial board member and peer reviewer of reputed international journals. He is also the recipient of the Maulana Azad National Fellowship awarded by the University Grants Commission (UGC) and has published a good number of research and review articles in reputed journals.

Sajad Ahmad Wani is currently working as Assistant Professor in the Department of Food Science & Technology, Government Degree College, Shopian, J&K, India. He has completed his Master's Degree in Food Technology from IUST, Awantipora, J&K, India and his PhD from Sant Longowal Institute of Engineering and Technology, Punjab, India. He passed the National Eligibility Test (NET) (2015), conducted by the Indian Council of Agricultural Research (ICAR) in December 2015. He has published more than 35 research/review articles, along with 10 book chapters and 1 book. He also writes articles for the popular magazines *Food & Beverage News* (India's first magazine for the Food & Beverage Industry). He has attended more than 40 international/national/conferences, seminars and workshops across the world. He serves as an editorial board member of a number of national and international journals. He is a potential peer reviewer of reputed international journals related to food science and technology issued by popular publishing houses. He is also working as a member of various associations like the Institute of Food Technologists, the Institute for Educational Research and Publication, the Association of Food Scientists and Technologists India and the Asian Council of Science Editors. He is the recipient of the "Maulana Azad National Fellowship" (MANF-2013-14) from the University Grant Commission, New Delhi, India.

Jasmeet Kour was awarded doctoral degree at Sant Longowal Institute of Engineering and Technology, Longowal, Sangrur, Punjab, India from the Department of Food Engineering and Technology. Since 2009, she has been serving as an Assistant Professor at the Department of Food Science and Technology, Government College for Women, Gandhi Nagar, Jammu, Jammu and Kashmir. She has conducted vast research into various prominent nutraceuticals derived from plant origin and her work has been published in reputed journals with high-impact factors issued by eminent publishing houses in the field of food science. She has also presented research and review paper at various national and international conferences. She has authored, as well as co-authored, numerous book chapters and scientific articles published in international books with prestigious publishing houses. She is also a participant in various international projects. As her research work left a great imprint, she was cordially invited to share her newest research findings at Tokyo University of Agriculture, Japan at the 6th International Conference on Agricultural and Biological Sciences (ABS 2020). She is currently working as an editorial board member and peer reviewer for various journals of repute.

Shiv Kumar works as an Associate Professor (Food Science and Technology) at Maharishi Markandeshwar Deemed to be University. He completed his MSc. in Food Technology, and obtained a PhD in Food Engineering and Technology from SLIET, Longowal, Punjab. In 2019, he was honored with the best PhD thesis award. He has taught a wide range of courses to undergraduate, postgraduate, and doctoral students in different areas of Food Science and Technology. He is actively involved in the supervision of PhD research scholars, MSc. students and undergraduate students in the Food Science and Technology discipline. He has filed a total of 7 utility patents, along with 2 copyrights; 6 of these patents have now been published. He has submitted multiple research projects on functional food to state and national funding agencies. He is actively publishing research and review articles as well as book chapters, and has been a reviewer for various journals of international repute. He has lifetime membership of the Society for Advancement of Human and Nature, the Akal Food Safety Quality Control Society, the Agricultural & Environmental Technology Development Society, and the Asia Society of Researchers.

Contributors

Mohammed Shafiq Alam
Department of Processing and Food Engineering
Punjab Agricultural University
Ludhiana, India

Zerina Begum
Department of Biotechnology
Shaheed Udham Singh College of Engineering and Technology
Mohali Punjab, India

Abida Bhat
Department of Immunology and Molecular Medicine
Sher-e-Kashmir Institute of Medical Sciences
Srinagar, India

C. G. Harshitha
Dairy Chemistry Division
National Dairy Research Institute
Karnal Haryana, India

Raees ul Haq
Department of Food Science & Technology
Govt. Degree College
J&K, India

Shreedevi Indapwar
School of Agriculture and Food Sciences
The University of Queensland
St. Lucia, Australia

Mahaldeep Kaur
Department of Microbial Biotechnology
Punjab University, India

Ramandeep Kaur
Sant Longowal Institute of Engineering and Technology Longowal, India

Maninder Kaur
Punjab Agricultural University
Ludhiana, India

Sandaldeep Kaur
Post Graduate Government College for Girls
Chandigarh, India

Jasmeet Kour
Sant Longowal Institute of Engineering and Technology
Longowal, India

Kawaljeet Kour
Department of Food Science and Technology
Faculty of Food Science and Technology
Government College for Women Gandhinagar
Jammu, India

Shiv Kumar
Maharishi Markandeshwar (Deemed To be University)
Haryana India

Anima Manhswita
School of Agriculture and Food Sciences
The University of Queensland
St. Lucia, Australia

Bharti Mittu
National Institute of Pharmaceutical Education and Research (NIPER)
Mohali, India

Santwana Palai
Department of Veterinary Pharmacology & Toxicology
College of Veterinary Science and Animal Husbandry
Odisha University of Agriculture and Technology
Bhubaneswar, India

Gareema Pandey
School of Science
Western Sydney University
NSW, Australia

Pratiksha
Department of Food Science & Technology, Amity University, Mohali Punjab, India

Sangeeta Prakash
School of Agriculture and Food Sciences
The University of Queensland
St. Lucia, Australia

Ritika Rathi
School of Agriculture and Food Sciences
The University of Queensland
St. Lucia, Australia

Kanchan Sharma
Department of Home Science
University of Jammu
Jammu and Kashmir, India

Neha Sharma
Center of Innovative and Applied Bioprocessing (CIAB)
Mohali Punjab, India

Sukhcharn Singh
Sant Longowal Institute of Engineering and Technology
Longowal, India

Divyang Solanki
School of Agriculture and Food Sciences
The University of Queensland
St. Lucia, Australia

Srishti
School of Agriculture and Food Sustainability
The University of Queensland
St. Lucia, Australia

Qiu Yi Tan
School of Agriculture and Food Sciences
The University of Queensland
St. Lucia, Australia

Elizabeth Thomas
Dairy Technology Division
National Dairy Research Institute
Karnal Haryana, India

Ilknur Ucak
Dr. Nigde Omer Halisdemir University
Faculty of Agricultural Sciences and Technologies

1 Overall Overview of the Medicinal Plants Found in Himalayan Region

Raees ul Haq and Shiv Kumar

1.1 INTRODUCTION

Plants are a source of numerous secondary metabolites that, either directly or indirectly, find their use in medicine. In general terms, medicinal plants are defined as any plant that is the source of therapeutic substances, irrespective of the plant part. The use of medicinal plants dates back to time immemorial as plants were the sole entities to humans attempting to find relief from a disease. The trials did not yield any substantive conclusion, but a few plants worked as a remedy to the problem. These plants find their use in traditional medicine throughout the world. Medicinal procedures that were concerned with the diseases and health of humans can be termed ethnomedicine; it addresses both well-being and healthcare practices. Figure 1.1 shows the different types of medicinal plants.

As suggested above, medicinal plants have been at the center of drug discovery (Nalawade et al., 2003; Chen et al., 2010; Chacko et al., 2010; Hamilton, 2004) and the history of this dates back through human history as driven by experience and taste that treated their illness. Medicinal plants are classified according to traditional systems as well as employing a more modern approach involving information of phytochemicals on classification. These plants have been the sources of numerous chemicals; it has been estimated that around 90% of medicinal plants were initially found in the wild. According to data provided by the International Union for Conservation of Nature (IUCN) and the Worldwide Fund for Nature (WWF), between 50,000 and 80,000 flowering plants have been used for medicinal purposes (Bentley, 2010; Ross, 2005). The estimates provided by the National Medicinal Plant Board India (NMPBI) report that India is home to 17,000 to 18,000 flowering plant species, of which around one-third have been used in traditional medicinal systems of Unani, Ayurveda, etc. Consumption of more than one hundred metric tons in case of 178 medicinal plants have been reported to be in trading out of 960 medicinal plants. Increased demand for medicinal plants, along with an increase in population, have led to a considerable decline in the level of plant species (Janmejaya, 2016).

The use of traditional medicinal plants can be a topic of discussion, particularly regarding the use of crude extracts or whole plants for treatment of a disease. The results obtained through the use of modern high-tech instrumentation has challenged the effectiveness of these medicinal plants as research has linked their use with serious side effects in children as well as adults (Sakurai, 2011; Posadzki et al., 2013; Basaran et al., 2022; Kulkarni et al., 2022; Philips & Abraham, 2022). One such example, and something that, in recent times, has been a topic of hot debate is *Tinospora cordifolia*, commonly known as giloy, which has been traditionally used throughout India as an immunity booster. The presence of clerodane furano-diterpenoids in *Tinospora cordifolia* is known to have harmful effects on the liver (Philips & Abraham, 2022; Kulkarni et al., 2022). A lot of chemical constituents is present in the herbs that are used as traditional medicines in one form or other. No doubt there is the presence of the drug or compound which tackles or treats a disease, but the same plant is source of numerous compounds that may have serious side effects for the human body.

DOI: 10.1201/9781003292340-1

FIGURE 1.1 Different types of medicinal plants (Sweet violet, Corn mint and Dandelion).

As part of a logical and safer approach, preference should always be provided to a known drug and selecting between herbal and a modern drug should be made cautiously or judiciously.

1.2 BRIEF OVERVIEW OF THE MEDICINAL PLANTS DISCUSSED IN THIS BOOK

Artemisia absinthium, generally recognized as wormwood, is a famous herb used as traditional community medicine around the world. The *Artemisia absinthium* is a plant from the family *Asteraceae*. All parts of the plant have a range of health benefits due to the presence of phytochemicals like polyphenols, flavonoids, tannins, etc. Extracts of the herb have been documented to possess neuroprotective, hepatoprotective, anti-inflammatory, cardiovascular, hypoglycemic, antioxidant, anticancer, immunomodulatory, and analgesic properties and also for its actions against infectious diseases (Ahamad et al., 2019; Ahameethunisa & Hopper, 2010).

Aquilegia pubiflora is a medicinally significant species belonging to the Ranunculaceae family and it can be found across the Himalaya region of India, Afghanistan and Pakistan. The plant, which is known locally as *Thandi buti* or *Dombaand*, is widely known as bristly flowered columbine or Himalayan columbine. This species has plentiful noteworthy pharmacological and therapeutic properties, including dyspepsia, astringent, antipyretic, cardiotonic, antiasthmatic, energizer, and anti-jaundice. The dried roots of plants have been known to fix eye problems, toothaches, homeopathy, aggravation, and snakebites and, above all, for its benefits to the sensory system. A new report has found that a methanolic form of *Aquilegia pubiflora* shows doable erythroid mobilization action, which is an indication that the plant may be a wellspring of fetal hemoglobin-creating phytocompounds and that it could be used in the treatment of β-thalassemia. In addition, *Aquilegia pubiflora* was primarily utilized in the treatment of influenza, the healing of wounds, gynecology, circulatory, jaundice, and cardiovascular illness (Munz, 1946; Calixto, 2016; Alsharif et al., 2020; Waris et al., 2021).

Atropa belladonna, often known as deadly nightshade, is a plant whose story is shrouded in myth, fear, and awe. There has been considerable use of *A. belladonna* in medicine and cosmetics. As detailed later in the book, atropine, hyoscine (scopolamine), and hyoscyamine, three tropane alkaloids, are among the active substances with anticholinergic effects which have been identified as present in belladonna. *A. belladonna's* medicinal plant species overview is conducted by covering practically all essential facts and providing a comprehensive database of references to researchers for their successful investigation and application in improving human welfare (Hazslinsky, 1956; Hartmann et al., 1986; Birch, 1993; Jouhikainen et al., 1999).

Redwood Pigweed (*Amaranthus retroflexus*), also termed tumbleweed, which is enriched with essential fatty acids, flavonoids, and other bioactive compounds. *Amaranthus* spp. is considered to be an underutilized species with the potential to become an alternative crop for future food security as it is highly nutritious and has an inherent tolerance of severe climate conditions. The seeds or grains of amaranth are rich in a wide range of nutrients, including starch, dietary fibres, proteins, and lipids, and can therefore be regarded as a healthy food. Conventional and novel green technologies were applied to extract bioactive compounds from this plant, which envisages its medicinal uses for the upcoming future. Overall, *A. retroflexus* has been reported to have antipyretic, anti-ulcer, and anti-inflammatory activities and has been found to be efficacious in the treatment of diseases like rheumatoid arthritis, asthma, and diseases of the urinary tract. Hence, all these properties of the plant make it an important plant from a medical point of view (Reza & Ferdousi, 2016; Arêas et al., 2016; Adegbola et al., 2020).

Mint (*Mentha arvensis*) is amongst one of the most common herbs cultivated in Asia and other parts of the world. Taxonomically, the plant has been placed in the *Lamiaceae* family and genus *Mentha*. It has an abundant concentration of bioactive compounds, such as essential oils, polyphenols, flavonoids, and flavonols, which have beneficial effects for human health. It is rich in menthol, for which reason it is sometimes also called menthol mint (Anca-Raluca et al., 2011; Alizadeh et al., 2013; Abdelli et al., 2016; Anwar et al., 2019; Al-Sabahi et al., 2016).

Medicinal plants have a high value to humans due to their medicinal constituents and potential biological consequences. The members of the *Nepeta* genus are largely distributed in the temperate to alpine zone. A total of 22 medicinal plants were recorded in the study area. These medicinal plants are used extensively by local communities for the treatment of nine major disease categories. The most frequently used medicinal remedies decoction (38.46%). Among the medicinal plants, herbs (100%) were most frequently used by the local populations of Kashmir, whereas leaves (44.4%) were used for the treatment of various ailments. The main focus is on the usage of *Nepeta* species in traditional medicine, with consideration also given to scientific studies on their biological effects and secondary metabolites. Essential oils extracted from the *Nepeta* genus are abundant in terpenoid and oxygenated terpenoid hydrocarbons, and the biological properties of these compounds are widely recognised. Terpenoid and phenolic components found in *Nepeta* species exhibit a range of beneficial properties, including antibacterial, insecticidal, antioxidant, anticonvulsant, analgesic, anti-inflammatory, and antidepressant actions. These characteristics illustrate *Nepeta's* importance in agriculture and medicine. The *Nepeta* genus exhibits therapeutic properties that are promising for several kinds of disorders; however, clinical research is required to confirm preclinical results. In order to evaluate the efficacy of active plant species and bioactive secondary metabolites, future guidelines for organising and performing extensive trials must be established. These trials can assess the benefits of using these drugs either alone, or in combination with, existing conventional medicines (Duke, 2002; Hawke, 2007; Formisano et al., 2011; Emami et al., 2017).

Basil (*Ocimum basilicum*) is enriched with essential oils, phenolic, flavonoids and other bioactive compounds. In the modern search for natural remedies, the plant has been increasingly employed in medicinal situations. It has both pharmaceutical and food applications, making it of interest plant in seeking out functional and biofunctional ingredients. Basil is a rich source of macro minerals, polyphenols, and anthocyanins. Conventionally, the herb has already been applied for the treatment of many diseases and health-related issues. Recent developments in technologies have increased its use in various drugs, medicines, and therapies, including cardiovascular- and cancer-related health problems. Utilization of novel extraction technologies followed by *in vitro, in silico* and *in vivo* analysis has raised its application in pharmacy. Overall, basil can be considered a plant with multiple functions and a variety of ingredients with both medicinal and non-medicinal applications (Al-Maskari et al., 2012; Falowo et al., 2019; Antonescu et al., 2021; Dhama et al., 2021).

Prunella vulgaris L. is an annual herb and a member of the Lamiaceae family which has been used for both medicinal and culinary purposes for thousands of years. *Prunella vulgaris* has provable clinical therapeutic benefits with regard to a variety of illnesses, including photophobia, vertigo, wound, goitre, and cancer. Several flavonoids, triterpenoids, polysaccharides, steroids, organic acids, volatile oils, and phenylpropanoids are identified as the important chemical elements purified from *Prunella vulgaris*. Modern pharmacologic researches have revealed that *Prunella vulgaris* possesses pharmacological activities, including antioxidant, anti-inflammatory, antihypertensive, hypoglycemic, hepatoprotective, immunoregulatory, antiviral, antibacterial, lipid-lowering and hypnotic effects. This chapter encompasses *Prunella vulgaris'* botany, phytochemistry, ethnopharmacology, traditional uses, analytical techniques, and pharmacological effects. *Prunella vulgaris* can be used for medical purposes and has significant scope for the development of new drug resources (Gazala et al., 2016; Kapoor et al., 2017; Chen et al., 2018; Ahmad et al., 2020).

The use of medicinal plants and herbs has a long history in traditional medicine. One such plant which attracts tremendous interest is rhubarb. Better known as Himalayan rhubarb, *Rheum emodi* belongs to the family polygonaceae. This is a herbaceous perennial plant with a thick rhizome. A native of Himalayan region, rhubarb's growth is favored by cold and sub-tropical regions. Anthroquinones such as rhein, physcion, chrysophanol, emodin, aloe-emodin, and stilbene compounds such as rhaponticin and rhapontigenin, are the main phytochemicals present in *Rheum*. Chrysophanol and physcion are particularly significant anthraquinones. The plant also shows therapeutic effects, along with antifungal and antimicrobial properties, and has been found effective in treating diabetes, anemia, gastric disorders, renal and hepatic disorders, menstrual disorders, and cancer (Goel et al., 1997; Agarwal et al., 2000; Babu et al., 2003; Cao et al., 2017).

Traditionally, *Saussurea costus* has been utilized for its properties against microbes, viruses, cancer, hepatotoxic agents, arthritis, and convulsions. Some of its active components could be developed into new medications to combat a variety of diseases. *Saussurea costus* is a critically endangered species which is limited to a small area of the Himalayas and hence needs to be protected by being cultivated extensively. Recent findings demonstrate that *Saussurea costus* roots contain an abundance of bioactive phytochemical substances. Furthermore, it is advantageous to isolate and characterize these bioactive compounds from the roots of *S. costus* in order to create novel, organic medicines which are efficient and effective (Gautam & Asrani, 2018; Abdallah et al., 2017; Al Otibi et al., 2020; Ali et al., 2021; Cai et al., 2021; Ali & Venkatesalu, 2022).

Dandelion (*Taraxacum officinale* L.) is used in herbal medicine in a number of countries, including Iran, Russia, India, and China. Traditional medicine has made extensive use of *T. officinale* owing to its anti-oxidant effects, anti-inflammatory activity, hepatoprotective properties, immunomodulatory action, anti-obesity effects, hypolipidemic effects, antidepressant effects, diuretic effects, and anti-urolithic effect. It is also effective against various cancers, Type-2 diabetes mellitus, neurodegenerative illnesses, cardiovascular diseases, gastrointestinal diseases, respiratory disease (COVID), and other ailments. The purpose of this chapter is to confirm the pharmacological potential of the phytoalimurgic plant by compiling all of the most recent information on phytochemical and also linked pharmacological activities. Previous studies has established the presence of bioactive metabolites, such as caffeoylquinic acids, sesquiterpenoids, and flavonoids, which contribute to its diuretic, antioxidant, antiviral, cytotoxic, antibacterial, anti-inflammatory, and hepatoprotective properties. As a result, dandelion is a medicinal plant which is proving of commercial interest in the nutraceutical and phytopharmaceutical industries (Alamgir, 2018; Grauso et al., 2019; Cheema & Singh, 2021; Dongare et al., 2021).

Sweet violet (*Viola odorata*) is one of the most common herbs, and has been cultivated over a long period in both Europe and Asia. It belongs to the family Violaceae and the genus Viola. It is known by various names across the world: as sweet violet, garden violet, banafsha, Gul-e-banafsha and many more in a range of different languages. It contains many bioactive compounds, such as essential oils, polyphenols, flavonoids, and flavonols, which have beneficial effects on human health. This chapter highlights the origin of Sweet violet (*Viola odorata*), along with its classification and composition. It emphasizes the different bioactive compounds, their beneficial effects, and their isolation and characterization. It also discusses the pharmacological applications of its different constituent components (Pranting et al., 2010; Akhbari et al., 2012; Gautam et al., 2012; Hamid et al., 2013; Feyzabadi et al., 2017; Drozdova and Bubenchikov, 2005).

1.3 CONCLUSION

Since ancient times, medicinal plants have served humans in the treatment of different ailments and people have relied upon these folk medicines which vary from region to region. The trust that people place in these medicinal plants or herbs requires thorough investigation through the employment of modern techniques for the elucidation of their chemical composition. At the time of writing, research has been going on for the past few decades, but the quantum of research is still in its infancy. The presence of the pharmaceutical compounds needs a detailed investigation regarding its efficacy, its side effects, and so on. Furthermore, the presence of other chemical constituents that make up the overall composition of the plant needs to be evaluated for safety, toxicity, drug interactions, and other safety procedures laid down by different regulatory authorities. As part of a logical and safer approach, preference should always be provided to a known and well-researched drug; whenever there is a choice to be made between a herbal and a modern drug, this should be made both cautiously and judiciously.

REFERENCES

Abdallah, E. M., Qureshi, K. A., Ali, A. M., & Elhassan, G. O. (2017). Evaluation of some biological properties of *Saussurea costus* crude root extract. *Bioscience Biotechnology Research Communications*, *10*(4), 601–611.

Abdelli, M., Moghrani, H., Aboun, A., & Maachi, R. (2016). Algerian *Mentha pulegium* L. leaves essential oil: Chemical composition, antimicrobial, insecticidal and antioxidant activities. *Industrial Crops and Products*, *94*, 197–205.

Adegbola, P. I., Adetutu, A., & Olaniyi, T. D. (2020). Antioxidant activity of *Amaranthus* species from the Amaranthaceae family—A review. *South African Journal of Botany*, *133*, 111–117.

Agarwal, S. K., Singh, S. S., Verma, S., & Kumar, S. (2000). Antifungal activity of anthraquinone derivatives from *Rheum emodi*. *Journal of Ethnopharmacology*, *72*(1–2), 43–46.

Ahamad, J., Mir, S. R., & Amin, S. (2019). A Pharmacognostic review on *Artemisia absinthium*. *International Research Journal of Pharmacy*, *10*(1), 25–31.

Ahameethunisa, A. R., & Hopper, W. (2010). Antibacterial activity of *Artemisia nilagirica* leaf extracts against clinical and phytopathogenic bacteria. *BMC Complementary and Alternative Medicine*. *10*, 1–6

Ahmad, G., Masoodi, M. H., Tabassum, N., & Mir, R. A. (2020). Phytochemical analysis and anti-inflammatory activity of various extracts obtained from floral spikes of *Prunella vulgaris* L. *Jordan Journal of Pharmaceutical Sciences*, *13*(1), 41–52.

Akhbari, M., Batooli, H., & Kashi, F. J. (2012). Composition of essential oil and biological activity of extracts of *Viola odorata* L. from central Iran. *Natural Product Research*, *26*(9), 802–809.

Al Otibi, F., Rizwana, H., Alharbi, R. I., Alshaikh, N., & Albasher, G. (2020). Antifungal effect of *Saussurea lappa* roots against phytopathogenic fungi and resulting morphological and ultrastructural changes. *Gesunde Pflanzen*, *72*(1), 57–67.

Alamgir, A. N. M. (2018). Phytoconstituents—Active and inert constituents, metabolic pathways, chemistry and application of phytoconstituents, primary metabolic products, and bioactive compounds of primary metabolic origin. In A. N. M. Alamgir (Ed.), *Therapeutic Use of Medicinal Plants and their Extracts: Volume 2* (pp. 25–164). Springer.

Ali, A., Alqaseer, K., Fatlawi, D., Shehab, S., Falah, M., Hassan, M., & Radhi, O. (2021). *Saussurea lappa*: An important medicinal plant for treatment different diseases: A review. *Kufa Journal for Nursing Sciences*, *11*(1), 1–19.

Ali, S. I., & Venkatesalu, V. (2022). Botany, traditional uses, phytochemistry and pharmacological properties of *Saussurea costus*–An endangered plant from Himalaya—A review. *Phytochemistry Letters*, *47*, 140–155.

Alizadeh, A., Alizadeh, O., Amari, G., & Zare, M. (2013). Essential oil composition, total phenolic content, antioxidant activity and antifungal properties of Iranian *Thymus daenensis* subsp. *digenesis* Celak. as in influenced by ontogenetical variation. *Journal of Essential Oil Bearing Plants*, *16*, 59–70.

Al-Maskari, M. Y., Hanif, M. A., Al-Maskri, A. Y., & Al-Adawi, S. (2012). Basil: A natural source of antioxidants and neutraceuticals. In M. Mohamed Essa, A. Manickavasagan, & E. Sukumar (Eds.), *Natural Products and Their Active Compounds on Disease Prevention* (pp. 463–471). Nova Science Publishers, Inc.

Al-Sabahi, J. N., Hanif, M. A., Al-Maskari, A. Y., Al Busaidi, M. S. M., Al Maskari, M. Y., & Al-Haddabi, M. H. (2016). Chemical report on wild growing *Mentha arvensis* and *Citrullus colocynthus* from Oman. *Journal of Essential Oil Bearing Plants*, *19*(3), 719–726

Alsharif, S. M., Salem, S. S., Abdel-Rahman, M. A., Fouda, A., Eid, A. M., Hassan, S. E., Awad, M. A., & Mohamed, A. A. (2020). Multifunctional properties of spherical silver nanoparticles fabricated by different microbial taxa. *Heliyon*, *6*(5), e03943.

Anca-Raluca, A., Boz, I., Padurariu, C., Atofani, D., Coisin, M., & Zamfirachie, M. M. (2011). Comparative biochemical and physiological research on taxa of *Mentha* L. genus. *Journal of Plant Development*, *18*, 41–45.

Antonescu, I. A., Antonescu, A., Miere, F., Fritea, L., Teuşdea, A. C., Vicaş, L., Vicaş, S. I., Brihan, I., Domuţa, M., Zdrinca, M., Zdrinca, M., & Cavalu, S. (2021). Evaluation of wound healing potential of novel hydrogel based on *Ocimum basilicum* and *Trifolium pratense* extracts. *Processes*, *9*(11), 2096.

Anwar, F., Abbas, A., Mehmood, T., Gilani, A.-H., & Rehman, N. (2019). Mentha: A genus rich in vital nutra-pharmaceuticals—A review. *Phytotherapy Research*, *33*, 2548–2570.

Arêas, J. A. G., Carlos-Menezes, A. C. C. C., & Soares, R. A. M. (2016). Amaranth. In B. Caballero, P. M. Finglas, & F. Toldrá (Eds.), *Encyclopedia of Food and Health* (pp. 135–140). Academic Press.

Babu, K. S., Srinivas, P. V., Praveen, B., Kishaore, K. H., Murthy, U. S. & Rao, J. M. (2003). Antimicrobial constituents from the rhizomes of *Rheum emodi*. *Phytochemistry*, *62*(2), 203–207.

Başaran, N., Paslı, D., & Başaran, A. A. (2022). Unpredictable adverse effects of herbal products. *Food and Chemical Toxicology*, *159*, 112762.

Bentley, R. (2010). *Medicinal plants*. Domville-Fife Press.

Birch, A. J. (1993). Investigating a scientific legend: The tropinone synthesis of Sir Robert Robinson F.R.S. *Notes and Records of the Royal Society of London*, *47*(2), 226–277.

Cai, Y., Gao, K., Peng, B., Xu, Z., Peng, J., Li, J., & Yan, Y. (2021). Alantolactone: A natural plant extract as a potential therapeutic agent for cancer. *Frontiers in Pharmacology*, *12*, 781033.

Calixto, G. M. F., Bernegossi, J., de Freitas, L. C., Fontana, S., & Chorilli, M. (2016). Nanotechnology-based drug delivery systems for photodynamic therapy of cancer: a review. *Molecules*, *21*(3), 342.

Cao, Y. J., Pu, Z. J., Tang, Y. P., Shen, J., Chen, Y. Y., Kang, A., Zhou, G. S. & Duan, J. A. (2017). Advances in bio-active constituents, pharmacology and clinical applications of rhubarb. *Chinese Medicine (United Kingdom)*, *12*(1).

Chacko, S. M., Thambi, P. T., Kuttan, R., & Nishigaki, I. (2010). Beneficial effects of green tea: A literature review. *Chinese Medicine*, *5*, 13.

Cheema, H. S., & Singh, M. P. (2021). The use of medicinal plants in digestive system related disorders—A systematic review. *Journal of Ayurvedic and Herbal Medicine*, *7*, 182–187.

Chen, S. L., Yao, H., Han, J. P., Liu, C., Song, J. Y., Shi, L. C., Zhu, Y. J., Ma, X. Y., Gao, T., Pang, X. H., Luo, K., Li, Y., Li, X. W., Jia, X. C., Lin, Y. L., & Leon, C. (2010). Validation of the ITS2 region as a novel DNA barcode for identifying medicinal plant species. *PLoS ONE*, *5*, e8613.

Chen, T. G., Li, L. Y., Wei, Y. R., & Zhang, L. W. (2018). Screening, identification and activity evaluation of pancreatic lipase inhibition in *Prunella vulgaris*. *China Journal of Chinese Materia Medica*, *43*(23), 4665–4671.

Dhama, K., Sharun, K., Gugjoo, M. B., Tiwari, R., Alagawany, M., Iqbal Yatoo, M., Thakur, P., Iqbal, H. M. N., Chaicumpa, W., Michalak, I., Elnesr, S. S., & Farag, M. R. (2021). A comprehensive review on chemical profile and pharmacological activities of *Ocimum basilicum*. *Food Reviews International*, *39*, 1–29.

Dongare, P. N., Kadu, T. B., Pohane, A. L., Dighade, S. J., & Patil, P. G. (2021). Review on pharmacognosy, phytochemistry and pharmacological activity of *Taraxacum officinale* (Dandelion plant). *World Journal of Advanced Research and Reviews*, *12*(02), 212–218

Drozdova, I., & Bubenchikov, R. (2005). Composition and anti-inflammatory activity of polysaccharide complexes extracted from sweet violet and low mallow. *Pharmaceutical Chemistry Journal*, *39*(4), 197–200.

Duke, J. A. (2002). *Handbook of medicinal herbs*. CRC Press.

Emami, S. A., Yazdian-Robati, R., Sadeghi, M., Baharara, J., Amini, E., Salek, F., & Tayarani-Najaran, Z. (2017). Inhibitory effects of different fractions of *Nepeta satureioides* on melanin synthesis through reducing oxidative stress. *Research in Pharmaceutical Sciences*, *12*(2), 160.

Falowo, A. B., Mukumbo, F. E., Idamokoro, E. M., Afolayan, A. J., & Muchenje, V. (2019). Phytochemical constituents and antioxidant activity of sweet basil (*Ocimum basilicum* L.) essential oil on ground beef from boran and nguni cattle. *International Journal of Food Science*, *2019*, 2628747.

Feyzabadi, Z., Ghorbani, F., Vazani, Y., & Zarshenas, M. M. (2017). A critical review on phytochemistry, pharmacology of *Viola odorata* L. and related multi-potential products in traditional Persian medicine. *Phytotherapy Research*, *2*, 1–8.

Formisano, C., Rigano, D., & Senatore, F. (2011). Chemical constituents and biological activities of *Nepeta* species. *Chemistry & Biodiversity*, *8*(10), 1783–1818.

Gautam, H., & Asrani, R. K. (2018). Phytochemical and pharmacological review of an ethno medicinal plant: *Saussurea lappa*. *Veterinary Research International*, *6*, 1–9.

Gautam, S.S., Navneet, & Kumar, S. (2012). The antibacterial and phytochemical aspects of *Viola odorata* Linn. extracts against respiratory tract pathogens. *Proceedings of the National Academy of Sciences, India Section B: Biological Sciences*, *82*(4), 567–572.

Gazala, Q., Murtaza, I., Ara, S., Qazi, H., Geelani, S. M., & Amir, S. (2016). Characterization and antimicrobial activity of some natural dye yielding plant species of Kashmir Valley. *Journal of Industrial Pollution Control*, *32*(2), 518–528.

Goel, V., Ooraikul, B., & Basu, T. K. (1997). Cholesterol lowering effects of rhubarb stalk fiber in hypercholesterolemic men. *Journal of the American College of Nutrition*, *16*(6), 600–604.

Grauso, L., Emrick, S., de Falco, B., Lanzotti, V., & Bonanomi, G. (2019). Common dandelion: A review of its botanical, phytochemical and pharmacological profiles. *Phytochemistry Reviews*, *18*(4), 1115–1132.

Hamid, K. S., Ranjbar, S. H., Esfehani, M., Mohammad, K., & Larijani, B. (2013). A systematic review of the antioxidant, anti-diabetic, and anti-obesity effects and safety of triphala herbal formulation. *Journal of Medicinal Plants Research*, *7*(14), 831–844.

Hamilton, A. C. (2004). Medicinal plants, conservation and livelihoods. *Biodiversity and Conservation*, *13*, 1477–1517.

Hartmann, T., Witte, L., Oprach, F., & Toppel, G. (1986). Reinvestigation of the alkaloid composition of *Atropa belladonna* plants, root cultures, and cell suspension cultures. *Planta Medica*, *52*(5), 390–395.

Hawke, R. (2007). A comparative study of cultivated catmints. *Chicago Botanical Garden*, *29*, 1–4.

Hazslinsky, B. (1956). Poisonous honey from deadly nightshade. *Zeitschrift für Bienenforschung*, *3*(3), 93–96.

Janmejaya, S. (2016). Medicinal plants and related developments in India; A peep into 5-year plans of India. *Indian Journal of Health Sciences and Biomedical Research (KLEU)*, *9*(1), 14–19.

Jouhikainen, K., Lindgren, L., Jokelainen, T., Hiltunen, R., Teeri, T. H., & Oksman-Caldentey, K. M. (1999). Enhancement of scopolamine production in *Hyoscyamus muticus* L. hairy root cultures by genetic engineering. *Planta*, *208*, 545–551.

Kapoor, R., Sharma, B., & Kanwar, S. S. (2017). Antiviral phytochemicals: An overview. *Biochemistry and. Physiology*, *6*(2), 220.

Kulkarni, A. V., Hanchanale, P., Prakash, V., Kalal, C., Sharma, M., Kumar, K., Bishnu, S., Kulkarni, A. V., Anand, L., Patwa, A. K., Kumbar, S., Kainth, S., Philips, C. A., & Liver Research Club India. (2022). *Tinospora Cordifolia* (Giloy)-induced liver injury during the COVID-19 pandemic-multicenter nationwide study from India. *Hepatology Communications*, *6*(6), 1289–1300.

Munz, P. A. (1946). *Aquilegia*: The cultivated and wild columbines. *Gentes Herbarum*, *7*, 1–150

Nalawade, S. M., Sagare, A. P., Lee, C. Y., Kao, C. L., & Tsay, H. S. (2003). Studies on tissue culture of Chinese medicinal plant resources in Taiwan and their sustainable utilization. *Botanical Bulletin of Academic Sinica*, *44*, 79–98.

Philips, C. A., & Abraham, L. (2022). *Tinospora cordifolia* (Giloy) and autoimmune-like liver injury—A classic case of Primum Non Nocere, "First, Do No Harm." *Journal of Clinical and Experimental Hepatology*, *12*, 245–246.

Posadzki, P., Watson, L. K., & Ernst, E. (2013). Adverse effects of herbal medicines: An overview of systematic reviews. *Clinical Medicine (London, England)*, *13*(1), 7–12.

Pranting, M., Loov, C., Burman, R., Goransson, U., & Andersson, D. I. (2010). The cyclotide Cycloviolacin O2 from *Viola odorata* has potent bactericidal activity against Gram-negative bacteria. *Journal of Antimicrobial Chemotherapy*, *65*(9), 1964–1971.

Reza M. A., & Ferdousi Z. (2016). Assessment of antioxidant, anticancer and antimicrobial activity of two vegetable species of *Amaranthus* in Bangladesh. *Complementary and Alternative Medicine*, *16*, 157.

Ross, I. A. (2005). *Medicinal plants of the world (volume 3): Chemical constituents, traditional and modern medicinal uses*. Humana Press Inc.

Sakurai, M. (2011). Perspective: Herbal dangers. *Nature*, *480*, S97.

Waris, A., Din, M., Ali, A., Ali, M., Afridi, S., Baset, A., & Khan, A. U. (2021). A comprehensive review of green synthesis of copper oxide nanoparticles and their diverse biomedical applications. *Inorganic Chemistry Communications*, *123*, 108369.

2 Wormwood (*Artemisia absinthium*)

Bharti Mittu, Mohammed Shafiq Alam, Maninder Kaur, Zerina Begum, and Jasmeet Kour

2.1 INTRODUCTION

Artemisia absinthium L. is more commonly termed wormwood; it is a shrubby, medicinal plant native to the countries of Eurasia, North Africa and the Middle East. The plant is a member of the largest and most extensively spread family of flowering plants, *Asteraceae*, which contains 500 different species and individual wild plants can be characterized as perennial, annual, or biennial. The parts above ground, such as the stem, leaves, petioles, flowers, fruit and seeds, have a pleasant and distinctive fragrance, whereas the root is aromatic in taste. Several common names exist for the plant: English names include green ginger, *Artemesia absinthium*, madewood, and wormwood; *Genepi* is the Latin name; *Vermouth* is the French term; Apsinthion in Greek; Majtari, Karmala, Majri and Mastiyarah in Hindi; Wermut, Bitterer Beifuss, Wermkraut in German; Nigayomogi in Japanese; Yang ai, Zhong ya ku hao in Chinese; Afsaneen and Damseeh in Arabic (Ahamad et al., 2019; Hussain et al., 2017; Szopa et al., 2020). *A. absinthium* is an important herb with antioxidant, anti-inflammatory, hypoglycemic, cardiovascular, hepatoprotective, anticancer, wound healing, antibacterial, antiviral, and insecticidal uses. Overdosing on it has been reported to have deleterious neurotoxic effects on the body, because of the presence of thujone and its analogs. *A. absinthium* causes disorders related to the central nervous system, headaches, nausea, convulsions, and intestinal muscle spasm if taken in high doses. The essential oil of *A. absinthium* is also prohibited for use in pregnant females, lactating mothers, and patients with digestive disorders, ulcers, and gastritis (Batiha et al., 2020). Hence, its use in individuals should be supported by thorough experimental research, safety profiles, and cautious risk–benefit analysis.

2.2 OCCURRENCE

Wormwood is distributed widely throughout the northern hemisphere of the world, including in Eurasia and North America. It is well acclimatized to the northern temperate climate of Europe and West Asia and was also recorded in North America for the first time in 1841 (Bora & Sharma, 2011). The plant is now naturalized across North America and in Canada. It is also widely found in England, Poland, and Scotland and also along the Australian coast. In India, the plant population is mainly concentrated in Kashmir and Himachal Pradesh. In Kashmir, the plant has been found to occupy sites up to an altitude of 2100 above sea level (Ahamad et al., 2019; Szopa et al., 2020).

A. absinthium can survive in arid or semiarid areas and swamp habitats (Wright, 2001; Yu & Zhong, 2001; Martín et al., 2003) and the species are recorded to have originated from Asia during the Cenozoic era, some 66 million years ago (Orege et al., 2021). These plants grow vigorously in the field that dominate a particular landscape by restricting the growth of other species. They grow naturally on uncultivated mountain slopes and wastelands; however, it can be cultivated in dry, moist or damp soil conditions.

DOI: 10.1201/9781003292340-2

TABLE 2.1
Systemic Classification of *Artemisia*

Kingdom	Plantae
Subkingdom	Tracheobionta
Division	Spermatophyta
Superdivision	Magnoliophhyta
Class	Magnoliopsida
Subclass	Asteridae
Order	Asterales
Family	Asteraceae
Genus	*Artemisia L.*
Species	*Artemisia absinthium L.*

2.3 SYSTEMIC CLASSIFICATION

The plant's morphology is very significant in distinguishing their taxonomic locus using the cladistics approach. The morphology of the plant is very difficult to find because of resemblances in its species that cause obstacles in their delimitation and identification. The genus is distributed into three different genera: *Artemisia*, *Abrotanum*, and *Absinthium*. Linnaeus placed all three different genera into a single genus called *Artemisia*. However, the genus is divided into four sections on the basis of their sterility/fertility of disc florets and the presence and absence of ray florets. *Absinthium* has been associated with the presence of pistillate fertile ray florets and they also have perfect fertile disc florets present on a hairy receptacle. *Abrotanum* has perfect fertile disc florets on smooth receptacles but does not contain ray florets. Furthermore, McArthur later classified the plant in a new group with a genus named *Tridantate*, although this is native to North America. He also extended the taxonomy on a classical basis such as *Anthemidae*. Bremer also revised the classification of the plant, recognized the phylogeny of the Asteraceae family (Hussain, 2020). The present-day classification of wormwood is given in Table 2.1 (Batiha et al., 2020; Ansari et al., 2019).

2.4 MORPHOLOGICAL FEATURES OF THE PLANT

Wormwood is an herbaceous plant that can grow up to a height of 150 cm. It has a characteristic and aromatic smell attributed to its essential oils. The plant blooms from the months of July to September, or even as a late October in some places. At the end of the growing period, the aerial parts die and they are harvested for their medicinal properties (Batiha et al., 2020).

2.4.1 Roots

The roots of the plant are relatively flat and perennial. The roots are well-developed and consist of a taproot, which occasionally reaches 2 inches in diameter with many anastomosing lateral branches spreading all around. The plant is rhizomatous; these rhizomes are strong, woody, and fibrous, from which numerous short and twiggy roots grow.

2.4.2 Stem

The stem is white in color and can grow up to a height of 150 cm. It is highly branched and leafy, and almost completely covered in fine silky hairs. Wormwood stems are firm and straight, and woody at the base, with very fine silky hairs covering the entire length of the stems and also on the leaves, giving the plant a grayish tint.

2.4.3 Leaves

The leaves of wormwood are long and differ from bottom to top in shape and size. They are arranged spirally and divided into deeply lobed leaflets, bearing essential oil-secreting hairs/glandular trichomes. The apical leaves are one or two times pinnate that are short-stalked in nature while the underlying leaves are abrupt pinnate and are long-petiolate. Leaves are alternate and silky with dense hairs on the leaf surface. These silky hairs act as a defensive sheet against extreme temperatures and also to help the plant to sustain a prolonged drought. The color of leaves is white to silver green, with the basal leaves having a length of 20 cm and a breadth of 3–7 cm. The tips of leaves are lanceolate to linear-lanceolate and the margins of the leaf-stalks are winged (Batiha et al., 2020).

2.4.4 Flowers

Flowering tops are branched with numerous flower heads. The flower heads or capitula (a type of flower head composed of numerous tiny florets) are short, spherical, suspended, flowery panicle, greenish-yellow flowers (small flowers are yellow in color). The flower heads are inclined and have a length of 3 mm and 4 mm in width. The bracts can be differentiated; the outer bracts are silky-pubescent with a rounded tip, while the inner ones are oval and broad with a membranous margin. The flower receptacle has only a few hairs. The disc florets are androgynous in nature, and the ray florets are female in nature with an extending stem. Its most noteworthy features are the silky hairy sepals with fertile flowers (Batiha et al., 2020).

2.4.5 Fruit

The ripe fruit is a cypsela (small achene with brown strips) and is uncrowned. It does not have any tuft of hairs or pappus to help in the dispersal of seeds. Up to the point of ripening, wormwood forms yellow to ocher-colored nut fruits, which are elongated in shape. These contain seeds which are greyish-brown and slightly long (up to 1 mm), as well as being cylindrical, flat, and urn-shaped, and are dispersed by gravity. A single plant produces up to 100,000 seeds in a growing season (Batiha et al., 2020).

2.5 NUTRITIONAL ATTRIBUTES OF THE PLANT

Wormwood (*Artemisia* species) is a herb found in many parts of the world. All of the above-ground parts of the plant have remarkable medicinal benefits, which can be used either fresh or dried. The whole plant has to be harvested during the flowering season and the aerial parts (the parts above ground) are used in their either fresh or dried forms for their medicinal properties. The formulation of the extracts of these dried parts lack calories, vitamins, or minerals but contain numerous bioactive components, the best known of which is thujone.

The nutritional attributes of wormwood showed that minerals such as zinc, potassium, sodium, magnesium, and low concentrations of vitamins like fat-soluble vitamin A, and water-soluble vitamin B and C are present. But the major components of the plant are characterized by the presence of a rich quantity of oils. These essential oils can be collected from plants throughout the growing season, but the maximum quantity of the oil is extracted at the starting of the flowering period and also when the leaves have begun to bud. The composition of wormwood essential oil varies on the basis of their geographical location and the soil and environmental conditions. The abundance of essential oil in the herb is found to be less than (0.2%) in dry climatic condition, whereas it extends to about 1.5% in humid environmental conditions (Beigh & Ganai, 2017). The oil content also depends on the type of extractant used for the purpose. Studies have proven that an ethanolic extract of the plant parts has a considerably increased content of flavonoids and phenols, and tannins with comparison to chloroform and other aqueous extracts (EFSA, 2014). The quantity and quality of essential oils is markedly affected due to the drying process. It has been observed that the organoleptic characteristics of the oil are also affected by even a slight rise in the air temperature. In addition, the collected herb should be kept, spread in thin layers, and dried in drying chambers or dried sheds at room temperature with the circulation of natural air (Lachenmeier et al., 2006).

2.6 PHYTOCHEMICAL OVERVIEW OF THE PLANT

A phytochemical overview of the wormwood plant shows that it is a good source of numerous secondary metabolites, namely, terpenoids (e.g., γ-terpinene trans-thujone, terpinen-4-ol, camphene, bornyl acetate, cadinene, guaiazulene, myrcene, linalool, and chamazulene, camphor, and trans-sabinyl acetate), alkaloids, lactones, organic acids and resins, tannins, phenol acetylenes, sterols, and essential oils, (Anibogwu et al., 2021; Ahameethunisa & Hopper, 2010). It contains flavonoids e.g., quercitin, flavonoid glycosides such as isorhamnetin-3-O-rhamnose glucoside, quercitin-3-O-D-glucoside, quercetin-3-O-rhamnoglucoside, isoquercitrin and isorhamnetin-3-O-glucoside, and phenolic acids as caffeic, ferulic, chlorogenic, coumaric, syringic, vanillic acids, and salicylic acid (Batiha et al., 2020). A summary of the important phytochemical constituents in wormwood is shown in Table 2.2.

Researchers have also reported the presence of many other compounds in lesser concentrations, such as chalcone—coumarins (herniarin, coumarin) (Msaada et al., 2015), cardamonin (Krebs et al., 2010a; Hatziieremia et al., 2006), fatty acids (Ahamad et al., 2019), carotenoids, tannins (Bora & Sharma, 2010; EFSA, 2014), lignans (Bora & Sharma, 2010), and resinous substances (EFSA, 2014).

Mamatova et al. studied the presence of β-pinene (23–24%) and β-thujone (18–19%) as the main components of essential oil of wormwood (Mamatova et al., 2019). The vital chemotypes of the essential oils of wormwood are: estragole (42–85%), sabinene (approx. 33–35%), methyl eugenol (approx. 20–25%), and elemicin (up to 50–57%), shown in Figure 2.1 (Ekiert et al., 2021). Terpinen-4-ol, β-ocimene, α-trans-ocimene, cis-ocimene, limonene and trans-anethole, α-phellandrene and β-phellandrene, capillene, and (Z)-artemidinare are among the other components of the oil in concentrations slightly bigger than 10% (Joshi et al., 2016; Bedini et al., 2017; Hassanzadeh et al., 2016; Sharopov et al., 2020).

Ahamad et al. (2019) characterized the isoflavone glycosides in methanolic extracts of the aerial parts of the plant. These are stigmast-5,22-dien-3β-ol-21-oic acid-3β glucopyranosyl-2′-octadec-9″-enoate; 3,11-dimethyldodecan-1,7-dioic acid-1- β-D-glucopyranosyl-6′-octadec-9″-enoate; tricosan-14-on-1,4-olide-5- eicos-9′-enoate and lanost-24-en-3β-ol-11-one-28-oic acid-21,23 α-olide-3β-D-glucopyranosyl-2′- dihydrocaffeoate-6′-decanoate. Ahamad et al. (2019) isolated two more new isoflavone glycosides from the same extract; *Artemisia* bis-isoflavonyl dirhamnoside and *Artemisia* isoflavonyl glucosyl diester.

TABLE 2.2
Some Important Constituents of Wormwood

Essential oil	ß-pinene, chamazulene, nuciferolbutanoate
Sesquiterpene lactones	Absinthin, artemetin, artemisinin, arabsin, artabsin, ketopelenolide, santonin-related lactones
Flavonoid	Artemitin, 3-rhamnoglucoside, rutin, spinacetin 3-glucoside, spinacetin 3-rhamnoglucoside, chlorogenic acid, caffeic acids and others
Ascorbic acid	Vitamin C
Phenolic compounds	Syringic acid, chlorogenic acid, caffeic acid
Glucosides	Absinthin, anabsinthin
Bitter principles	Artabsin, absinthin, artamarininquebrachitol, and others

Source: Hashimi et al., 2019.

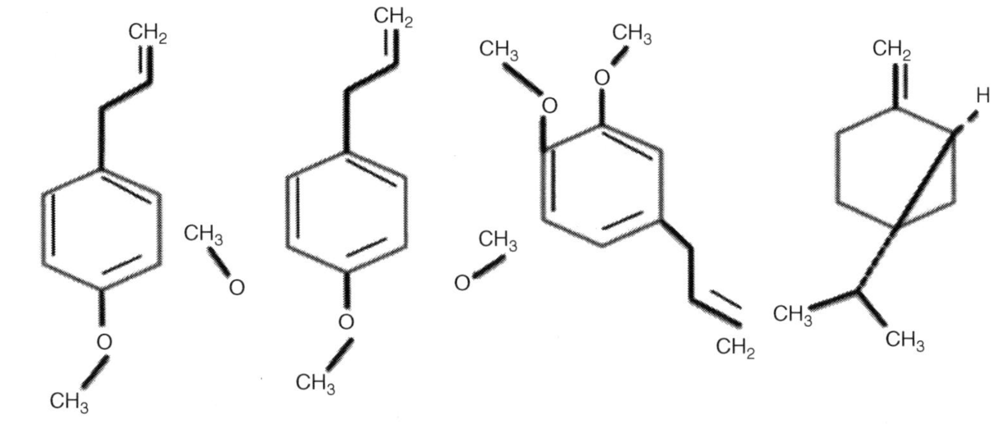

FIGURE 2.1 Chemical structure of volatile oils of *Artemisia*.

Two diastereomeric homo-diterpeneperoxides from the above-ground parts of *A. absinthium* have been reported. The combination of these two complexes exhibited antimalarial activity *in vivo* with an EC50 value of 1 µg/ml. The homoditerpenes, 1-(E)-8-isopropyl-1,5-dimethylnona-4,8-dienyl-4-methyl2,3-dioxa-bicyclo[2,2,2] oct-5-ene and iso-1-(E)-8-isopropyl-1,5-dimethyl-nona-4,8-dienyl-4-methyl2,3-dioxa-bicyclo[2,2,2]oct-5-ene,were reported by different researchers (Ahamad et al., 2013; Ahamad et al., 2014; Ahamad et al., 2019).

A. *absinthium* also contains bitter sesquiterpenoid lactones, of which absinthin (a guaianolide dimer) is the main metabolite (0.2–0.28%). Isomers of absinthin (anabsin, anabsinthin, artabsin (0.04–0.16%), and absintholide) are also present. Artamaridinin, artamarin, artamarinin and artamaridin are all present in low concentrations. The bitter components are best extracted from plants harvested in September (Bora & Sharma, 2010).

The medicinal applications of the plant depends on the presence of these bioactive components. The fresh plants are considered to yield about 40 to 70% of azulene; an important bioactive constituent isolated from *A. absinthium* (Batiha et al., 2020).

2.7 STRUCTURAL CHARACTERIZATION OF BIOACTIVE COMPONENTS

Gas chromatography-mass spectrometry technique assessed the essential oil components of aerial parts of plant *A. absinthium*. The results pointed out that the wormwood essential oils consist of 73.5% phenylpropanoids, 24.3% monoterpene hydrocarbons (mainly chamazulene), and only 0.2% of sesquiterpenoids (Bedini et al., 2017).

Studies on Iranian wormwood essential oil showed a predominance of β-pinene and β-thujone, whereas Moroccan *A. absinthium L.* essential oil is primarily represented by 39.69% α-thujone, 10.96% sabinyl acetate, and β-thujone (7.25%) (Msaada et al., 2015). The chemical structure of main bioactive constituents of plant *A. absinthium* are tabulated in Table 2.3.

2.8 EXTRACTION AND ISOLATION TECHNIQUES OF THE BIOACTIVE COMPOUNDS

The plant species exhibit a comprehensive variety of bioactive compounds which are found to be effective in a number of health disorders. Many kinds of research have been performed to extract these bioactive compounds that can conserve the variety and magnificence of the species.

TABLE 2.3
Bioactive Compounds Isolated from *A. absinthium*

Name of the Bioactive Molecule	Class of the Bioactive Molecule	Chemical Structure of the Bioactive Molecule
Artemisinin	Endoperoxide having sesquiterpene lactone	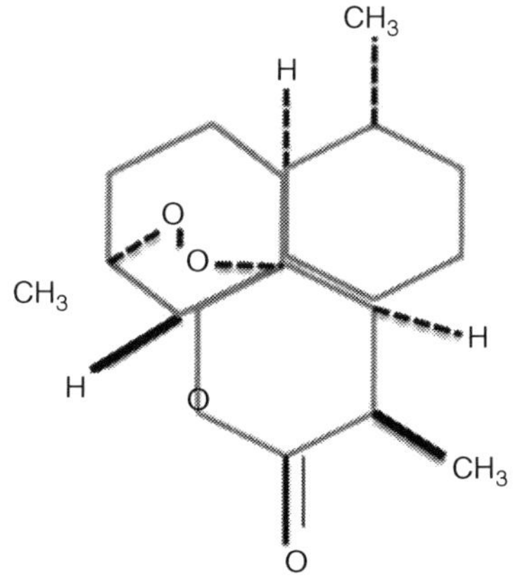
β-pinene	Bicyclic monoterpene	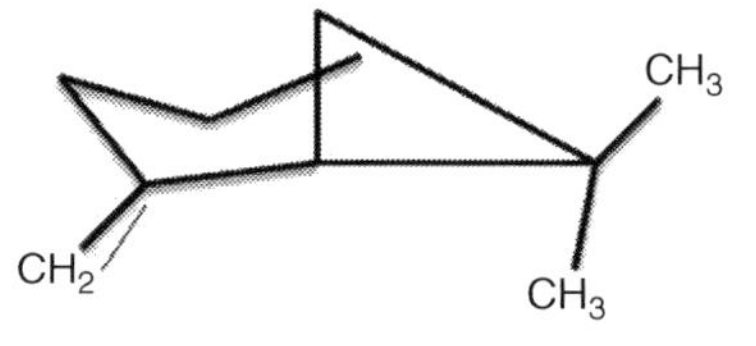
α-thujone	Bicyclic monoterpene ketone	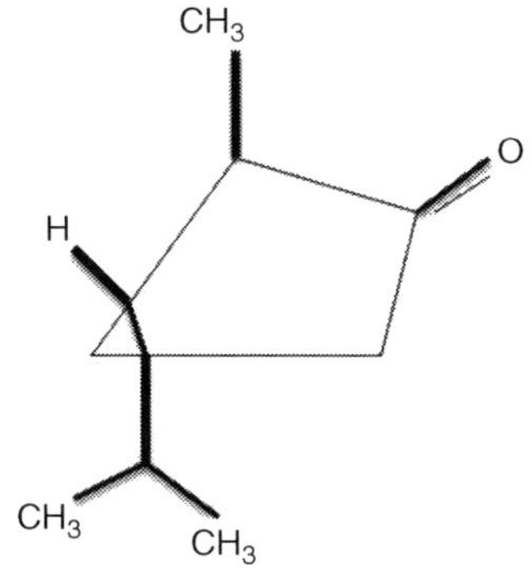
β-thujone	Bicyclic monoterpene ketone	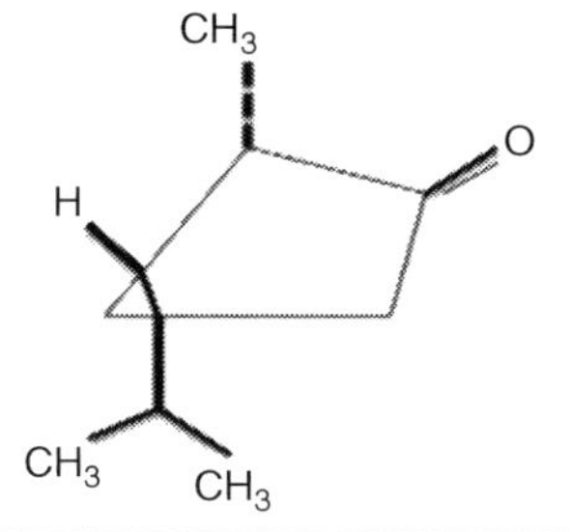

(Continued)

TABLE 2.3 (Continued)

Name of the Bioactive Molecule	Class of the Bioactive Molecule	Chemical Structure of the Bioactive Molecule
(E)-sabinyl acetate	Monoterpenoid	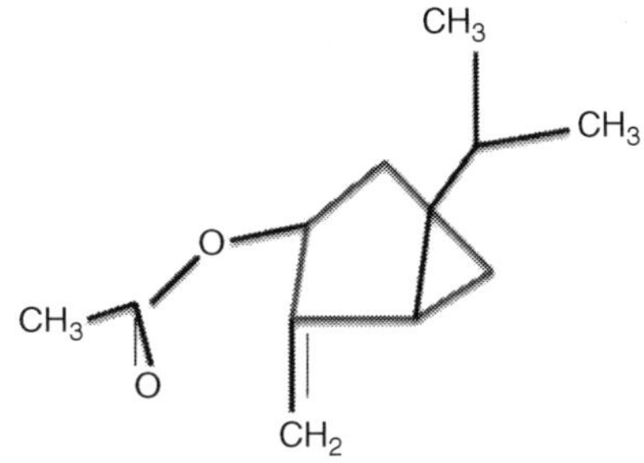
Bornyl acetate	A bicyclic monoterpene and Acetate ester of borneol	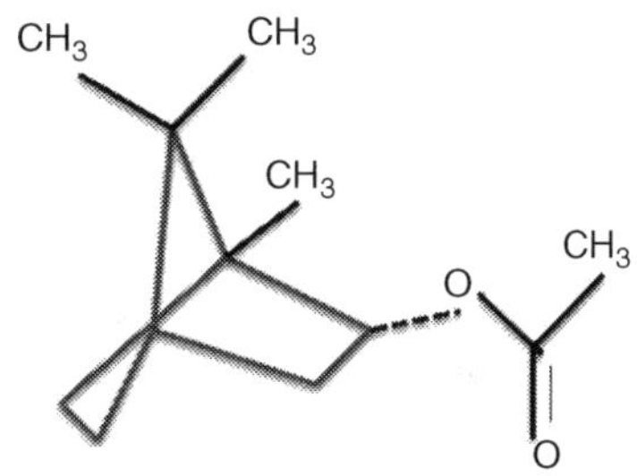
Terpinen-4-ol	An isomer of the terpineol, monoterpene alcohol	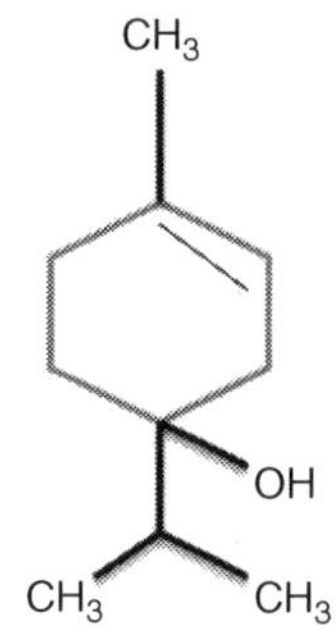
Camphene	Bicyclic monoterpene	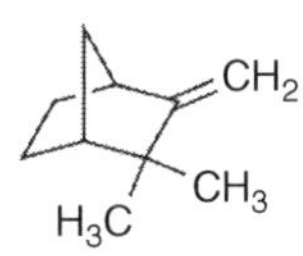
Cadinene	Bicyclic sesquiterpenes	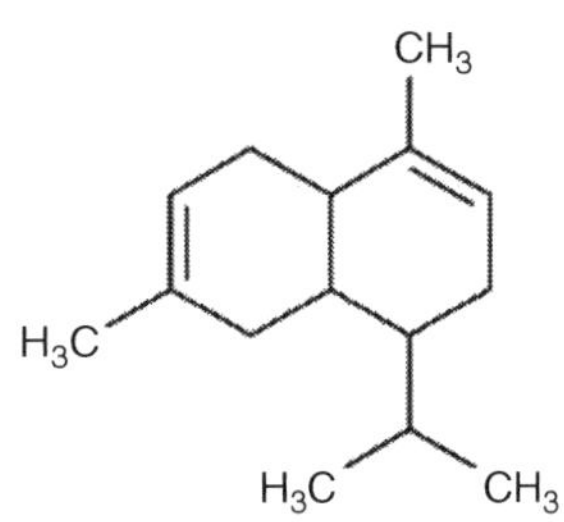
Myrcene	Monoterpene; alkene hydrocarbon	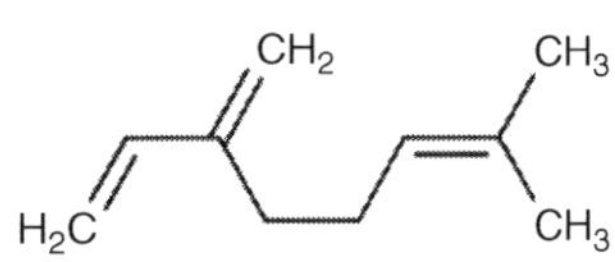

(*Continued*)

TABLE 2.3 (Continued)

Name of the Bioactive Molecule	Class of the Bioactive Molecule	Chemical Structure of the Bioactive Molecule
Guaiazulene	Bicyclic sesquiterpene	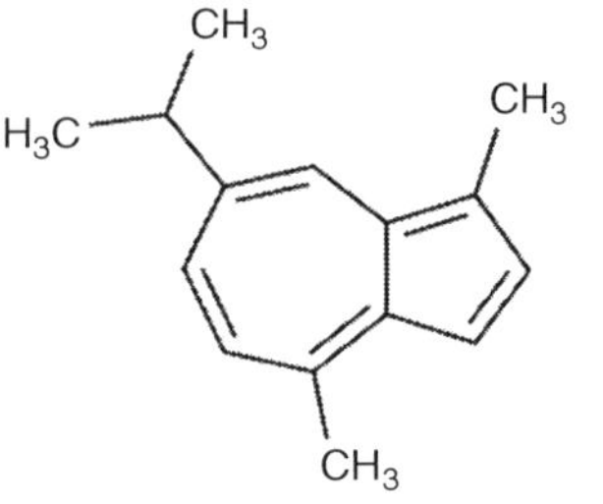
γ-Terpinene	Monoterpene	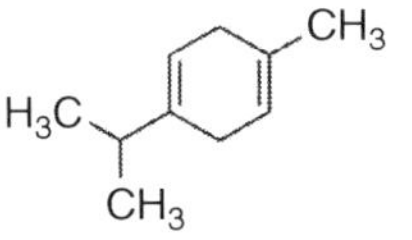
Linalool	Acyclic monoterpene alcohol	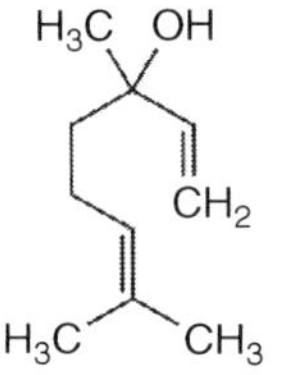
Camphor	Terpenoid	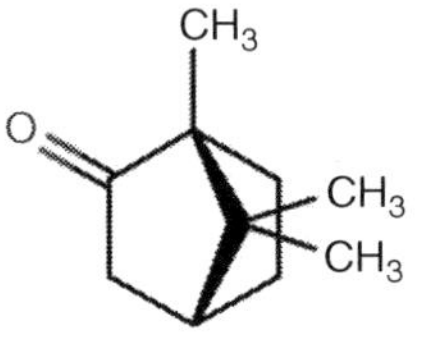

Source: Batiha et al., 2020.

Traditional methods of extraction utilize large quantity of organic solvents, but result in minimum yields (Capuzzo et al., 2013). Thus, it is vital and valuable to explore novel green extraction techniques with methodologies emphasizing on maximum extraction output with minimum energy requirement. At present, the two most prevalent high–pressure methods with high extraction efficiencies, namely supercritical fluid extraction (S.F.E.) and pressurized liquid extraction (PLE), are used for the extraction of phytochemicals compounds of natural plant sources (Capuzzo et al., 2013).

Supercritical fluid extraction (SFE) is an advanced process that separates the extractant from the medium through the use of an extracting solvent. These fluids perform at predetermined conditions of temperature and pressure, known as the supercritical region, i.e. when these supercritical fluids are elevated beyond their critical point, they exhibit physicochemical properties like liquid–liquid solvent power, minimum surface tension, and gas-like transport properties. Supercritical carbon dioxide (CO_2) is the preferable gas in bioactive extraction as it is highly pure, nonflammable, odorless, colorless, cost-effective, non-poisonous, and reusable permitting effectiveness at 72.9 atm; 31.3 °C (Capuzzo et al., 2013). Hence, the use of CO_2 is observed to be ecofriendly since it is being recycled. The supercritical fluid extraction assembly includes a compression pump and filter with a 1 l extraction vessel and two separators of 1.8 l capacity. Separators are used for the regaining of heavy compounds and volatile oils separately (Martín et al., 2011).

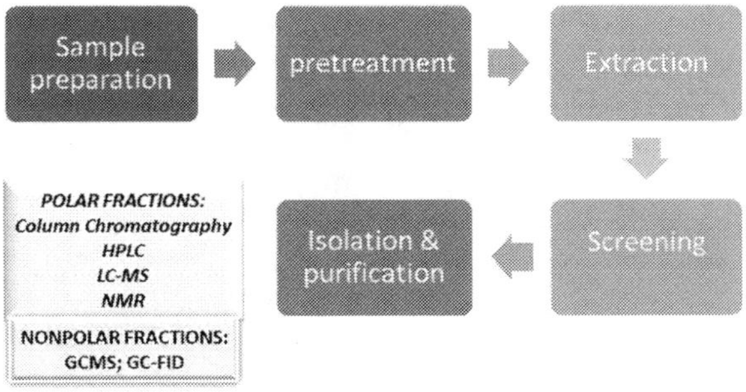

FIGURE 2.2 Flowchart showing various stages of extraction as well as isolation of different phytochemicals compounds of *Artemisia* (Anibogwu et al., 2021).

PLE is a "green" technology for the mining of nutraceuticals from foods and herbal plants. This technique is actually an addition of the SFE process where solvents (organic to weak acids) are used instead of CO_2. This technique is made distinctive by the use of suitable pressures (500–3000 psi), to withstand the solvent in its liquid phase and high temperature (25–200 °C). A small amount of solvent is required to perform this technique, which results in an enhanced extraction because of an improved mass transfer rate. Reduced viscosity and improved solvent solubility under increased temperature helps in mass transfer and matrix penetration (Cao et al., 2017).

In *Artemisia* species, the low concentrations of bioactive compounds lead to an abundant task that can affect the improvement of their extraction as well as analysis using conventional methods, something which is almost unachievable. The green technique of hydrophilic deep eutectic solvents (DESs) has extended the attention as a suitable replacement for the effective withdrawal of bioactive components. In this method, two or more low flammable, non-toxic, biodegradable, negligible water-miscible and cost-effective components bind to each other by hydrogen bonding. They can be simply prepared by mixing the constituents in an appropriate molar ratio where one component is a hydrogen bond acceptor while the other is a hydrogen bond donor, which makes them applicable for varied pharmaceutical applications. For the extraction of artemisinin from wormwood, a hydrophobic DES known as N81Cl-NBA is used with the highest extraction yield. It is prepared by mixing 1-butanol (1:4) and methyl trioctyl ammonium chloride. Results demonstrated that DESs are known as "designer solvents" and that they can be used as eco-friendly extraction solvents for extracting different bioactive compounds of plant components (Gallego et al., 2018; Cao et al., 2017).

Anibogwu and his co-researchers have reported the use of high-performance liquid chromatography (HPLC), gas chromatography mass spectrometry, high-performance thin layer chromatography (HPTLC), and nuclear magnetic resonance for the characterization and authentication of many secondary metabolites such as terpenoids, flavonoids, saponins, and polysaccharides from *Artemisia* spp. The flow chart of bioactive compound extraction and isolation from wormwood is shown in Figure 2.2 (Anibogwu et al., 2021).

2.9 APPLICATIONS OF THE ACTIVE INGREDIENTS

In traditional and classical medicinal practices, wormwood had been used all over the globe from time immemorial. Several researchers have conducted studies in this genus and documented their huge applications in promoting the health of millions globally.

Studies demonstrated the wide use of the wormwood essential oil for its antioxidant activity, neuro-protective, hepatoprotective, antitumor, anti-inflammatory, antidepressant, antimicrobial, antifungal, anthelminthic, antimalarial and insecticidal properties.

In Europe, wormwood is a popular plant which is used as a spice, while in many countries in Asia, the plant extracts are in use for the treatment of gastrointestinal ailments, insomnia, epileptic seizures, and as an anesthetic and hypnotic. It has been found to be very effective in the treatment of inflammation, helminthiasis, and many types of fever (Bora & Sharma, 2011). The Korean community uses the leaf and stem extracts for the treatment of liver disorders, hyperlipemia, and infected cholecystitis. Chinese folk medicine uses the dried parts of the herb for the prevention and treatment of chronic inflammations of the liver. This may be due to the inhibitory response of lipid peroxidation down-regulation of apoptosis mediated by TNF-alpha. In recent microbial studies, chlorogenic acid was found to destroy the cell integrity by disrupting the cell membrane, which ultimately led to the death of the bacterial cell. In some studies, the ethyl acetate, n-hexane, and menthol extracts displayed greater microbial activity due to a high concentration of phenolics, terpenoids, and volatile oils. Thujone, a neurotoxin, excites the brain by blocking gamma aminobutyric acid (GABA), a neurotransmitter that has a calming impact on the central nervous system (CNS) (Mamatova et al., 2019 Batiha et al., 2020; Hashimi et al., 2019; Fahd et al., 2020; Msaada et al., 2015; Merajuddin et al., 2012). Moreover, Guarrera reported the use of the plant as an insect repellent; as an anthelminthic agent, and also as an additive source for ruminants, particularly to promote ruminal fermentation for the efficient use of diets (Guarrera, 1999).

In traditional Arabic herbal medicine, the herb had been in use for the treatment of gum infections (gingivitis), sleep disorders, and as a masking agent against the bitter taste of medicines. It was also used for the treatment of foot and mouth disease in cattle and sheep. In Central Asian countries, including Russia, it was used to cure indigestion and gastric inflammation; and also dermal issues such as allergic rashes and itching (Sharopov et al., 2020). The Indian ayurvedic form of medicine relates the species with the effective treatment of various types of fever, parasitic intestinal worms, intestinal smooth muscle cramps, an immunostimulant, and also as an emmenagogue i.e. to stimulate or increase menstrual flow (Obolskiy et al., 2011). In Kashmir and Himachal, a paste formulated from the leaves was used to treat wounds on the legs of their livestock (particularly yaks and donkeys). Moreover, they use the extract as an anesthetic for toothache, stomach ache, fever, dysentery, and intestinal parasitic worms (Joshi et al., 2016).

Artemisinin, a sesquiterpene lactone, demonstrated excellent anticancer activities (Ferreira et al., 2010), as an antibactericidal and fungicidal agent (Appalasamy et al., 2014), and confirmed anti-inflammatory and immunomodulatory functions; these were probably due to its capability to maintain equilibrium between free radicals (reactive oxygen species and reactive nitrogen species) and antioxidant resistance within the body, which can cause various oxidative stress conditions, thereby leading to diseases. As such, essential oils of absinthe established its position for vital applications in the food, pharma, and cosmetic sector. *A. absinthium* L. is an essential oil from France containing (Z)-epoxyocimene and chrysanthenyl acetate as vital components which can inhibit the growth of *Saccharomyces cerevisiae var. chevalieri* and *Candida albicans*. Turkish wormwood essential oils have proven effective against a large number of fungi and primarily consists of 1,8 -cineole, chamazulene, and camphor. Furthermore, *Absinthium* essential oil from Uruguay, which is rich in thujone, displayed antifungal properties (Msaada et al., 2015). In German customary herbal medicine, the plant parts are used to repel insects such as flies, moths, lice, and bugs. The plant had remained associated with rituals in funeral processions and to ward off evil spirits (Bäumler, 2007).

The French Pharmacopoeia and the European Pharmacopoeia classified the fresh flowering herb part of *A. absinthium* as a raw material in some homeopathic forms of medicine. The formulated potion should comprise a lowest level of 0.04%–0.05% w/w of hydroxycinnamic acid derivatives, which is shown in the form of chlorogenic acid. As a homeopathic medicine, the plant is prescribed to improve conditions such as hallucinations, dizziness, nervousness, insomnia, and convulsions (Lockie, 2006).

In addition to these applications, raw materials obtained from wormwood have been used in cosmetics for the production of various skin care agents, and applications applied to the scalp, body, and face. These include moisturizing creams with SPF filter, lotions, shampoos, face masks and serums, tonics, cleansing milk, and under-eye patches. These products are principally obtained from essential oils by a process of steam distillation. The extracts are also used as a component in perfumes and in aromatherapy during body massages and baths (Szopa et al., 2020).

In the food industry, the plant has been investigated for its safety profiles and recommended as a resource for the manufacture of alcoholic beverages, spirits, and bitters (Basta et al., 2007). In America, wormwood extracts have been used as a flavoring agent in eggs, chicken, and seafood; in France, it is used to improve the taste of eggs, mayonnaise, and sour cream (Anibogwu et al., 2021). Dried above-ground parts of the plant can be traditionally used for the production of an alcoholic beverage, either as a single component or in combination with other herbs. The European Authority of Food Safety (EFSA) have decreed that the percent content of thujone (α- and β-forms) in alcoholic beverages should not exceed 35 mg/kg and 10 mg/kg in bitter and plain sprits, respectively.

2.10 PHARMACOLOGICAL ACTIONS AND USES OF *A. ABSINTHIUM*

A. absinthium have been documented to treat various ailments such as liver inflammation, hepatitis infection, stomach ache, dyspepsia, gastritis and flatulence, loss of appetite, enlargement of the spleen, esophageal bowel syndrome, depression, epileptic seizures, chronic fever, dermatitis, and the healing of wounds. It has also been reported for its use as an anthelminthic and insect repellant, as well as an additive source to improve the rumen fermentation for the optimal utilization of food. This can help with labor pains as well as the treatment of sclerosis and leukemia. In the food industry, it's widely used to manufacture aperitifs, spirits, and bitters. The plant ointment is used extensively to reduce muscle and joint stiffness, as well as to help in the healing of injuries. Wormwood is also used to alleviate childbirth pain, as well as menstrual cycle symptoms, as well as heart problems and hypertension (Sharopov et al., 2012; Jaleel et al., 2016; Nibret & Wink, 2010; Manganelli, et al. 2000).

2.10.1 Antioxidant Activity

A. absinthium has been associated with antioxidant activities. The presence of numerous phenolic and flavonoid chemicals is responsible for this effect. The methanolic extract of the herb has exhibited neuroprotection abilities. *A. absinthium* causes the decreased peroxidation of lipids by reducing levels of thio-barbituric acid substances and releases different endogenous antioxidants (for example glutathione peroxide, superoxide dismutase levels) which clearly implies its role in preventing oxidative stress-related disorders. Another study demonstrated the antioxidant properties of ethanolic extracts *A. absinthium* on fibroblast-like cells and its protection from oxidative stress. Study was conducted to investigate the ability of the plant extracts to suppress 1,1diphenyl-2-picryl-hydrazyl (DPPH) free radical and reactive hydroxyl during the fenton reaction is confined by 5,5dimethyl-1-pyrroline- N-oxide (Batiha et al., 2020; Bora & Sharma, 2011; Craciunescu et al., 2012; Canadanovic-Brunet et al., 2005).

2.10.2 Antitumor Activity

A. absinthium showed anti-tumor activity against human (HCT-116) cancer cell line by using MTT assay (3-(4,5-dimethylthiazol-2-yl)-2,5-diphenyltetrazolium). The methanolic extract was used for the study and the results show a marked reduction in the viability of the HCT-116 cancer cells. The protein levels and mRNA of BAX (Bcl2-Associated X protein) increased, but those of BCL-2 (B-cell lymphoma 2) fell in extract-treated cells. The extract induced arrest at the cell-cycle stage and apoptosis in the cell line; the protein level, as well as the activity of caspase 3 and mitochondria membrane potential damage, also improved significantly. Activation of the mitochondrial pathway predominantly encouraged apoptosis in the HCT-116 cells (Nazeri et al., 2020). Ahamad and coworkers studied the anti-cancerous activity of *A. absinthium* against human breast cancer cell lines MDA-MB-231 (estrogen-insensitive) as well as MCF-7 cancer cell lines (estrogen-responsive). Results demonstrated that the methanolic extract concentration of 25 g/ml can cause a 50% reduction in MDA-MB- 231 and MCF-7 cell proliferation in comparison to the control (Ahamad et al., 2019). The mitogen-activated protein kinase (M.E.K.)/extracellular signal-regulated kinase (ERK) signaling pathway which activated the mitochondrial pathway of caspase activation due to extract. This pathway monitors Bad and Bcl2 family proteins, resulting in the apoptotic death of MCF-7 and MDA-MB-231 cells. Chlorogenic acid has been shown to have a lethal carcinogenic effect in the large intestine, tongue, and liver. Artesunate, a derivative of artemisinin, has also shown anticancer activity both in vitro as well as in vivo. Another chemical (artemisetin) isolated from *A. absinthium* demonstrated a substantial anticancer impact against melanoma B16, but did not affect Pliss lymph sarcoma growth retardation. It is hypothesized that artemisinin bioactivation occurs in the endosome; this is followed by the release of iron by internalized transferrin, which is triggered by different pH. Carbon-centered radicals are subsequently generated by iron-activated artemisinin, leading to lysosome rupture and ROS. formation, and resulting in mitochondrial damage and cell death. Many studies have related artemisinin toxicity to decreased cytokinesis, tumor invasion inhibition, metastasis, and ROS (Shafi et al., 2012; Maobe et al., 2013; Ferreira et al., 2010).

Previous literature reported the anti-cancerous property of *A. absinthium* against extensive cancer cell lines such as ovarian cancer (HO-8910), chronic myeloid leukemia (KML-562), and cervical cancer (HeLa) (Parekh et al., 2009; Taleghania et al., 2020).

The anticancer property of the plant extract was investigated on endometrium (ECC-1) cancer cells, human colon (DLD-1), and (HEK-293) embryonic kidney cells. Cytotoxic effects were studied with MTT, apoptotic activity with DNA fragmentation ELISA and AO/EB fluorescent staining. *A. absinthium* extract showed cytotoxic activity on the human colon and endometrium cancer cells. In contrast, only a low effect was found on embryonic kidney cells. Increased free radicals led to DNA damage of cancer cells and, ultimately, cell death (Koyuncu, 2018). Recently, Sultan et al. (2020) produced similar results using hot methanolic extracts of *A. absinthium* leaves. In their study, the cytotoxicity effect of various concentrations of the extract was determined against the breast cancer cell line (MCF-7 ATCC) and the human colon cancer cell line (HCT 116 ATCC) by performing the MTT assay. The results showed that the extract effectively inhibited the propagation of breast cancer cells at a concentration of 80.0 ± 3.92 µg/ml as IC-50 value, but that they do not suppress the growth of the treated cancer cells of the human colon (Sultan et al., 2020).

2.10.3 Neuroprotective and Antidepressant Effects

A. absinthium has neuroprotective effects and plays an essential part in the synthesis of endogenous antioxidants such as glutathione peroxidase and superoxide dismutase (Bora & Sharma, 2010). Additionally, the plant's nicotinic and muscarinic effects increase cognitive function. The ethanol extract of the plant has shown anticholinesterase activity consistently and prevented lead-induced neurotoxicity. It also restored altered Ach E and monoamine oxidase (MAO) enzyme activity to near-normal levels in rats. The plant improved the changes induced by excessive lead poisoning in the neurons and glial cells of adult rats, implying that *A. absinthium* may also help prevent mercuric chloride-driven neurodegeneration. Furthermore, the natural sesquiterpene dimer and caruifolin are produced from *A. absinthium* L. that significantly prevented the organization of intracellular free reactive oxygen species, suggesting that this mechanism of action could be related to neuroprotective activity. Mahmoudi et al. evaluated the antidepressant effect of *A. absinthium* and showed that the extract reduced the immobility duration in the same way as imipramine (Mahmoudi et al., 2009). This antidepressant effect could be the result of various mechanisms, including MAO inhibition, depression relief, and selective serotonin re-uptake inhibition, and, possibly, the amalgamation of bioactive substances found in the extract (Hallal et al., 2016; Zeng et al., 2015).

Rashidi and others studied the role of *A. absinthium* ethanolic extracts on Parkinson's disease. They indicated that the extracts have neuroprotective effects against 6-OHDA (6-hydroxydopamine—a usual neurotoxin)-induced oxidative (human neuroblastoma) SH-SY5Ycells death through the reduction of malon-di-aldehyde (MDA), intracellular ROS and apoptosis. The novel therapeutic potential of *A. absinthium* for protection and for the treatment of neurodegenerative diseases such as Parkinson's disease was confirmed (Rashidi et al., 2021).

2.10.4 Immuno-modulatory and Wound Healing Activity

In medicine, *A. absinthium* has been shown to be effective against immune-mediated disorders. Studies reported that the extracts of plants suppress the tumor necrosis factor alpha (TNF-α) in Crohn's disease, thereby assisting in healing (Krebs et al., 2010a). The ethanolic extracts of the plant showed a potent immune-modulatory property and the maturation of dendritic cells by increasing the expression of CD40. Polysaccharides derived from *A. absinthium* exhibited immunomodulatory properties through triggering the Th1 response and activating NO production. Hoseinian et al. demonstrated wound-healing ability due to *A. absinthium* by causing the repair of an Achilles tendon wound of a rabbit. This could be attributed to thujone and pinene, both scavenging free radicals (Hoseinian et al., 2018; Shahnazi et al., 2015).

Sultan et al. reported the *in vivo* wound-healing properties of a reaction combination of hot methanolic extract of wormwood in male rats, with an effectiveness which matched that of standard povidone iodine cream. The reaction mixture of the methanolic extract was prepared in milli-Q water. The efficiency of the extract was tested using various biomarkers such as IL-6, IL-1β, TNF-α, caspase-3, and caspase-9 levels. Results demonstrated that the extract reaction mixture displayed better wound-healing properties than standard povidone iodine cream; the modulation of the cytokine network and biomarkers level demonstrated wound healing (Sultan et al., 2020).

Boudjelal et al. (2020) reported the wound-healing activity of *A. absinthium* methanolic fraction on rats and two ointments of different concentrations (5% and 10%) were formulated for the study. Cicatryl-Bio was taken as the reference drug and results showed that the ointment with 10% concentration of methanol extract significantly increased the contraction in wound, similar to that shown by the reference drug. However, there was the absence of any oral or topical adverse side effects of the experimental animal, which also established the safety of the ointments (Boudjelal et al., 2020).

2.10.5 Hepatoprotective Effects

According to Mohammadian and others administration of *A. absinthium* hydroalcoholic extract improves liver function and decreases the concentration of oxidative stress indicators. The extract stimulates, and also conserves, the integrity of the hepatocellular membrane, resulting in lowered activity of serum aspartate andalanine aminotransferase. The plant decreased hexobarbitone-induced sleep time and showed cholagogic (inducing a flow of bile) and excretory activity, as well as stimulating the secretion of bile acid. Similarly, *A. absinthium* exhibited antioxidant activity by scavenging free radicals such as hydrogen peroxide (H_2O_2), DPPH. As a result, the plant may be utilized to reduce hepatic injury and as an alternative for synthetic medications used to treat liver illness. The hepatoprotective effects of this herb are likely at dosages less than 200 mg/kg, whereas, at large concentrations, the herb may be toxic to the liver. The hepatoprotective effect of *A. absinthium* extracts on acute liver toxicity is demonstrated by lowering lipid peroxidation and tetra-chloro-methane toxicity. Hepatic problems, such as hepatitis and hepatobiliary diseases, were also improved. The decrease of liver microsomal drug metabolizing enzymes, calcium channel blocking and free radical scavenging activity, are all conceivable mechanisms based on its hepatoprotective qualities (Mohammadian et al., 2016; Grycyk et al., 2019).

2.10.6 Anti-inflammation and Anti-snake Venom Activity

Absinthium extracts have demonstrated anti-inflammatory properties in several studies, a finding which is linked to flavonoids and sesquiterpene-type substances. The regulators of inflammation such as serotonin, histamine, prostaglandins, and bradykinins are inhibited by these substances, resulting in anti-inflammatory effects. Furthermore, Ahamad and coworkers discovered that when methanolic *A. absinthium* extract was supplied at 300, 500, and 1000 mg/kg doses, and varied levels of anti-inflammatory action were seen in experimental models. The delayed anti-inflammatory action of methanolic *A. absinthium* extract could be due to the delay in the absorption of the plant extracts (Ahmad et al., 1992). Lee et al. have shown the anti-inflammatory effects of 5,6,3′,5′ tetramethoxy 7,4′hydroxyflavone derived from *A. absinthium in vitro* and *in vivo* in cells stimulated with lipopolysaccharide. It was found that 5,6,3′,5′ tetramethoxy 7,4′ hydroxyflavone suppressed the expression of pro-inflammatory mediators such as inducible NO synthase (iNOS), nitric oxide (NO), prostaglandin E(2) (PGE (2)), cyclooxygenase 2 (COX2), and nuclear factor kappaB (NFkB). In collagen-treated animals, it also reduced the level of tumor necrosis factor (TNF). Cardomonin, derived from *A. absinthium* extracts, employed a direct influence on the transcription factor binding to DNA by consistently suppressing the release of NO and the expression of iNOS (Lee et al., 2004). Zeng et al. have discovered that a natural sesquiterpene dimer identified in *A. absinthium*, caruifolin D, has anti-inflammatory potential and could be a source for future medicines to treat neuro-inflammation (Zeng et al., 2015). Choi and others found that flavone extracted from *A. absinthium* suppressed IL10 production and had anti-inflammatory effects on the cytokine, decreasing arthritis in experimental mice generated by collagen (Choi et al., 2006; Beigh & Ganai, 2017; Batiha et al., 2020).

A study conducted by Nalbantsoy and others reported that *A. absinthium* reduces the swelling induced by the venom or carrageenan of viper species (*Montivipera xanthina*) in rodents. Before administrating the snake venom or carrageenan rodents were intraperitoneally administered with methanolic extract of the plant. The results indicated a significant reduction in the rodent's paw edema with doses of 25 and 50 mg/kg of the plant extract; by contrast, carrageenan-induced inflammation were effective with doses from 12.5 to 25 and 50 mg/kg (Nalbantsoy et al., 2013).

2.10.7 Antipyretic and Analgesic Activities

Khattak et al. demonstrated the antipyretic activity by different extracts (chloroform, hexane, aqueous) of *A. absinthium* in experimental rabbits with subcutaneous yeast inoculations (Khattak et al., 1985). Furthermore, up to a dose like 1.6 g/kg of A. *absinthium*, no deleterious effects were found in the animal. *A. absinthium* extracts exerts topical antinociceptive effects in animal model, according to Zeraati et al. Additionally, applying an ointment containing the plant extract topically to patients with osteoarthritis of the knee improves clinical symptoms (Zeraati et al., 2014; Batiha et al., 2020).

2.10.8 Cardiovascular Activity

A. absinthium extracts had hypolipidemic, antiatherosclerotic, and hypocholesterolemic action, according to Hurrell and his team (Hurrell et al., 2015). Daradka et al. considered the lipid-lowering effects of *A. absinthium* in hypercholesterolemic rabbits. They discovered that an ethanolic extract of *A. absinthium* lowers blood cholesterol from 143mg/dl to 60.08 mg/dl marking a reduction by 58 times. The plant's cholestatic activity in the liver, which involves the abrogation of lipoproteins or the removal of lipid hydrolytic enzymes in the lysosomes produced by hepatocytes, is thought to be responsible for some of the plant's lipid-lowering effects. In another study, Daradka et al. displayed the efficiency of wormwood in lowering total cholesterol levels in diabetic rats by inhibiting the enzyme activities involved in producing cholesterol and reducing insulin-induced lipolysis (Daradka et al., 2014). Furthermore, Khori et al. studied that *A. Absinthium* extracts exhibit antiarrhythmic properties when used to treat supraventricular tachyarrhythmia in experimental models (the Langendorff model of isolated heart of rats). The study showed that various concentrations of the methanolic extracts of the plant had depressed effects on the Wenchebach Cycle Length (WBCL), AV Conduction Time (AVCT), and Functional and Effective Refractory Periods of the isolated heart (Khori & Nayebpour, 2007; Hurrell et al., 2015; Batiha et al., 2020).

2.10.9 Hypoglycemic Effects and Renal Properties

Farzaneh and his team reported the efficiency of wormwood extract in decreasing the harmful impact of an immunosuppressant, and azathioprine on kidney tissues (especially glomerulus, renal corpuscle, and proximal and distal urine collecting ducts) (Farzaneh et al., 2015). Further, Kharoubi et al. demonstrated that the plant's aqueous extract showed antioxidant activity in renal and hepatic tissues and restores the membrane sodium–potassium pump, calcium ATPase as well as magnesium ATPase to the normal level. High concentration of phenols and flavonoids in the extract is linked to this action. These phytochemicals also improved the renal tissue of rats treated with azathioprine and also improved oxidative damage in the liver and kidneys (Kharoubi et al., 2008). Another study in alloxan-administered diabetic rats reported the ability of different extracts of *A. absinthium* in lowering the level of kidney dysfunction by elevating the serum protein levels and lowering the range of urea and creatinine. The study further demonstrated that ethanolic extract of the plant in different concentrations displayed hypoglycemic activity in alloxan-administered diabetic rat models. 500 mg/kg and 1000 mg/kg doses resulted in higher hypoglycemic action in comparison to 250 mg/kg dose. The action may be due to phytochemicals such as thujyl alcohol, α- and β-thujones, cadinene, sabinene, azulenes, bisabolene, phellandrene, and pinene. In particular, thujone has been effective in improving insulin-activated glucose transporter by enhancing adenosine monophosphate-activated protein kinase (AMPK), which principally translocated insulin-stimulated glucose transporter type 4 (GLUT4) to the cell surface (Daradka et al., 2014).

Krebs et al. studied that wormwood extracts free from thujone can be used for the treatment of patients with immunoglobulin A (IgA) nephropathy for the management of protein levels in their urine (Krebs et al., 2010b). Hassan and his coworkers documented the mode of action of the plant extracts in lowering the level of lipids and glucose in the bloodstream in patients with type-2 diabetic mellitus with overly high cholesterol and triglycerides levels (Hassan et al., 2018).

2.10.10 Anti-Ulcer and Digestive Effects

The ethanolic, methanolic, hexane, and chloroform extracts of wormwood have been evaluated for its anti-ulcer properties in experimental ulcerogenic rats (acetylsalicylic acid induced). The results may be due to the decreased volume of gastric juice secreted and the reduced activity of the peptidase enzyme (Shafi et al., 2004). Azizi et al. supported the anti-ulcer nature of the plant extract in mice. The oral administration of the ethanolic extract was found to enhance palatability and thus affected feed intake. The extracts had no effect on the secretion and activity of mucin, but resulted in a significant elevation in the secretion of pepsin enzyme and gastric acid decreased the amount of gastric juice and lowered the rate of digestion (Azizi et al., 2016). In contrast, clinical studies by Palomino and Koch reported that oral administration of the ethanolic extracts of *A. absinthium* increases the secretion of gastric juices, bile, and intestinal juices in humans (Palomino & Koch, 2017).

Kim et al. documented the higher rate of digestibility in sheep fed with wormwood extract in comparison to control animals (Kim et al., 2002; Kim et al., 2006). Furthermore, the bitter substances in the extracts of plant were used in the treatment of ailments associated with the gastrointestinal tract. These bitter substances increased the production and secretion of bile (Kreitmair, 1951).

Hbika et al. (2022) described the effect of *A. absinthium* aqueous extracts on the pancreatic α-amylase enzyme and the intestinal α-glucosidase enzyme *in vitro* and *in vivo*. Three different methods, namely, the 2, 2-diphenyl-1picrylhydrazyl (D.P.P.H.) free radical scavenging, iron reduction, and the β-carotene bleaching method, were used to check the antioxidant property of the extract. The aqueous extract significantly suppressed the levels of pancreatic α-amylase as well as intestinal α-glucosidase, *in vitro*, while in *in vivo* experimentation in rats (normal and STZ-diabetic), there was significant attenuation of starch- and sucrose-induced hyperglycemia. Naringenin and caffeic acid are responsible for the inhibitory effect of this extract (Hbika et al., 2022).

2.10.11 Activities Related to Infectious Diseases

2.10.11.1 Antibacterial Activities

Essential oils of the *A. absinthium* plant has demonstrated a wide spectrum of inhibitory effects against, in particular, Gram-positive bacteria and other microorganisms (Fiamegos et al., 2011) *A. absinthium* ethanolic extracts showed inhibitory effects against *Staphylococcus aureus* (ATCC 29213) strain (Dülger et al., 1999). By contrast, its methanolic extract demonstrated antimicrobial effects against species such as *Bacillus subtilis, B. cereus, Salmonella typhimurium, B. cereus, S. thermophilus, and Pseudomonas putida 1617* (Sengul et al., 2011). The topical application of its hydroalcoholic extract showed wound-healing activities as it inhibited the growth of *S. aureus* at the wound site. This antibacterial property of the plant extract is well attributed to its bioactive components, as these biomolecules have been reported to increase the permeability of the biological membranes by interfering with the phospholipid composition of the membranes (Moslemi et al., 2012). Juteau and others demonstrated that this activity of *A. absinthium* was significantly higher when compared to gentamicin in a number of bacteria, viz., *S. aureus, Salmonella. typhi, E. coli, Proteus vulgaris, Klebsiella pneumoniae*, and *P. aeruginosa*. (Juteau et al., 2003; Asili et al., 2015). In addition, essential oils from *A. absinthium* were found to suppress the growth of *Listeria monocytogenes, S. aureus, and B. cereus* by decreasing the biosynthesis of DNA, proteins, RNA, and polysaccharides in the bacterial cells (Belay et al., 2011).

Two bioactive components, chlorogenic acid and 4,5-di-O-caffeoylquinic acid, have been studied to control antibiotic resistance in individual strains of various Gram-positive bacteria by inhibiting the activity of membrane efflux pumps. This study establishes that *A. absinthium* may act as a source of future drugs capable of decreasing the antibiotic resistance against pathogenic bacteria (Fiamegos et al., 2011).

Such antibacterial activities of *A. absinthium* have made it an important candidate for its use as a natural preservative for food and pharmaceutical industries.

2.10.11.2 Antiviral Activities

Antiviral action of the plant extracts have been demonstrated against the hepatitis virus and human deficiency virus-1 (HIV-1). In a study, Ansari et al. observed that after the oral administration of *A. absinthium* extract to patients with HBeAg-negative or -positive chronic hepatitis B showed antiviral activity against hepatitis B virus. The mechanism of action may significantly reduce and suppress the HBV DNA levels, and normalize alanine aminotransferase activity in a remarkable fashion with no detrimental side effects. The normalization of the inflammation and functioning of the liver was restored significantly in such patients. Further, the extracts inhibited the integrase enzyme in HIV-1 patients, thereby blocking the integration of the reversibly transcribed viral DNA into that of the host cell (Anwar et al., 1998; Ansari et al., 2018; Ahamad et al., 2019).

2.10.11.3 Antiprotozoal Activities

Antiprotozoal activities of the plant extracts have been demonstrated with regard to a number of human and livestock parasites such as *Trypanosoma cruzi, T. brucei, Leishmania infantum, Eimeria, Plasmodium, Toxoplasma, Babesia*, and *Theileria* (Fernández-Calienes Valdés et al., 2008; Nozari et al., 2016). *Eimeria* species is responsible for the occurrence of Coccidiosis in ruminants and in poultry. Aqueous extracts of wormwood have been found to be significantly effective in reducing the number of oocytes in *Eimeria*-infected broiler (Kostadinovic et al., 2012; Habibi et al., 2016). The antileishmanial activity of the plant extract was reported from a study on *Leishmania strains* (*L. donovani* and *L. aethiopica*) and also on *L infantum* (Tariku et al., 2011). Studies also show the role of essential oil against *T. cruzi* under experimental conditions (Bailen et al., 2013). The leishmanicidal and trypanocidal activities of the extracts are chiefly attributed to its flavonoids and essential oil composition (Martínez-Díaz et al., 2015). Monzote et al. reported leishmanicidal activity in a

BALB/c mouse model of experimental cutaneous leishmaniasis (Monzote et al., 2014). Intraperitoneal, subcutaneous, and oral administration of aqueous and alcoholic leaf extracts suppressed the number of parasites in the body of the mice. The aqueous extract abrogated *P. falciparum* activity; and the dilution of 1:35 showed a maximum inhibition percentage (89.9%) (Fernández-Calienes Valdés et al., 2008; Ahamad et al., 2019). Alcohol and aqueous *A. absinthium* leaf extracts also displayed antimalarial action (Nahrevanian et al., 2012; Zafar et al., 1990). This antimalarial action is displayed by artemisinin, which works by stimulating both the heme- and mitochondrial-mediated degradation cascade (Antoine et al., 2014). The plant stimulates scavenging of the free radical in the heme-mediated cascade (O'Neill et al., 2010). Additionally, artemsinin act on *P. falciparum* translationally controlled tumor protein (PfTCTP), and impairs its function by covalently binding with it (O'Neill et al., 1998; Eichhorn et al., 2013).

In a clinical trial comprising patients with amoebic dysentery, *A. absinthium* extract was administered orally in a capsular form of 500 mg concentration. Results showed relief in various clinical stages with complete eradication of the pathogen, *Entamoeba histolytica*, in 72% of cases without any side effects (Ahamad et al., 2019).

2.10.11.4 Antifungal Activities

Antifungal activities of *A. absinthium* could be linked mainly to the existence of thujone in its essential oil composition. Reports suggest a significant role of thujone, as is evident from a number of studies on extracts obtained from different areas. The Croatian *A. absinthium* essential oil exhibited increased fungicidal activity against *Candida albicans* and *Saccharomyces cerevisiae var. chevalieri* when compared with the French one, which is free from thujone (Juteau et al., 2003). Similarly, Uruguay *A. absinthium* extract, being rich in thujone, exhibited antifungal ability against *Botrytis cinerea* and *Alternaria* sp., while the fungicidal activity of *A. absinthium* essential oil from Turkey was due to camphor, 1,8-cineole, and chamazulene as main constituents (Msaada et al., 2015). More reports of potent antifungal effects leading to significant increase in the inhibition zones have been documented in studies on *Aspergillus flavus, Trichophyton mentagrophytes, Microsporum canis, A. niger, Epidermophyton floccosum, Candida albicans*, and *C. neoformans* (Ewais et al., 2014). Joshi et al. demonstrated the role of essential oils against *Micrococcus luteus, M. flavus, Bacillus subtilis, Penicillium chrysogenum*, and *A. fumigatus* (Joshi, 2013). Recently, *A. absinthium*–silver nanoparticles have been designed, experimented with, and approved for its fungicidal properties against three pathogenic microbes belonging to the genus Candida (del-Pilar Rodríguez-Torres et al., 2019).

2.10.11.5 Antihelminthic Activities

Antihelminthic activity of the plant extract is evident from a number of studies; and the essential oil constituent thujone (both α- and β- form) are reported to display such actions (Meschler & Howlett, 1999). Aqueous and ethanol extracts of the above-ground parts of *A. absinthium* displayed elevated action on gastrointestinal nematodes in sheep in comparison to known synthetic antihelminthic agents. Another study supports this finding, revealing that the ethanolic extracts of the plant showed significant activity against juvenile larvae of *Trichostrongylus colubriformis* (parasite of rabbits) and the eggs of *Ascaris suum* (porcine roundworm) (Tamargo et al., 2017; Urban et al., 2008). The methanolic extracts also display lethal action against the larvae of *Trichinella spiralis* and significantly decreases its number in various muscles in rats (Caner et al., 2008). Beshay showed the antihelminthic activity of the plant extract in *Hymenolepis* infection; the extract in a concentration-dependent manner induced damage in the integument of the worm, the accumulation of lipids, the destruction of the nephridial canal, the paralysis of the worm, and even death (Beshay, 2018).

Because of this potential, the plant could be a future replacement for synthetic agents in the management of pathogenic diseases of humans as well as animals.

2.10.11.6 Insecticidal Effects

A. absinthium has been reported to exhibit insecticidal activities. The characteristic odor of the plant makes it a natural repellent to insect and this action is due to the secretion of absinthin (sesquiterpene lactone), which reduces the development of neighboring plants (Maw et al. 1985). In addition, it was reported that *A. absinthium* essential oil possesses activity against Ascaris (Chiasson et al., 2001), insecticidal (Kordali et al., 2006), and repellent properties against insects like flies, mosquitoes, fleas and ticks (Jaenson et al., 2005; Parveen et al., 2014). Ethanolic extract displayed complete inhibitory effects against cattle tick (*Rhipicephalus microplus*) eggs hatching *in vitro* and also provided antifeedant ability on *Leptinotarsa decemnlineata* and *Rhopalosiphum padi* (Ertürk & Uslu, 2007). Govindarajan and Benelli registered that *A. absinthium* essential oils and its major constituents exhibit larvicidal activity against *Culex quinquefasciatus, Culex tritaeniorhynchus, Anopheles subpictus, Anopheles stephensi, Aedes albopictus, and Aedes aegypti* larvae (Govindarajan & Benelli, 2016). Leaf extracts in solvents such as methane, acetone, and petroleum ether had a marked toxic effect against the first instar larvae of *Chrysomya albiceps*. However, the lethal action of these extracts against the larva, pupa, and adult varied according to the solvent used in the extraction and the concentration of the extract. Larval mortality was achieved only by the petroleum ether extract at a dose of 0.4 g/ml (Lc50 dose of 0.11 g/ml). Acetone and the methanolic extract had showed delayed toxic effects on pupae resulting from the treated larvae stage. In contrast, the toxicity of all the extracts are extended to the adult insect, which resulted from the treated larval stage. The period of the larval and pupal stages also increased significantly (Alhuraysi et al., 2021) (Table 2.4).

TABLE 2.4
Pharmacological Applications of Different Bioactive Compounds as Reported by Different Studies

Pharmacological Applications	Bioactive Compound	Mechanism of Action
Antioxidant activity	Phenolic compounds (gallic acid, coumaric acid, vanillic acid, syringic acid, chlorogenic salicylic acid) and flavonoids including quercetin and rutin	Decreased lipid peroxidation level associated with decreasing thiobarbituric acid reactive substances (TBARS.) level and the recovery of endogenous antioxidant (e.g., superoxide dismutase (SOD)) glutathione (GSH) (Bora & Sharma, 2011)
Antitumor activity	Chlorogenic acid, artesunate, dihydroartesunate, artemisenin	Activation of the MEK/ERK pathway, activates the mitochondrial pathway of caspase activation, antiproliferatic activity; stimulate cell death by apoptosis (Seo et al., 2003)
Neuroprotective and antidepressant effects	Several phytochemical compounds	Inhibition of acetyl cholinesterase enzymes activity (Hoseinian et al., 2018)
Immunomodulatory and wound healing activity	Polysaccharides; β-thujone and β-pinene	Activation of NO synthesis; Free radical scavenging activity (Shanazi et al., 2015)
Hepatoprotective effects	Thujone	Suppression of liver microsomal drug-metabolizing enzymes, free radical scavenging activity, calcium channels blocking (Amat et al., 2010)
Anti-inflammatory effects	hydroxyflavone, cardomonin, caruifolin D	Suppressing the expression of the proinflammatory mediators; inhibit Tumor necrosis factor-alpha expression (Hatziieremia et al., 2006)
Antipyretic and analgesic effects	22-dien-3 Bat, 24-β-ethyl p-cholesta-7	Nicotinic and muscarinic action (Khattak et al., 1985)
Cardiovascular activity	Essential oil	Inhibiting the activities of the enzymes involved in the cholesterol biosynthesis leads to lowered cholesterol level in blood; lipolysis under the influence of insulin is also decreased

(Continued)

TABLE 2.4 (Continued)

Pharmacological Applications		Bioactive Compound	Mechanism of Action
Renal and hypoglycemic effects		α- and β-thujones, azulenes, cadinene, bisabolene, sabinene, phellandrene, pinene	High contents of total phenolic compounds and flavonoids as well as antioxidant effect of wormwood extract (Farzaneh et al., 2015; Xia et al., 2020)
Antiulcer and digestive effects		Essential oils	production and secretion of bile enhanced; insulin-activated glucose transporter activity elevated (Kim et al., 2006)
Activities against infectious diseases	Antibacterial activity	Essential oil; Chamazulene	The biosynthesis of proteins, RNA, DNA and polysaccharide in the bacterial cells are suppressed (Zanousi et al., 2016)
	Antiviral activity	Essential oil and other bioactive components	Inhibits integrase enzyme in HIV infection (Abad et al., 2012)
	Antiprotozoal activity	Essential oil, flavonoids, artemisinin	Stimulation of both heme and mitochondrial-mediated degradation cascade, inhibit PfATP6, inhibiting LDH, SAMS, PyrK, SpdSyn, OAT enzyme activities.
	Antifungal activity	Essential oil	total phenolic compounds and flavonoids present in higher concentrations (Sengul et al., 2011)
	Antihelminthic activity	α-and β-thujones	Decrease juvenile (L3) larval motility and development of egg of *Ascarissuum* in an *in vitro* model (Rodríguez-Torres et al., 2019)
	Insecticidal activity	Essential oil	Stimulation of both heme and mitochondrial-mediated degradation cascade, inhibit PfATP6, inhibiting LDH, SAMS, PyrK, SpdSyn, OAT enzyme activities (Kordali et al., 2006)

2.11 CONCLUSION

A. absinthium, also known as wormwood, belongs to genus *Artemisia*. Almost all parts of the plant are of great importance in pharmaceuticals and had been associated with traditional applications in many parts of the world. To date, phytochemical and pharmacological research demonstrates that the plant extracts shows a remarkable efficiency to be used as a neuro-protective, hepatoprotective, antioxidant, antidepressant, anticancer, antimicrobial, antifungal, antihelminthic, acaricidal, antimalarial, and insecticidal agent.

Artemisia effortlessly acclimatizes to an extensive range of temperatures, but it flourishes best in wet and damp soils. This species adapts and survives in both hot and cold temperature environments; hence, it is found in most parts of the globe. Its potential for adaptability provides the plant with an extremely valued defensive mechanism.

The occurrence of thujone and its analogs exerts neurotoxic effects upon the long-term use of *A. absinthium*. Reports further demonstrate that an intake of a greater amount of *A. absinthium* has been associated with major central nervous system disorders, and show symptoms such as intestinal cramps, vomiting, headaches, and dizziness.

A. absinthium is indexed into the category of critically endangered species, and therefore coherent efforts should be made to protect the species. Efforts should also be put forward to improve and modernize the cultivation of the plant on a large scale.

This plant has vast potential for scientists and investigations will lead to the discovery of many more novel components which could be used as potent drugs for the prevention and treatment of critical illnesses.

REFERENCES

Abad, M. J., Bedoya, L. M., Apaza, L., & Bermejo, P. (2012). The *Artemisia* L. Genus: a review of bioactive essential oils. *Molecules*, *17*(3), 2542–2566.

Ahamad, J., Ali, M., & Mir, S. R. (2013). New glycoside esters from the aerial parts of *Artemisia absinthium* Linn. *The Natural Products Journal*, *3*(4), 260–267.

Ahamad, J., Ali, M., & Mir, S. R. (2014). Isoflavone glycosides from aerial parts of *Artemisia absinthium* Linn. *Chemistry of Natural Compounds*, *49*(6), 696–700.

Ahamad, J., Mir, S. R., & Amin, S. (2019). A Pharmacognostic review on *Artemisia absinthium*. *International Research Journal of Pharmacy*, *10*(1), 25–31.

Ahameethunisa, A. R., & Hopper, W. (2010). Antibacterial activity of *Artemisia nilagirica* leaf extracts against clinical and phytopathogenic bacteria. *BMC Complementary and Alternative Medicine*. *10*, 1–6

Ahmad, F., Khan, R. A., & Rasheed, S. (1992). Study of analgesic and anti-inflammatory activity from plant extract of *Lactuca scariola* and *Artemisia absinthium*. *Journal of Islamic Academy of Sciences*, *5*(2), 111–114.

Alhuraysi, A. M. S., Elsheikh, T. M. Y., & Abu El-Ghiet, U. M. (2021). The insecticidal activity of *Artemisia absinthium* leaves Extracts against Blowfly, *Chrysomya albiceps* (Calliphoridae). *Egyptian Academic Journal of Biological Sciences, F. Toxicology and Pest Control*, *13*(1), 115–127.

Amat, N., Upur, H., & Blažeković, B. (2010). In vivo hepatoprotective activity of the aqueous extract of *Artemisia absinthium* L. against chemically and immunologically induced liver injuries in mice. *Journal of Ethnopharmacology*, *131*(2), 478–484.

Anibogwu, R., Jesus, K. D., Pradhan, S., Pashikanti, S., Mateen, S., & Sharma, K. (2021). Extraction, isolation and characterization of bioactive compounds from *Artemisia* and their biological significance: A review. *Molecules*; *26*(22), 6995.

Ansari, S., Shamshi, Y., & Khan, Q. A. (2019). A review of *Artemisia absinthium*, Linn.(afsanteen) with special reference of Unani medicine. *Journal of Pharmaceutical and Scientific Innovation*, *8*(1), 11–18.

Ansari, S., Siddiqui, M. A., Malhotra, S., & Maaz, M. (2018). Antiviral efficacy of qust (*Saussurea lappa*) and afsanteen (*Artemisia absinthium*) for chronic Hepatitis B: A prospective single-arm pilot clinical trial. *Pharmacognosy Research*, *10*, 282.

Antoine, T., Fisher, N., Amewu, R., O'Neill, P. M., Ward, S. A., & Biagini, G. A. (2014). Rapid kill of malaria parasites by artemisinin and semi-synthetic endoperoxides involves ROS-dependent depolarization of the membrane potential. *The Journal of antimicrobial chemotherapy*, *69*, 1005–1016.

Anwar, M., Hakim, M., & Siddiqui, M. (1998). Clinical efficacy of *Artemisia absinthium* Linn. viral Hepatitis with specific reference to ejection fraction of heart. *Hamdard Medicus*, *41*, 93–95.

Appalasamy, S., Lo, K. Y., Ch'ng, S. J., Nornadia, K., Othman, A. S., & Lai-Keng, C. (2014). Antimicrobial activity of artemisinin and precursor derived from in vitro plantlets of *Artemisia* annua L. *BioMed Research International*, *2014*, 215872.

Asili, J., Emami, S. A., Eynolghozat, R., Noghab, Z. S., Bazzaz, B. S. F., & Sahebkar, A. (2015). Chemical composition and in vitro efficacy of essential oil of seven *Artemisia* species against ESBL producing multidrug-resistant Escherichia coli. *Journal of Essential Oil Bearing Plants*, *18*, 124–145.

Azizi, K., Moemenbellah, F. S., Fard, Q. A., & Mohammadi-Samani, S. (2016). Antiulcer activity after oral administration of the wormwood ethanol extract on lesions due to Leishmania major parasites in BALB/C mice. *Asian Journal of Pharmaceutical Research and Health Care*, *8*, 33–41.

Bailen, M., Julio, L. F., Diaz, C. E., Sanz, J., Martínez-Díaz, R. A., Cabrera, R., Burillo, J., & Gonzalez-Coloma, A. (2013). Chemical composition and biological effects of essential oils from *Artemisia absinthium* L. cultivated under different environmental conditions. *Industrial Crops and Products*, *49*, 102–107.

Basta, A., Tzakou, O., Couladis, M., & Pavlović, M. (2007). Chemical composition of *Artemisia absinthium* L. from Greece. *Journal of Essential Oil Research*, *19*, 316–318.

Batiha, G. E. S., Olatunde, A., El-Mleeh, A., Hetta, H. F., Al-Rejaie, S., Alghamdi, S., & Rivero-Perez, N. (2020). Bioactive compounds, pharmacological actions, and pharmacokinetics of wormwood (*Artemisia absinthium*). *Antibiotics*, *9*(6), 353.

Bäumler, S. (2007). *Heilpflanzen Praxis Heute*. Urban and Fischer.

Bedini, S., Flamini, G., Cosci, F., Ascrizzi, R., Echeverria, M. C., Guidi, L., et al. (2017). *Artemisia* spp. essential oils against the disease-carrying blowfly *Calliphora vomitoria*. *Parasites Vectors*, *10*(1), 1–10.

Beigh, Y. A., & Ganai, A. M. (2017). Potential of wormwood (*Artemisia absinthium* Linn.) herb for use as additive in livestock feeding: A review. *The Pharma Innovation Journal*, *6*, 176–187.

Belay, G., Tariku, Y., Kebede, T., Hymete, A., & Mekonnen, Y. (2011). Ethnopharmacological investigations of essential oils isolated from five Ethiopian medicinal plants against eleven pathogenic bacterial strains. *Phytopharmacology*, *1*, 133–143.

Beshay, E. V. N. (2018). Therapeutic efficacy of *Artemisia absinthium* against Hymenolepis nana: In vitro and in vivo studies in comparison with the anthelmintic praziquantel. *Journal of Helminthology*, *92*(3), 298–308

Bora, K. S., & Sharma, A. (2010). Neuroprotective effect of *Artemisia absinthium* L. on focal ischemia and reperfusion-induced cerebral injury. *Journal of Ethnopharmacology*, *129*, 403–409.

Bora, K. S., & Sharma, A. (2011). Evaluation of antioxidant and free-radical scavenging potential of *Artemisia absinthium*. *Pharmaceutical Biology*, *49*(12), 1216–1223.

Boudjelal, A., Smeriglio, A., Ginestra, G., Denaro, M., & Trombetta, D. (2020). Phytochemical Profile, Safety Assessment and Wound Healing Activity of *Artemisia absinthium* L. *Plants*, *9*(12), 1744.

Canadanovic-Brunet, J. M., Djilas, S. M., Cetkovic, G. S., & Tumbas, V. T. (2005). Free-radical scavenging activity of wormwood (*Artemisia absinthium* L) extracts. *Journal of the Science of Food and Agriculture*, *85*(2), 265–272.

Caner, A., Doskaya, M., Degirmenci, A., Can, H., Baykan, S., Uner, A., Basdemir, G., Zeybek, U., & Guruz, Y. (2008). Comparison of the effects of *Artemisia vulgaris* and *Artemisia absinthium* growing in western Anatolia against trichinellosis (*Trichinella spiralis*) in rats. *Experimental Parasitology*, *7*, 81–85.

Cao, J., Yang, M., Cao, F., Wang, J., & Su, E. (2017). Well-designed hydrophobic deep eutectic solvents as green and efficient media for the extraction of artemisinin from *Artemisia annua* leaves. *ACS Sustainable Chemistry and Engineering*, *5*, 3270–3278

Capuzzo, A., Maffei, M. E., & Occhipinti, A. (2013). Supercritical fluid extraction of plant flavors and fragrances. *Molecules*, *18*(6), 7194–7238.

Chiasson, H., Bélanger, A., Bostania, N., Vincent, C., & Poliquin, A. (2001). Acaricidal properties of *Artemisia absinthium* and *Tanacetum vulgare* (Asteraceae) essential oils obtained by three methods of extraction. *Journal of Economic Entomology*, *94*, 167–171.

Choi, S. C., Choi, E. J., Oh, H. M., Lee, S., Lee, J. K., Lee, M. S., Shin, Y. I., Choi, S. J., Chae, J. R., & Lee, K. M. (2006). DA-9601, a standardized extract of *Artemisia* asiatica, blocks TNF-α-induced IL-8 and CCL20 production by inhibiting p38 kinase and NF-κB pathways in human gastric epithelial cells. *World Journal of Gastroenterology*, *12*, 4850.

Craciunescu, O., Constantin, D., Gaspar, A., Toma, L., Utoiu, E., & Moldovan, L. (2012). Evaluation of antioxidant and cytoprotective activities of *Arnica montana* L. and *Artemisia absinthium* L. ethanolic extracts. *Chemistry Central Journal*, *6*(1), 1–11.

Daradka, H. M., Abas, M. M., Mohammad, M. A., & Jaffar, M. M. (2014). Antidiabetic effect of *Artemisia absinthium* extracts on alloxan-induced diabetic rats. *Comparative Clinical Pathology*, *23*, 1733–1742.

del-Pilar Rodríguez-Torres, M., Acosta-Torres, L. S., Díaz-Torres, L. A., Padrón, G. H., García-Contreras, R., & Millán-Chiu, B. E. (2019). *Artemisia absinthium*-based silver nanoparticles antifungal evaluation against three Candida species. *Materials Research Express*, *6*(8), 085408.

Dülger, B., Ceylan, M., Alitsaous, M., & Uğurlu, E. (1999). Antimicrobial Activity of *Artemisia absinthium* L. *Turkish Journal of Biology*, *23*, 377–384.

Eichhorn, T., Winter, D., Büchele, B., Dirdjaja, N., Frank, M., Lehmann, W. D., Mertens, R., Krauth-Siegel, R. L., Simmet, T., & Granzin, J. (2013). Molecular interaction of artemisinin with translationally controlled tumor protein (TCTP) of *Plasmodium falciparum*. *Biochemical Pharmacology*, *85*, 38–45.

Ekiert, H., Swiątkowska, J., Knut, E., Klin, P., Rzepiela, A., Tomczyk, M., & Szopa, A. (2021). *Artemisia dracunculus* (Tarragon): A review of its traditional uses, phytochemistry and pharmacology. *Frontiers of Pharmacology*, *12*, 653993.

Ertürk, Ö., & Uslu, U. (2007). Antifeedant, growth and toxic effects of some plant extracts on *Leptinotarsa decemlineata* (Say.) (*Coleoptera, Chrysomelidae*). *Fresenius Environmental Bulletin*, *16*(6), 602–608.

European Food Safety Authority. (2014). Outcome of the consultation with Member States and EFSA on the basic substance application for *Artemisia absinthium* for use in plant protection as fungicide in wheat and as nematicide and insecticide in vegetables. *EFSA Supporting Publications*, *11*(10), https://doi.org/10.2903/sp.efsa.2014.en-665

Ewais, E., Aly, M. M., Ismail, M., Abdel Shakour, E., & Hassanin, M. (2014). Antibacterial, antifungal, antitumor and toxicity of essential oils of *Salvia officionalis*, *Thymus vulgaris, Eugenia caryophyllata* and *Artemisia absinthium*. *Scientific Journal of Flowers and Ornamental Plants*, *1*, 265–274.

Fahd, A. N., Omar, M. N., Ramzi, A. M., Ali, S. A., & Abdullah, A. A. (2020). Cytotoxic, antimicrobial and antioxidant activities and phytochemical analysis of *Artemisia judaica* and *A. sieberi* in Saudi Arabia. *African Journal of Pharmacy and Pharmacology*, *14*(8), 278–284

Farzaneh, F., Ebrahim, H. S., & Akbar, V. (2015). Investigating on effect of wormwood extract on reduction of renal toxicity in treated rats by azathioprine. *Biomedical and Pharmacology Journal*, *8*(1), 291–299.

Fernández-Calienes Valdés, A., Mendiola Martínez, J., Scull Lizama, R., Vermeersch, M., Cos, P., & Maes, L. (2008). In vitro anti-microbial activity of the Cuban medicinal plants *Simarouba glauca* DC, *Melaleuca leucadendron* L and *Artemisia absinthium* L. *Memórias do Instituto Oswaldo Cruz*, *103*(6), 615–618.

Ferreira, J. F., Luthria, D. L., Sasaki, T., & Heyerick, A. (2010). Flavonoids from *Artemisia annua* L. as antioxidants and their potential synergism with artemisinin against malaria and cancer. *Molecules*, *15*(5), 3135–3170.

Fiamegos, Y. C., Kastritis, P. L., Exarchou, V., Han, H., Bonvin, A. M., Vervoort, J., Lewis, K., Hamblin, M. R., & Tegos, G. P. (2011). Antimicrobial and efflux pump inhibitory activity of caffeoylquinic acids from *Artemisia absinthium* against gram-positive pathogenic bacteria. *Public Library of Science ONE*, *6*, e18127

Gallego, R., Montero, L., Cifuentes, A., Ibáñez, E., & Herrero, M. (2018). Green extraction of bioactive compounds from microalgae. *Journal of Analysis and Testing*, *2*, 109–123.

Govindarajan, M., & Benelli, G. (2016). *Artemisia absinthium*-borne compounds as novel larvicides: Effectiveness against six mosquito vectors and acute toxicity on non-target aquatic organisms. *Parasitology Research*, *115*, 4649–4661.

Grycyk, R. A., Kireev, I. V., Struk, O. A., & Ivanochko, V. M. (2019). Investigation of acute toxicity and hepatoprotective effect of extracts of *Artemisia vulgaris* L. and *Artemisia absinthium* L. herbs. *Medical and Clinical Chemistry*, *4*, 147–155.

Guarrera, P. M. (1999). Traditional antihelmintic, antiparasitic and repellent uses of plants in Central Italy. *Journal of Ethnopharmacology*, *68*, 183–192.

Habibi, H., Firouzi, S., Nili, H., Razavi, M., Asadi, S. L., & Daneshi, S. (2016). Anticoccidial effects of herbal extracts on *Eimeria tenella* infection in broiler chickens: In vitro and in vivo study. *Journal of Parasitic Diseases*, *40*, 401–407.

Hallal, N., Kharoubi, O., Benyettou, I., Tair, K., Ozaslan, M., & Aoues, A. (2016). In vivo amelioration of oxidative stress by *Artemisia absinthium* L. administration on mercuric chloride toxicity in brain regions. *Journal of Biological Sciences*, *16*, 167–177.

Hashimi, A., Siraj, M. B., Ahmed, Y., Siddiqui, M. A., & Jahangir, U. (2019). One for All-*Artemisia absinthium* (Afsanteen) "A potent unani drug". *TANG Humanitas Medicine*, *9*(4), 51–59.

Hassan, M., Niazi, A. T., Khan, S., & Gul, F. (2018). Antidiabetic and antihyperlipidemic effects of *Artemisia absinthium* L., *Citrullus colocynthis* (L.) Schrad. and *Gymnema sylvestre* (Retz.) R. Br. ex Sm. on type II diabetes hyperlipidemic patients. *Indian Journal of Traditional Knowledge*, *17*, 233–239.

Hassanzadeh, M. K., Tayarani Najaran, Z., Nasery, M., & Emami, S. A. (2016). Tarragon (*Artemisia dracunculus* L.) oils. In V. R. Preedy (Ed.), *Essential Oils in Food Preservation, Flavor and Safety* (pp. 813–817). Elsevier.

Hatziieremia, S., Gray, A. I., Ferro, V. A., Paul, A., & Plevin, R. (2006). The effects of cardamonin on lipopolysaccharide-induced inflammatory protein production and MAP kinase and NFκB signalling pathways in monocytes/macrophages. *British Journal of Pharmacology*, *149*, 188–198.

Hbika, A., Daoudi, N. E., Bouyanzer, A. H., Bouhrim, M., Mohti, H., Loukili, E. H., Mechchate, H., Al-Salahi, R., Fahd, A. N., Bnouham, M., & Zaid, A. (2022). *Artemisia absinthium* L. aqueous and ethyl acetate extracts: Antioxidant effect and potential activity in vitro and in vivo against pancreatic α-amylase and intestinal α-glucosidase. *Pharmaceutics*, *14*(3), 481.

Hoseinian, A., Moslemi, H. R., & Sedaghat, R. (2018). Antioxidant properties of *Artemisia absinthium* accelerate healing of experimental Achilles tendon injury in rabbits. *Herba Polonica*, *64*(1), 36–43

Hurrell, J. A., Puentes, J. P., & Arenas, P. M. (2015). Medicinal plants with cholesterol-lowering effect marketed in the Buenos Aires–La Plata conurbation, Argentina: An urban ethnobotany study. *Ethnobiology and Conservation*, *4*, https://doi.org/10.15451/ec2015-9-4.7-1-19

Hussain, A. (2020). The genus *Artemisia* (Asteraceae): A review on its ethnomedicinal prominence and taxonomy with emphasis on foliar anatomy, morphology, and molecular phylogeny. *Proceedings of the Pakistan Academy of Sciences: B. Life and Environmental Sciences*, *57*(1), 1–28

Hussain, M., Raja, N. I., Akram, A., Iftikhar, A., Ashfaq, D., Yasmeen, F., Mazhar, R., Imran, M., & Iqbal, M. (2017). A status review on the pharmacological implications of *Artemisia absinthium*: A critically endangered Plant. *Asian Pacific Journal of Tropical Disease*, *7*(3), 185–192

Jaenson, T. G. T., Pålsson, K., & Borg-Karlson, A. K. (2005). Evaluation of extracts and oils of tick-repellent plants from Sweden. *Medical and Veterinary Entomology*, *19*, 345–352.

Jaleel, G. A. R. A., Abdallah, H. M. I., & Gomaa, N. E. S. (2016). Pharmacological effects of ethanol extract of Egyptian *Artemisia* herba-alba in rats and mice. *Asian Pacific Journal of Tropical Biomedicine*, *6*(1), 44–49.

Joshi, R. K. (2013). Volatile composition and antimicrobial activity of the essential oil of *Artemisia absinthium* growing in Western Ghats region of North West Karnataka, India. *Pharmaceutical Biology*, *51*, 888–892.

Joshi, R. K., Satyal, P., & Setzer, W. (2016). Himalayan aromatic medicinal plants: A review of their ethnopharmacology, volatile phytochemistry, and biological activities. *Medicines*, *3*(1), 6.

Juteau, F., Jerkovic, I., Masotti, V., Milos, M., Mastelic, J., Bessiere, J. M., & Viano, J. (2003). Composition and antimicrobial activity of the essential oil of *Artemisia absinthium* from Croatia and France. *Planta Medica*, 69, 158–161.

Kharoubi, O., Slimani, M., Aoues, A., & Seddik, L. (2008). Prophylactic effects of wormwood on lipid peroxidation in an animal model of lead intoxication. *Indian Journal of Nephrology*, *18*, 51.

Khattak, S. G., Gilani, S. N., & Ikram, M. (1985). Antipyretic studies on some indigenous Pakistani medicinal plants. *Journal of Ethnopharmacology*, *14*, 45–51.

Khori, V. & Nayebpour, M. (2007). Effect of *Artemisia absinthium* on electrophysiological properties of isolated heart of rats. *Journal of Physiology and Pharmacology*, *10*; 303–311.

Kim, J., Kim, C. H., & Ko, Y. D. (2002). Influence of dietary addition of dried wormwood (*Artemisia* sp.) on the performance and carcass characteristics of Hanwoo steers and the nutrient digestibility of sheep. *Asian Australasian Association of Animal Production Societies*, *15*, 390–395.

Kim, S. C., Adesogan, A. T., Kim, J. H., & Ko, Y. D. (2006). Influence of replacing rice straw with wormwood *(Artemisia montana)* silage on feed intake, digestibility and ruminal fermentation characteristics of sheep. *Animal Feed Science and Technology*, *128*, 1–13.

Kordali, S., Aslan, I., Çalmaşur, O., & Cakir, A. (2006). Toxicity of essential oils isolated from three *Artemisia* species and some of their major components to granary weevil, *Sitophilus granarius* (L.) (*Coleoptera: Curculionidae*). *Industrial Crops and Products*, *23*, 162–170.

Kostadinovic, L., Levic, J., Galonja-Coghill, T., & Ruzicic, L. (2012). Anticoccidian effects of the *Artemisia absinthium* L. extracts in broiler chickens. *Archivos de Zootecnia*, *15*, 69.

Koyuncu, I. (2018). Evaluation of anticancer, antioxidant activity and phenolic compounds of *Artemisia absinthium* L. Extract. *Cellular and Molecular Biology*, *64*(3), 25–34.

Krebs, S., Omer, B., Omer, T. N., and Fliser, D. (2010a). Wormwood (*Artemisia absinthium*) for poorly responsive early-stage IgA nephropathy: A pilot uncontrolled trial. *American Journal of Kidney Diseases*, *56*, 1095–1099.

Krebs, S., Omer, T. N., & Omer, B. (2010b). Wormwood (*Artemisia absinthium*) suppresses tumour necrosis factor alpha and accelerates healing in patients with Crohn's disease—A controlled clinical trial. *Phytomedicine*, *17*(5), 305–309

Kreitmair, H. (1951). *Artemisia absinthium* L., the true wormwood. *Die Pharmazie*, *6*, 27–28.

Lachenmeier, D. W., Walch, S. G., Padosch, S. A., & Kröner, L. U. (2006). Absinthe—A review. *Critical Reviews in Food Science and Nutrition*, *46*, 365–377.

Lee, H. G., Kim, H., Oh, W. K., Yu, K. A., Choe, Y. K., Ahn, J. S., Kim, D. S., Kim, S. H., Dinarello, C. A., & Kim, K. (2004). Tetramethoxy hydroxyflavone p7F downregulates inflammatory mediators via the inhibition of Nuclear Factor κB. *Annals of the New York Academy of Sciences*, *1030*, 555–568.

Lockie, A. (2006). *Encyclopedia of Homeopathy*. DK Publishing New York.

Mahmoudi, M., Ebrahimzadeh, M., Ansaroudi, F., Nabavi, S., & Nabavi, S. (2009). Antidepressant and antioxidant activities of *Artemisia absinthium* L. at flowering stage. *African Journal of Biotechnology*, *8*, 7170–7175.

Mamatova, A. S., Korona-Glowniak, I., Skalicka-Woźniak, K., Józefczyk, A., Wojtanowski, K. K., Baj, T., Sakipova, Z. B., & Malm, A. (2019). Phytochemical composition of wormwood (*Artemisia gmelinii*) extracts in respect of their antimicrobial activity. *BMC Complementary and Alternative Medicine*, *19*(1), 288.

Manganelli, R. E. U., Chericoni, S., & Baragatti, B. (2000). Ethnopharmacobotany in Tuscany: Plants used as antihypertensives. *Fitoterapia*, *71*, S95–S100.

Maobe, M. A., Gatebe, E., Gitu, L., & Rotich, H. (2013). Preliminary phytochemical screening of eight selected medicinal herbs used for the treatment of diabetes, malaria and pneumonia in Kisii region, southwest Kenya. *European Journal of Applied Sciences*, *5*(10), 01–06.

Martín, J., Torrell, M., Korobkov, A. A., & Vallès, J. (2003). Palynological features as a systematic marker in *Artemisia* L. and related genera (*Asteraceae, Anthemideae*)-II: Implications for Subtribe Artemisiinae delimitation. *Plant Biology*, *5*, 85–93.

Martín, L., Julio, L., Burillo, J., Sanz, J., & Mainar, A. (2011). Comparative chemistry and insect antifeedant action of traditional (Clevenger and Soxhlet) and supercritical extracts (CO2) of two cultivated wormwood (*Artemisia absinthium* L.) populations. *Industrial Crops and Products*, *34*(3), 1615–1621

Martínez-Díaz, R. A., Ibáñez-Escribano, A., Burillo, J., Heras, L. D. L., Prado, G. D., Agulló-Ortuño, M. T., Julio, L. F., & González-Coloma, A. (2015). Trypanocidal, trichomonacidal and cytotoxic components of cultivated *Artemisia absinthium* Linnaeus (*Asteraceae*) essential oil. *Memórias do Instituto Oswaldo Cruz*, *110*, 693–699.

Maw, M., Thomas, A., & Stahevitch, A. (1985). The biology of Canadian weeds: 66. *Artemisia absinthium* L. *Canadian Journal of Plant Science*, *65*, 389–400.

Merajuddin, K., Mousaa, A. A., Syamasundarb, K. V., & Alkhathlan, H. Z. (2012). Determination of chemical constituents of leaf and stem essential oils of *Artemisia monosperma* from Central Saudi Arabia. *Natural Product Communications*, *7*(8), 1079–1082.

Meschler, J. P., & Howlett, A. C. (1999). Thujone exhibits low affinity for cannabinoid receptors but fails to evoke cannabimimetic responses. *Pharmacology Biochemistry and Behavior*, *62*, 473–480.

Mohammadian, A., Moradkhani, S., Ataei, S., Shayesteh, T. H., Sedaghat, M., Kheiripour Grycyk, N., & Ranjbar, A. (2016). Antioxidative and hepatoprotective effects of hydroalcoholic extract of *Artemisia absinthium* L. in rat. *Journal of Herbmed Pharmacology*, *5*(1), 29–32.

Monzote, L., Piñón, A., Scull, R., & Setzer, W. N. (2014). Chemistry and leishmanicidal activity of the essential oil from *Artemisia absinthium* from Cuba. *Natural Product Communications*, *9*, 1799–1804.

Moslemi, H. R., Hoseinzadeh, H., Badouei, M. A., Kafshdouzan, K., & Fard, R. M. N. (2012). Antimicrobial activity of *Artemisia absinthium* against surgical wounds infected by Staphylococcus aureus in a rat model. *Indian Journal of Microbiology*, *52*, 601–604.

Msaada, K., Salem, N., Bachrouch, O., Bousselmi, S., Tammar, S., Alfaify, A., Al Sane, K., Ben Ammar, W., Azeiz, S., & Marzouk, B. (2015). Essential oils and phenolics. *Journal of Chemistry*, *2015*, 1–12.

Nahrevanian, H., Sheykhkanlooye Milan, B., Kazemi, M., Hajhosseini, R., Soleymani Mashhadi, S., & Nahrevanian, S. (2012). Antimalarial effects of Iranian flora *Artemisia sieberi* on *Plasmodium berghei* In vivo in mice and phytochemistry analysis of its herbal extracts. *Malaria Research and Treatment*, *2012*, 1–8.

Nalbantsoy, A., Erel, Ş. B., Köksal, Ç., Göçmen, B., Yildiz, M. Z., & Karabay Yavaşoĝlu, N. Ü. (2013). Viper venom induced inflammation with *Montivipera xanthina* (Gray, 1849) and the anti-snake venom activities of *Artemisia absinthium* L. in rat. *Toxicon*, *65*, 34–40.

Nazeri, M., Mirzaie-Asl, A., Saidijam, M., & Moradi, M. (2020). Methanolic extract of *Artemisia absinthium* prompts apoptosis, enhancing expression of Bax/Bcl-2 ratio, cell cycle arrest, caspase-3 activation and mitochondrial membrane potential destruction in human colorectal cancer HCT-116 cells. *Molecular Biology Reports*, *47*, 8831–8840.

Nibret, E. & Wink, M. (2010). Volatile components of four Ethiopian *Artemisia* species extracts and their in vitro antitrypanosomal and cytotoxic activities. *Phytomedicine*, *17*, 369–374.

Nozari, S., Azadmehr, A., Jahanihashemi, H., Nassiri, A. M., Adine, M., Javadi, F., Hajiaghaee, R., Shahnazi, M., & Saraei, M. (2016). In vitro Anti-toxoplasma effects of ethanolic extracts of *Artemisia absinthium* L., *Carum copticum* L. and *Gossypium hirsutum*. *Journal of Medicinal Plants*, *2*, 72–79,

O'Neill, P. M., Barton, V. E., & Ward, S. A. (2010). The molecular mechanism of action of artemisinin—The debate continues. *Molecules*, *15*, 1705–1721.

O'Neill, P. M., Pan, X. Q., Hossler, P. A., Walker, D. J., Yowell, C. A., Carlton, J., Dame, J. B., & Meshnick, S. R. (1998). The *Plasmodium falciparum* translationally controlled tumor protein homolog and its reaction with the antimalarial drug artemisinin. *Journal of Biological Chemistry*, *273*, 16192–16198.

Obolskiy, D., Pischel, I., Feistel, B., Glotov, N., & Heinrich, M. (2011). *Artemisia dracunculus* L. (tarragon): A critical review of its traditional use, chemical composition, pharmacology, and safety. *Journal of Agricultural and Food Chemistry*, *59*(21), 11367–11384.

Orege, J. I., Adeyemi, S. B., Tiamiyu, B. B., Akinyemi, T. O., Ibrahim, Y. A., & Orege, O. B. (2021). *Artemisia* and *Artemisia*-based products for COVID-19 management: Current state and future perspective. *Advances in Traditional Medicine*, *23*, 85–96.

Palomino, O., & Koch, J. (2017). *Assessment Report on Artemisia absinthium L., herba*. European Medicines Agency. https://docplayer.net/90203895-Assessment-report-on-*Artemisia-absinthium*-l-herba.html

Parekh, H. S., Liu, G., & Wei, M. Q. (2009). A new dawn for the use of traditional Chinese medicine in cancer therapy. *Molecular Cancer*, *8*, 21

Parveen, S., Godara, R., Katoch, R., Yadav, A., Verma, P., Katoch, M., & Singh, N. (2014). In vitro evaluation of ethanolic extracts of *Ageratum conyzoides* and *Artemisia absinthium* against cattle tick, *Rhipicephalus microplus*. *The Scientific World Journal*, *2014*, 858973.

Rashidi, R., Ghorbani, A., Rakhshandeh, H., & Mousavi, S. H. (2021). Protective effect of *Artemisia absinthium* on 6-hydroxydopamine-induced toxicity in SH-SY5Y cell line. *Avicenna Journal of Phytomedicine*, *11*(3), 238–246.

Sengul, M., Ercisli, S., Yildiz, H., Gungor, N., Kavaz, A., & Çetin, B. (2011). Antioxidant, antimicrobial activity and total phenolic content within the aerial parts of *Artemisia absinthium*, *Artemisia santonicum* and *Saponaria officinalis*. *Iranian Journal of Pharmaceutical Research*, *10*, 49.

Seo, J. M., Kang, H. M., Son, K. H., Kim, J. H., Lee, C. W., Kim, H. M., Chang, S. I., & Kwon, B. M. (2003). Antitumor activity of flavones isolated from *Artemisia argyi*. *Planta medica*, *69*(3), 218–222.

Shafi, G., Hasan, T. N., Syed, N. A., Al-Hazzani, A. A., Alshatwi, A. A., Jyothi, A., & Munshi, A. (2012). *Artemisia absinthium* (AA): A novel potential complementary and alternative medicine for breast cancer. *Molecular biology reports*, *39*(7), 7373–7379.

Shafi, N., Khan, G. A., & Ghauri, E. G. (2004). Antiulcer effect of *Artemisia absinthium* L. in rats. *Pakistan Journal of Scientific and Industrial Research*, *47*, 130–134.

Shahnazi, M., Azadmehr, A., Hajiaghaee, R., Mosalla, S., & Latifi, R. (2015). Effects of *Artemisia absinthium* L. extract on the maturation and function of dendritic cells. *Jundishapur Journal of Natural Pharmaceutical Products*, *10*, e20163.

Sharopov, F. S., Salimov, A., Numonov, S., Bakri, M., Sangov, Z., Habasi, M., Aisa. H. A., & Setzer, W. N. (2020). Phytochemical study on the essential oils of tarragon (*Artemisia dracunculus* L.) growing in Tajikistan and its comparison with the essential oil of the species in the rest of the world. *Natural Product Communications*, *15*, 1–7.

Sharopov, F. S., Sulaimonova, V. A., & Setzer, W. N. (2012). Composition of the Essential oil of *Artemisia absinthium* from Tajikistan. *Records of Natural Products*, *6*, 127–134.

Sultan, M. H., Zuwaiel, A. A., Moni, S. S., Alshahrani, S., Alqahtani, S. S., Madkhali, O., & Elmobark, M. E. (2020). Bioactive principles and potentiality of hot methanolic extract of the leaves from *Artemisia absinthium* L″ in vitro cytotoxicity against Human MCF-7 breast cancer cells, antibacterial study and wound healing activity. *Current Pharmaceutical Biotechnology*, *21*(15), 1711–1721.

Szopa, A., Pajor, J., Klin, P., Rzepiela, A., Elansary, H. O., Fahed, A. Al-Mana., Mattar, M. A., & Ekiert, H. (2020). *Artemisia absinthium* L.—Importance in the history of medicine, the latest advances in phytochemistry and therapeutical, cosmetological and culinary uses. *Plants*, *9*(9), 1063.

Taleghania, A., Emamib, S. A., & Tayarani-Najaran, Z. (2020). *Artemisia*: A promising plant for the treatment of cancer. *Bioorganic and Medicinal Chemistry*, *28*(1), 115180.

Tamargo, B., Monzote, L., Piñón, A., Machín, L., García, M., Scull, R., & Setzer, W. N. (2017). In vitro and in vivo evaluation of essential oil from *Artemisia absinthium* L. formulated in nanocochleates against cutaneous leishmaniasis. *Medicines*, *4*, 38.

Tariku, Y., Hymete, A., Hailu, A., & Rohloff, J. (2011). In vitro evaluation of antileishmanial activity and toxicity of essential oils of *Artemisia absinthium* and *Echinops kebericho*. *Chemistry and Biodiversity*, *8*, 614–623.

Urban, J., Kokoska, L., Langrova, I., & Matejkova, J. (2008). In vitro anthelmintic effects of medicinal plants used in Czech Republic. *Pharmaceutical Biology*, *46*, 808–813.

Wright, C. W. (2001). *Artemisia. Medicinal and Aromatic Plant—industrial profile*. CRC Press.

Xia, M., Liu, D., Liu, Y., & Liu, H. (2020). The therapeutic effect of *Artemisinin* and its derivatives in kidney disease. *Frontiers in pharmacology*, *11*, 380.

Yu, H., & Zhong, S. (2001). *Artemisia species in traditional Chinese medicine and the discovery of artemisinin*. CRC Press.

Zafar, M., Hamdard, M., & Hameed, A. (1990). Screening of *Artemisia absinthium* for antimalarial effects on *Plasmodium berghei* in mice: A preliminary report. *Journal of Ethnopharmacology*, *30*, 223–226.

Zanousi, M. B. P., Nekoei, M., & Mohammadhosseini, M. (2016). Composition of the essential oils and volatile fractions of *Artemisia absinthium* by three different extraction methods: hydrodistillation, solvent-free microwave extraction and headspace solid-phase microextraction combined with a novel QSRR Evaluation. *Journal of Essential Oil Bearing Plants*, *19*(7), 1561–1581.

Zeng, K. W., Liao, L. X., Song, X. M., Lv, H. N., Song, F. J., Yu, Q., Dong, X., Jiang, Y., & Tu, P. F. (2015). Caruifolin D from *Artemisia absinthium* L. inhibits neuroinflammation via reactive oxygen species-dependent c-jun N-terminal kinase and protein kinase c/NF-κB signaling pathways. *European Journal of Pharmacology*, *767*, 82–93.

Zeraati, F., Esna-Ashari, F., Araghchian, M., Emam, A. H., Rad, M. V., Seif, S., & Razaghi, K. (2014). Evaluation of topical antinociceptive effect of *Artemisia absinthium* extract in mice and possible mechanisms. *African Journal of Pharmacy and Pharmacology*, *8*(19), 492–496.

3 Atis (*Annona squamosa*)

Bharti Mittu, Zerina Begum, Sandaldeep Kaur, Mahaldeep Kaur, Abida Bhat, Neha Sharma, and Raees ul Haq

3.1 INTRODUCTION

The custard apple tree, *Annona squamosa* L. (Annonaceae), commonly called "Atis", is a tropical fruit tree. However, because of the edible character of the crop, it is grown all across India. As a result, it has been widely cultivated across the tropics. Because of the more outstanding sugar content, approximately 58% of dry mass in the fruit pulp, it has a very sweet flavor, and it is evident that the pulp of fruit pulp has a greater calorific value. The *Annona* genus contains 70 species, three of which bear fruit: *Annona squamosa, Annona reticulata, and Annona muricata* (Figure 3.1). Morphological features were used to distinguish the three species (Kumar et al., 2021; Hosseinabadi, 2021).

Annona squamosa is the major class of flowering plants belonging to the Annonaceae family, and is native to the West Indies and America. The fruit is modest and evergreen, medium in size and cultivated in many areas of India. According to the World Health Organization, traditional medicines are used by 80% of the population in developing nations to treat ailments and provide primary health care. Researchers in the literature reveal that every portion of the plant has traditional therapeutic uses. It's good for heart disease, diabetes, hyperthyroidism, cancer, ulcers, and wounds, among other things. This plant is known to have a variety of medical benefits.

Furthermore, the crumpled leaves were sniffed to alleviate hysteria and fainting spells and smeared to ulcers and wounds as healing properties (Ma et al., 2017; Narwade and Aher, 2019). The sugar apple, *Annona squamosa* L., which is also a member of the Annonaceae family, is native to the tropics. The tree is between 3 and 6 meters tall, with an open crown of uneven branches and zigzagging twigs. The fruit is dark brown in color and has depressions, which give it a quilted appearance; the pulp is very soft, sweet and reddish-yellow in color. The seeds contain considerable amounts of proteins, lipids, and carbohydrates, making them suitable for use in food and feed. More ever the presence of phenolic compounds within the seeds provide is related to their antioxidant potential. The oil seeds are rich in unsaturated fatty acids, principally oleic and linoleic, and a good source of total tocopherol. Apple seed oil could be used as a raw material in the manufacturing process of biodiesel. It is a tiny tropical tree, and a well-known Annonaceae cultivar. It has a number of different names: Custard apple, sugar apple, and fruta do conde. It contains poisonous seeds and have a variety of uses, most of which are traditional. They have a greater amount of annonaceous acetogenins (ACGs). The most effective inhibitors of mitochondrial complex I are related to this group of polyketides (Bermejo et al., 2005).

3.2 OCCURRENCE

Herbal medicine has a long history, dating back to the dawn of civilization. According to ancient sources, plants were traditionally employed for medical resolutions in Greece, China, Egypt and India long before the Christian era (Sandeep and Mittal, 2017). Its popular name, Sitaphal, highlights how the fruit has a fascinating history. According to legend, Sita, the wife of Lord Rama, is reported to have eaten this fruit during her vanvaas (period spent in the jungle). Another legend recounts that when Ravana abducted Sita during the Vanvaas period, the tear drops from her eyes to

 DOI: 10.1201/9781003292340-3

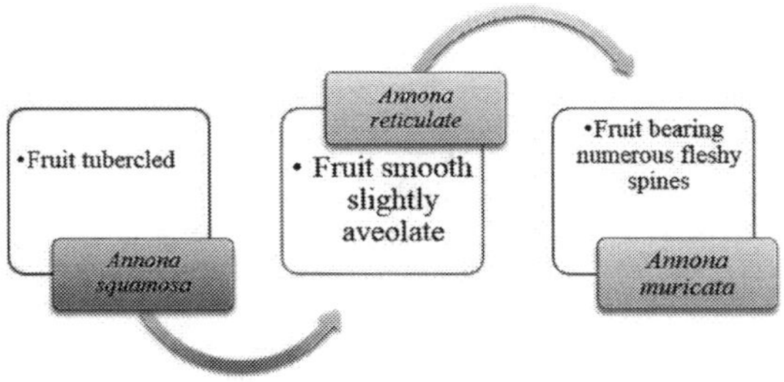

FIGURE 3.1 Some of the common species of *Annona* genus.

the ground, leading to the development of Stephan trees in the woods. This name derives from Sanskrit, where "sheet" means cold and "phal" means fruit; excess eating can make you an individual feel cold, and as it has a cooling impact on the bodies, therefore the term "Sitaphal" (Kaur et al., 2015; Narwade and Aher, 2019). The plant is a semi-deciduous, small tree with a height of 3–7 m, having broad, irregularly spreading branches, an open crown, and light brown bark with scars. The inner color of bark is light yellow and is slightly bitter. The twigs were brown in color, mottled with light brown dots. The leaves of the plant occur singly and are around 11 x 5cm long; with pale green color on both surfaces. The soft aggregate fruit generated from a flower's multiple loosely joined pistils is distinct from other members of the genus. Each pistil has its own tubercle, usually 1.3–1.9 cm in length and 0.6–1.3 cm wide. The fruit is round, heart-shaped, ovate or conical, and 5–10 cm in diameter. Flowers are yellowish-green, having fragrance with slender hairy stalks, produced singly or in short lateral clusters about 2.5 cm in length, and 2–4 flowers. The pulp is white and delicious with a powerful scent, the bloom is white and powdery, and the seeds, oblong, shiny and smooth, and either blackish or white, are inserted in each carpel.

3.3 SYSTEMIC CLASSIFICATION

The taxonomic classification of *Annona squamosa* is shown in Figure 3.2.

3.4 MORPHOLOGICAL FEATURES OF THE PLANT

The height of atis is dependent on the age of the plant; fully grown, mature ones can grow to as tall as 8 to 10 meters in height (Chopra et al., 1998; Pinto et al., 2005; Handique et al., 2022). Evident and notable differences in the major dimensions of the leaf and the length of petiole are present.

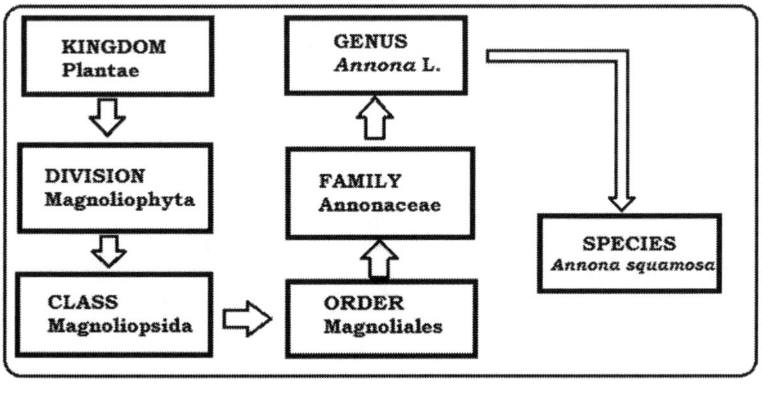

FIGURE 3.2 Taxonomic classification of *Annona squamosa*.

The plant has a lanceolate leaf blade, an acute leaf base, and acumlinate apex with an entire leaf margin. The length of the petiole also varies according to their species and age; it can be roughly up to 1.5 cm, whereas the length of the leaf may reach up to 25 cm (Kumar et al., 2018; Ghawade et al., 2018). The fruit shape is described as being cordate; that is, roughly spherical to oval. The range of fruit color is from greenish to yellowish. Just as with the variations in leaf size, the fruit size also varies considerably. As reported by different studies, it ranges up to 10 cm in length. Although this can be influenced by several factors, including the number of fruits, soil fertility, and the hours of sunlight, it can also be dependent on hormonal changes. Gibberellins released by the seeds are also known to contribute, or control, the final fruiting size (Thakur and Singh, 1967; Bakane et al., 2015; Chandel et al., 2018). In addition to environmental and external factors, fruit weight shows direct significance with the primary dimensions. More than the external factors, the genetic makeup of the plant is known to dominate the pulp color, which can be either light or dark creamy white (Bhatnagar et al., 2012). The colors of then seeds can be black or brown, or a combination of the two, and their number within the fruit can reach to around 90 per fruit; the seed dimensions are totally dependent of the fruit dimensions (Wongs-Aree and Noichinda, 2011; Satyanarayana et al., 2013; Handique et al., 2022).

3.5 NUTRITIONAL ATTRIBUTES OF THE PLANT

The nutritional content of *Annona squamosa* is shown in Table 3.1.

3.5.1 Carbohydrates

A. squamosa has the highest concentration of glucose in its leaves, seeds, and fruits. The fructose content of Annona in seeds, fruits and leaves is reported as being highest at Menofia and Alexandria, whereas in Giza and Mansoura the fructose content is highest in fruits, followed by in the seeds and the leaves. The leaves contains the lowest amounts of sucrose. At Giza and Mansoura, the highest concentration of total soluble solids (TSS) was detected in fruits, while the lowest concentration was observed in the leaves. The highest amount was detected in seeds in Menofia, while TSS content was distributed in ascending order in the seeds, leaves, and fruits at Alexandria. Increased glucose concentration in the plant's growing portions shows metabolic regulation linked to increased enzyme activity, which helps plants to survive harsh environmental circumstances and also promotes growth. The greatest TSS levels were found in plant sections growing in sand-rich soils subjected to water stress due to inadequate capacity of the soil to hold water. The stress-induced accumulation of high soluble sugar levels has also been observed in the shoots of various plant species. Fruits had the highest level of polysaccharides, while the leaves found in Giza and Alexandria had the lowest,

TABLE 3.1
Nutritional content (per 100 g fresh fruit) of *Annona squamosa* (Suchitra and Shrinivasan, 2015)

Energy	104 k Cal
Protein	1.6 g
Fat	0.4 g
Fibre	3.1 g
Carbohydrate	23.5 g
Phosphorous	47 g
Calcium	17 g
Iron	4.37 g
Vitamin C	37 mg

whereas seeds had the lowest levels in Mansoura and Menofia. At Giza and Mansoura, the total levels of carbohydrates were found in the following order: fruits > seeds > leaves. However, in terms of order Menofia showed the pattern leaves > fruits > seeds, while Alexandria had seed > fruits > leaves. TSS levels in the fruits in Giza and Menofia were often higher than in Alexandria and Mansoura. Meanwhile, Mansoura and Giza had a higher level of total carbohydrates in their fruits than in Menofia and Alexandria. The fibres contain cellulose, hemicellulose, and pectic substances.

3.5.2 Vitamins

The highest level of vitamin C was found in seeds from *A. squamosa* growing in Menofia, Alexandria, and Mansoura, followed by leaves and fruits. By contrast, plants growing in Giza had the largest content in leaves, followed by fruits, and, finally, the seeds. Menofia's fruits have a higher vitamin C concentration than encountered in other locations. Squamosa has a high content of vitamin C. The custard apple has been observed to have a high ascorbic acid concentration of 34–44 mg per 100 g. Folic acid was detected in the leaves of *A. squamosa*, followed by fruits found in each of the four zones. By contrast, Mansoura's fruits have a higher folic acid level than the others.

3.5.3 Phenolics

The highest levels of flavonoids and total antioxidants were found in *A. squamosa* growing in Menofia, Alexandria, and Mansoura. Fruits contain the most phenols, followed by leaves and seeds. On the other hand, plants growing near Giza have the highest seed content, followed by leaves and fruits. Menofia had the greatest total flavonoid and phenol levels in its fruits (Kalidindi et al., 2015).

3.5.4 Protein

Annona squamosa is a tropical shrub with great pharmaceutical abilities for the treatment of cardiac disease, thyroid disorders, cancer, and diabetes. The extracts of ASL (Annona squamosal leaf) showed many phytochemical activities, such as carbohydrates, proteins, alkaloids, flavonoids, phenolics, glycosides and saponins. Previous literature reported that ASLs contain the highest protein content compared to the fruit and seeds. ASL from Menofia contains 13.47 mg/g fresh weight of protein content; however, 17.26 mg/g FW protein content was noted in ASLs from Mansoura. Thus, a higher amount of protein in an extract of ASL added the nutritive importance of the food that is used by humans as well as animals. Furthermore, aqueous and methanolic fractions of ASLs demonstrated a higher content of proteins and amino acids. The biuret test confirms the amino acid and protein in the aqueous extract of ASLs. However, Millon's test confirms proteins and amino acids in the methanolic extracts of ASL. However, limited studies have been carried to quantify the protein and amino acid in ASLs (Kalyani et al., 2019; Shukry et al., 2019).

3.5.5 Minerals and Vitamins

Annona's leaves contain many minerals, such as vitamins, namely, A, C, E, B_1, B_2, B_3, B_9, magnesium, copper, iron, phosphorus, calcium, potassium, zinc, sodium, and selenium. These minerals are essential to the maintenance of a healthy body, as they assist in the performance of numerous actions, such as the regulation of healthy bones and teeth, relaxation and muscle contraction and blood clotting, the regulation of blood pressure, nerve function, a healthy immune system, the metabolism of energy, and many enzymes. Macro- and microminerals, namely, Se (in trace amounts), Zn (0.455 mg/100 g), Mg (65.65 mg/100 g), Cu (0.312 mg/100 g), and P (43.1 mg/100 g), were found in extracts of ASLs (Akram et al., 2020; Varadharaj et al., 2014).

ASL were found to contain vitamins such as vitamin B and C in a proportion of 8.12–11.98 and 12–17 g/g FW, respectively. It also found to contains various other vitamins, such as vitamin B1

(0.065 mg/100 g), B_2 (0.155 mg/100 g), B_3 (1.105 mg/100 g), C (34.4 mg/100 g), and A and E (in trace amounts). ASLs were reported to contain around 3.5% of vitamin C. These nutrients help to regulated numerous processes of the body, such as bone development, the maintenance of strong skin, the development of epithelial tissue, improves the sense of sight, immune response, the healing of wounds, and the strength of connective tissues. Several enzymes, such as vitamins B_1, B_2, B_3, and B_9 are the cofactors leading to oxidation reduction as well as the metabolism of carbohydrates (Himesh et al., 2012; Varadharaj et al., 2014; Akram et al., 2020; Kumar et al., 2021).

3.6 PHYTOCHEMICAL OVERVIEW OF PLANT

3.6.1 Profile of Essential Oil

In India, *Annona squamosal* has been cultivated for its roots, leaves, stems, and fruits because they have been associated with numerous nutraceutical and pharmaceutical applications. In addition, ASLs extract contain an essential oil that possess excellent antimalarial and antiparasitic properties. Previous studies reported the chemical composition of the essential oil present in ASLs. Using the hydro distillation method, a yield of nearly 0.13% is collected from shade-dried leaves. Approximately 88.6% of the total amount of oil is collected from the leaves of the plant. Studies demonstrated the presence of sesquiterpenoids, followed by monoterpenes extracted from ASL essential oil after investigation using gas chromatography (GC)–mass spectrometry (GC-MS) and GC–flame ionization detection. Cadinol (9.4%), cadinene (11.2%) and (E)-caryophyllene (15.9%) were the main components among the sesquiterpenes. The composition of ASLs essential oil (ASLEO) was found to be varied across the globe. For example, Brazilian ASLEO contains germacrene D (17.1%) and (E)-caryophyllene (27.4%) as the predominant components, whereas in Egyptian ASLEO, linalool (7.7%), diacetyl (9.3%) and carvone (24.9%) were found to be the major constituents. Similarly, caryophyllene (11.6–24.5%), limonene (0.8–11.7%), and pinene (1.0–11.9%) were reported to be the main components found in ASLEO from Vietnam. Around 0.12% (v/w) essential oil has been extracted from fresh leaves of *Annona squamosa* using the hydrodistillation method. GC and GC/MS analyses of ASLEO represented 86% of the total oil extracted. In addition, germacrene D (21.3%) and caryophyllene (23%) were reported to be the main constituents from ASLEO of the northern region of India, whereas caryophyllene (14.1%) and cedrene (23.3%) were more prevalent in the southern region of India (Verma et al., 2016; Meira et al., 2015; Kumar et al., 2021)

3.6.2 Secondary Metabolite Profile

The leaves of *Annona squamous* are considered to be insignificant agricultural waste products, which is, why they are either discarded or burnt. However, they have been used in many countries, such as Laos, India, Malaysia, and Vietnam, as a traditional treatment for the curing of problems related to heart, dysuria, fever, dysentery, fainting, worm infections, hemorrhage, and constipation. ASL extracts consist of many phytochemicals that are associated with various medicinal applications, such as phenols, antidiabetic, antioxidant, antiviral, antidiarrheal, antimicrobial, antiobesity, and antitumor activities. Acetogenins, fatty acids, alkaloids, quinones, flavonoids, terpenoids, saponins, glycosides, tannins, diterpenes, sesquiterpenes, anthocyanins, steroids, and phenols were the main phytochemicals extracted from ASLs. Studies reported the applications of polyphenolics against many chronic diseases such as cancer, diabetes, neurodegenerative and cardiovascular diseases. These polyphenolics helps in maintaining many biochemical and physiological procedures, such as enzymatic activity, cell proliferation, signal transduction pathways, and the potential for cellular redox (Nguyen et al., 2020; Yadav, 2019; Katole et al., 2018; Soni et al., 2018).

3.7 STRUCTURAL CHARACTERIZATION OF THE BIOACTIVE COMPONENTS

The *Annona* plant as a whole is a pool of bioactive components, but the maximum contents are extracted from the leaves and seeds. The major constituents include amino acids, phenolic compounds, acetogenins (ACGs), cyclopeptides (CPs), alkaloids (ALKs), essential oils (EO), and carotenoids (Madona et al., 2021; Bhardwaj et al., 2020). Essential oil composition was examined using gas chromatography-flame ionization detector (GC-FID) and GC-mass spectrometry (GC-MS). These two techniques were employed to isolate the volatile compounds of sugar apple fruit (Hemalatha et al., 2017; Adesanwo et al., 2021).

3.7.1 Acetogenins

These natural products basically consist of long aliphatic chains of fatty acids with one or more rings of tetrahydrofuran (THF), oxygenated (ketone, epoxides, hydroxyls) moieties, double bonds, and butyrolactone at the end of the hydrocarbon chain (Bhardwaj et al., 2020; Okhale et al., 2016). These can be classified further based on the spatial arrangement of the tetrahydrofuran rings presented in Figure 3.3.

3.7.2 Alkaloids

These secondary metabolites containing nitrogen have low molecular weights and are categorized related to the carbon–nitrogen ring, such as pyrrole, isoquinoline, piperidine, pyridine, and

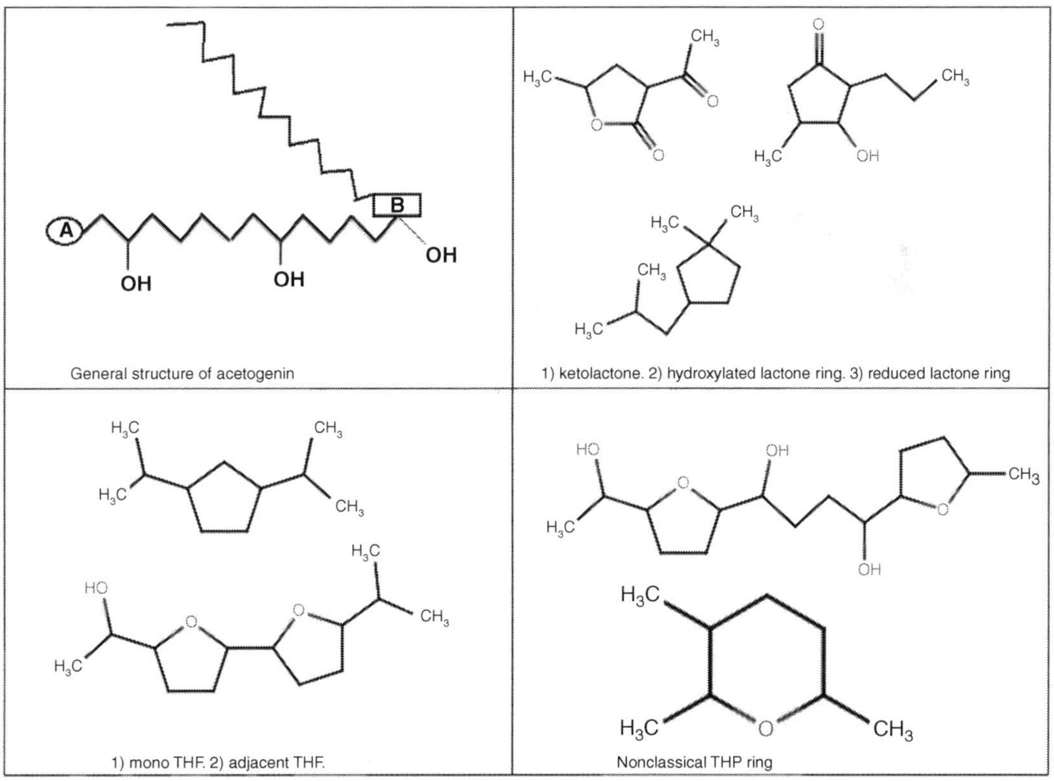

FIGURE 3.3 General structure of acetogenin (The A in acetogenin may contain ketolactone, or hydroxylated lactone ring or reduced lactone ring and the B in structure may contain mono THF or adjacent THF or nonadjacent THF or non THF or nonclassical THP ring).

aporphine alkaloids and indole, followed by derivatives of oxoapophine, are majorly present in *Annona* species (Nugraha et al., 2019). The structures of some of the essential alkaloids extracted from *Annona* spp. are given in Figure 3.4, 3.5 and 3.6.

3.7.3 Essential Oils

Several portions of the *Annona* plant, principally the leaves, have shown the presence of essential oils. Terpenes and phenylpropenes are the two major constituents of a mixture of oils in *Annona*, and sesquiterpenes derivatives are among the major essential oils shown in Figure 3.7 (Bhardwaj et al., 2020).

3.7.4 Other Compounds

Phenolic compounds are the other critical secondary metabolites found in various species of *Annona* (Hemalatha et al., 2017). These compounds consist of at least one unit of phenol and, on the basis of their structures, they can be classified into different subcategories, such as the tannins, stilbenes,

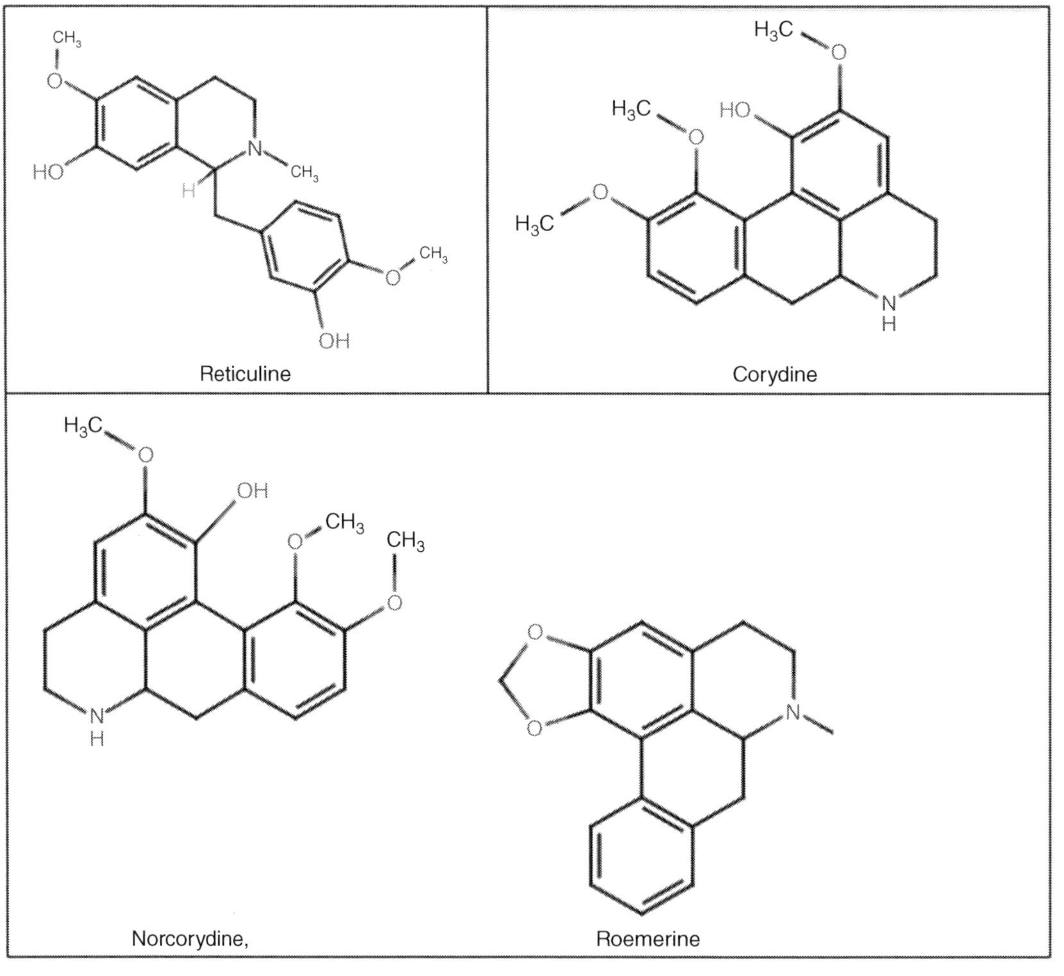

FIGURE 3.4 Reticuline, corydine, norcorydine and roemerine alkaloids from *Annona* spp.

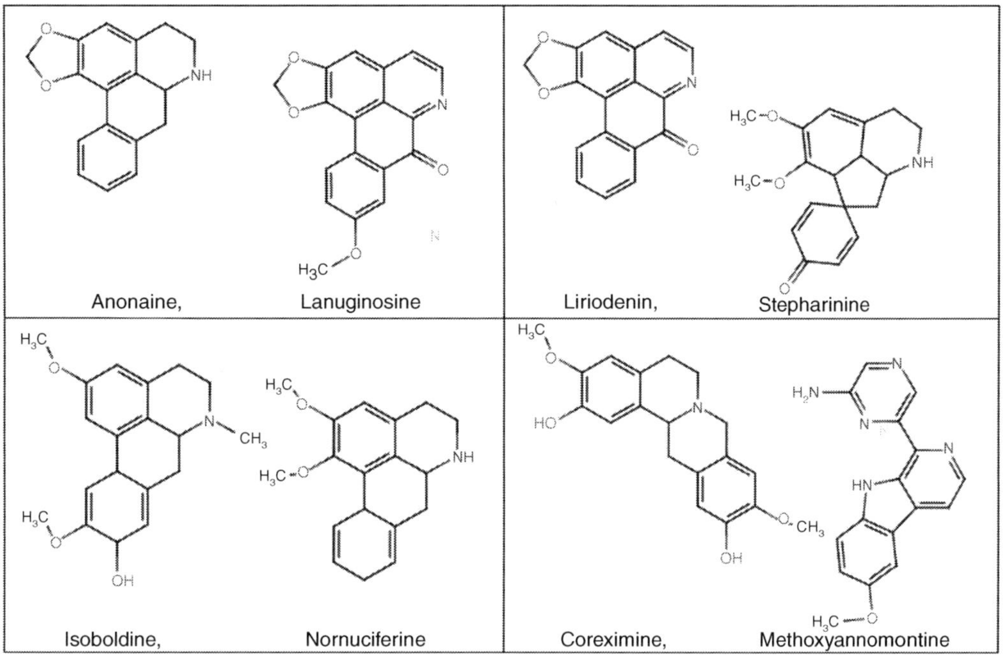

FIGURE 3.5 Structure of anonaine, lanuginosine, liriodenin, stepharinine, isoboldine, nornuciferine, coreximine and methoxyannomontine alkaloids found in *Annona* spp.

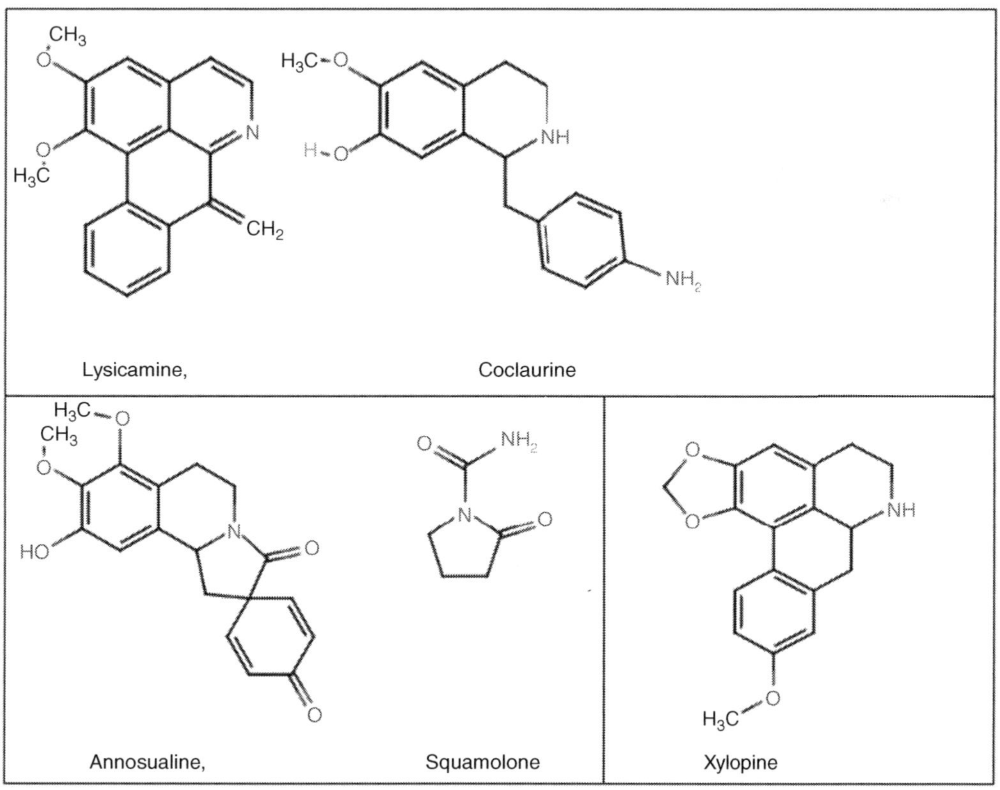

FIGURE 3.6 Structural skelton of lysicamine, coclaurine, annosqualine, squamolone and xylopine alkaloids of *Annona*.

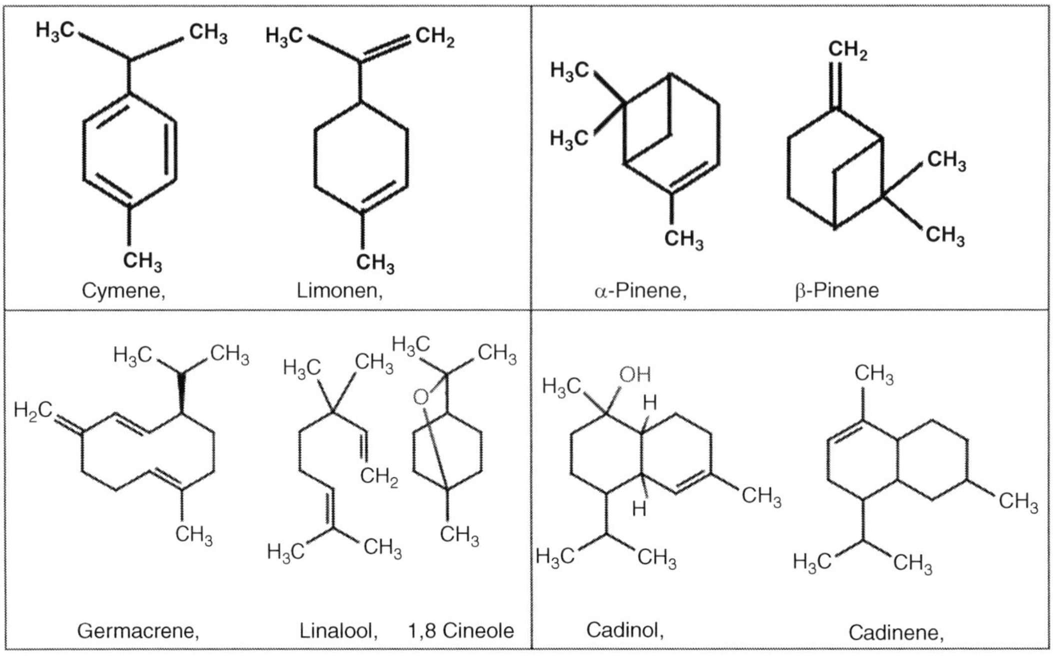

FIGURE 3.7 Structures of essential oil components found in the *Annona* species (cymene, limonen, α-pinene, β-pinene, germacrene, linalool, 1,8 cineole, cadinol, and cadinen).

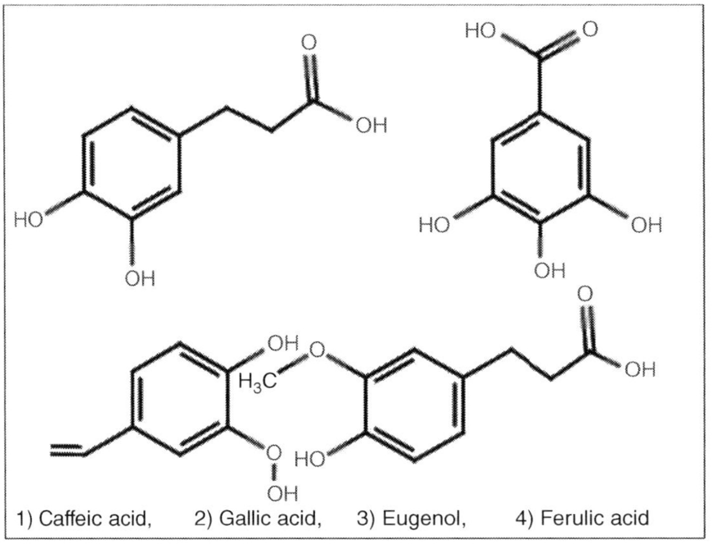

FIGURE 3.8 Structure of caffeic acid, gallic acid, eugenol, and ferulic acid and other compounds from *Annona* species

lignans, phenolic acids, flavonoids, quinines, and so on, depicted in Figure 3.8 (Kumar et al., 2021). The other nutrients found in the *Annona* species are vitamins, carotenoids and amides. The leaves, seeds, and fruit pulp are key sites for the buildup of vitamins and carotenoids (Correa-Gordillo et al., 2012; Vijayameena et al., 2013). The plant contains concentrations of various minerals, such as vitamins, potassium (K), zinc (Zn), iron (Fe), sodium (Na), calcium (Ca), selenium (Se), copper

(Cu), magnesium (Mg), A, C (ascorbic acid), E, B_1 (thiamine), B_2 (riboflavin), B_3 (niacin), and B_9 (folic acid) (Kumar et al., 2021). Amino acids, steroids, glycosides, anthocyanins, carbohydrates, and minerals are also found in *A. reticulata* and *A. senegalensis* (Jamkhande and Wattamwar, 2015). Amides, purines, β-cryptoxanthin, sterols, lutein, β-carotene, phytosterols, and phospholipids are present in *A. cherimoya* (Guillopé et al., 2011).

3.8 EXTRACTION AND IDENTIFICATION OF BIOACTIVE COMPONENTS

Studies reported the influence of various solvents, such as ethyl alcohol, hexane and methanol, for the extraction of antioxidants from various plants. Solvents of different polarities have been used to isolate phenolic compounds from plants accurately.

Air-dried powder of ASL extracts was macerated with 70:30 hydroethanol and kept in cold conditions for 72 h. The extracts were then passed through a Whattman filter paper No. 42 (125 mm) and the organic layer was collected and evaporated. The collected extracts were then concentrated with a water bath temperature of 40 °C, using a rotary evaporator, followed by the lyophilization of crude extracts into paste and analyzed further (Bhardwaj et al., 2020).

3.8.1 Gas Chromatography-Mass Spectrum Analysis (GC-MS)

Literature reported that the phytochemicals in the *Annona squamosa* leaf extracts were identified using the GC-MS technique. Extracts were analyzed using gas chromatography-mass spectrometer (GC-MS) with Elite-1 fused silica as a capillary column and consisting of 100% dimethyl poly siloxane. A 70eV ionization energy was used for the detection, and helium gas with a flow rate of 1.51ml/min. was used as carrier gas. Temperature programming of 70 °C (isothermal for 2 min.), with an increase of 300 °C for 10 min. was used; there was a total GC running time of 35 min. The concentration was quantified by comparing its average peak area to the total areas. National Institute of Standards and Technology (NIST) database was used for the identification of bioactive components. The mass spectrum of unknown compounds was compared with the known compounds to identify bioactive components (Hemlatha and Satyanarayana, 2015).

3.9 APPLICATION OF THE ACTIVE INGREDIENTS

3.9.1 Anti-Tumor Activity

Cancer is the leading cause of death on the planet. Researchers have recently focused on the anti-tumor capabilities of herbs' bark, pericarp, and seeds and active plant compounds have been identified for their anticancer properties. In H22 xenograft-bearing mice, the seed oil has an anti-tumor effect, with a maximal inhibitory rate (MIR) of 53.54% when administered orally. The key chemical elements of the anti-tumor action, unsaturated fatty acids, downregulate the Interleukin-6/Jak/Stat3 signalling pathway. Annonaceous acetogens from *Annona squamosa* have been identified as anticancer agents. Seed oil included the anticancer compounds 12, 15-cissquamostatin-A and bullatacin, which were found to have strong anticancer effects in human cancer cell lines and in mouse H22 cells.

3.9.2 Antimalarial Activity

Roemerolidine, N-nitroxylopine, and duguevalline are the alkaloids extracted from *Annona* which have been associated with antimalarial activity (Saha, 2011). The entire components which are present in atis showed modest antiplasmodial activity against a chloroquine-sensitive strain (D_{10}). And a chloroquine-resistant strain (Dd2) and *Annona squamosa* signifies a main source of a novel antimalarial compound as proved by the antimalarial agent quinine and artemisinin, which were discovered in chinchona species (Kumar et al., 2021; Tylor et al., 2011).

3.9.3 Antidiabetic Action

The previous literature revealed that the antihyperglycemic effect of extracts of atis root was investigated by inducing streptozocin (STZ) in Albino wistar rats with diabetes mellitus and an insufficiency of insulin. Previous literature reported that giving an aqueous extract of *Annona squamosa* to diabetic rats for 30 days helps to reduce the glucose level in blood, urea, uric acid, and creatinine levels, but that it also increased insulin peptide activity, the globulin albumin ratio, and the restoration of all marker enzymes to near-control levels (Kumar et al., 2021; Pandey and Barve, 2011)

3.9.4 Antiulcer Activity

The active ingredients of *Annona squamosa* twigs, like N-methylcorydaldine, (+)-O-methylarmepavine and isocorydine, have antisecretory properties that protect against peptic ulcers. It inhibits the H+-k+ ATPase pump and lowers gastric acidity, pepsin, and gastrin levels. Similarly, leaf extract protects mice from developing ulcers caused by aspirin and pyloric ligation (Win and Myat, 2017).

3.9.5 Antioxidant Activity

In Taiwan, ferric reducing antioxidant powder (FRAP) assay evaluated the antioxidant activity in the mature fruit of different varieties. Many research studies in India found that *Annona squamosa, Annona cherimola*, and *Annona muricata* extracts have a great level of antioxidant properties (Kaur et al., 2015; Okhale et al., 2016).

3.9.6 Insecticidal Activity

The presence of active principles identified in *Annona squamosa* leaf also appears responsible for some insecticidal activities. Because this plant's extract is a contact toxin for insects, it can easily penetrate into the body wall and tracheal system, causing insect death, and thereby showing insecticidal activity against *Sitophilusoryzae*. The use of insecticides from a plant origin may reduce the health risks associated with synthetic pesticides, and the herb's great level of insecticidal action appears to back up its traditional claims (Kumar et al., 2010).

3.9.7 Vasorelaxant Activity

Vasodilators are drugs that help to treat cerebral vasospasm and hypertension as well as helping in the improvement of peripheral circulation. A cyclic octapeptide, cyclosquamosin B, isolated from the seeds of *Annona squamosal*, was discovered to have a possible vasorelaxant effect on the rat aorta. The suppression of calcium influx via voltage-dependent calcium channels from the extracellular space may be a major contributor to cyclosquamosin's vasorelaxant activities (Kumar et al., 2021).

3.9.8 Hepatoprotective Activity

The hepatoprotective efficacy of *Annona squamosa* alcoholic and water extracts was investigated in hepatotoxic rats in order to see if it may be used to treat hepatotoxicity in humans. The ASLs extracts were unable to reverse the hepatic injury caused by isoniazid and rifampicin in the experimental investigation. However, they were able to help to reduce the influence of these medications in the liver. The antioxidative action of the flavonoids found in *Annona squamosal* leaf extracts may be responsible for its protective role (Kumar et al., 2021; Okhale et al., 2016).

3.10 PHARMACOLOGICAL APPLICATIONS

3.10.1 Antimicrobial Activity

Antimicrobials collected from plants have significant prophylactic qualities and are thought to be economical and safe replacements to synthetic antimicrobials, which have more noticeable adverse effects. *Annona squamosal* also contains a wide range of pharmacological qualities, including antibacterial activity, due to the presence of alkaloids, flavonoids, phytosterols, oils, phenols saponins, and glycosides. *Annona squamosal* leaf extracts have been found to exhibit exceptional antimicrobial activity in various studies. An active acetogenin component called annotemoyin extracted from leaf extract of chloroform and some flavonoid compounds separated from aqueous leaf extract of the plant both showed significant levels of antibacterial activity. Acetogenins, such as squamostatin, cholesteryl glucopyranoside, and squamocin, inhibited the growth of gram-positive bacteria, like *Staphylococcus aureus*, *Bacillus megaterium*, *Bacillus cereus*, and *Bacillus subtilis* (Gleye et al., 1998; Okhale et al., 2016).

3.10.2 Anticancer Activity

Human tumors can avoid apoptosis that ultimately results in development of different types of cancer and tumor in the body. One key reason for the failure of therapy failure is cancer cells' strong resistance to apoptosis in response to a relevant stimulation. As a result, various cancer treatment techniques, such as radiation therapy and chemotherapy, rely heavily on the apoptosis of cancer. Natural pharmaceutical pharmaceuticals have been developed due to the extended past of employing natural items such as ethnomedicines, which are affordable and have fewer side effects that some expensive synthetic treatments (Madona et al., 2021). In the field of anticancer medicines, naturally generated bioactive chemicals with apoptosis-inducing properties have attracted a lot of attention in recent decades (Okhale et al., 2016).

3.10.3 Molluscicidal Properties

Schistosomiasis is an endemic, parasitic disease which primarily affects tropical and subtropical regions; it results from the activities of the Schistosomamansoni worm in the liver of infected individuals. Molluscicides were used in prophylactic treatment to disrupt the worm's evolutionary cycle while simultaneously destroying the snail *B. glabrata* (intermediate host). Biomphalariaglabrata adult forms and egg masses were tested using ethanolic extracts of *Annona squamosa*. This has been utilized as a poisonous agent against the snail since traditional times, and experimental tests of the seed, root, stem, bark, and leaf's ethanolic extract have also shown molluscicidal action against the adult snail.

3.10.4 Genotoxic Effect

Because chemical pesticides pose a risk to humans and the environment, biopesticides have increasingly taken their place. In this case, an *Annona squamosa* seed extract yielded the chemical isosquamocin, which could be employed as a good insecticide for plant protection. The comet assay and other relevant investigations were used to assess the compound's genotoxicity, which demonstrated that *A. squamosa*'s genotoxicity and biochemical effects may not pose a significant risk to humans. However, doses must be further confirmed by the development of various mutagenic assays in order to ensure that humans are not exposed to unnecessary health risks (Katole et al., 2018).

3.10.5 Larvicidal Activity

Mosquitoes pose a significant hazard to human health due to their ability to transmit deadly diseases. The development of resistance, cross-resistance, rising costs, and potential toxicity problems resulting from the use of synthetic insecticides are just a few of the factors that have sparked attention in the innovation of plant-based solutions in recent years. *Annona squamosa* has been shown to have both larvicidal and growth-regulating properties against *Anstephensi* and other mosquitos. The great efficacy of Atis as a mosquito larvicide has been examined; however, the active ingredient that has a poisonous substance against larval species has yet to be identified (Hemlatha & Satyanarayana, 2015).

3.10.6 Antihelmintic Activity

Among the gastronomic delights *Haemonchuscontortus* is the most common and pathogenic species of gastrointestinal nematode, causing a significant death rate in young animals, especially during the rainy season. Artificial anthelmintics are frequently used to decrease the damages mentioned above. Because of the presence of active compounds with parasiticidal effects, *Annona squamosa* seed powder is employed against insects in Northeast Brazil. The antihelmintic action of extracts and isolated components from *A. squamosa* seeds was tested on *H. contortus* egg hatching. The structure of compound isolated from the ethyl acetate extract revealed by spectroscopic analysis as a C_{37} tri-hydroxy adjacent bistetrahydrofuran acetogenin and it reduced the hatching of *H. contortus* eggs at a concentration of about 25 mg/ml. The anthelmintic efficacy of seed extract of atis against the adult earthworm *Pheritimaposthuma* was also studied, and it was discovered that the methanolic extract had an anthelmintic effect, resulting in the death of the earthworm (Okhale et al., 2016) (Figure 3.9).

3.11 CONCLUSION

Annona squamosa is widely called the sugar or custard apple or, alternatively, the atis plant. *Annona squamosa* is mostly cultivated in both tropical & subtropical areas of the globe. Atis has been associated with many pharmacological applications and values. The plant is traditionally used for treating various diseases and disorders. Various pharmacological applications of this plant are draw on its antioxidants, insecticidal, anti-ulcerative, analgesic, antitumor, anti-inflammatory, and antimalarial properties. In recent times, the phytochemicals present in this plant have been the subject of major attention, with a view to it being applied to various challenges.

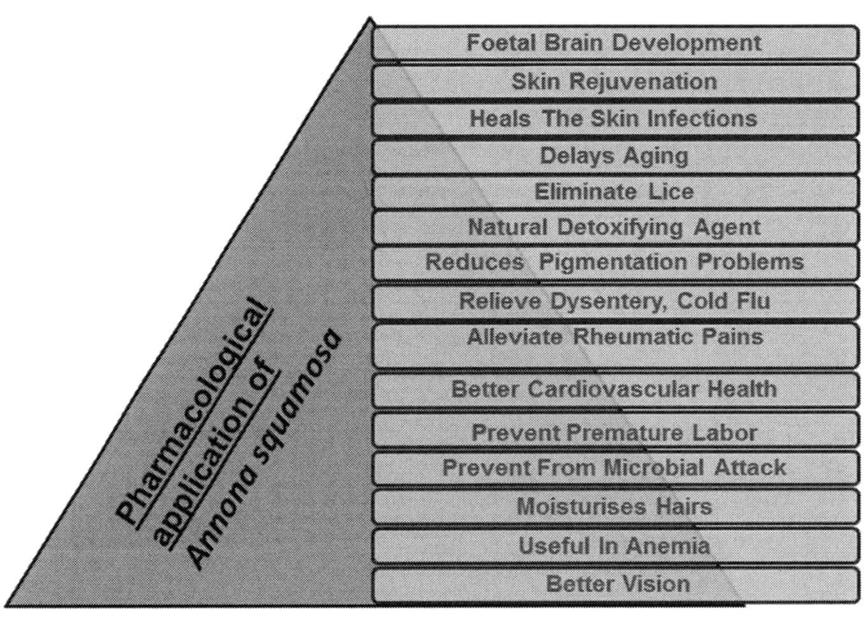

FIGURE 3.9 Health benefits of Atis (*Annona squamosa*).

REFERENCES

Adesanwo, J. K., Akinloye, A. A., Otemuyiwa, I. O., & Akinpelu, D. A. (2021). Chemical characteristics and biological activities of *Annona squamosal* fruit pod and seed extracts; *Journal of Exploratory Research in Pharmacology*, *6*(1), 5–15.

Akram, M., Munir, N., Daniyal, M., Egbuna, C., Găman, M. A., Onyekere, P. F., & Olatunde, A. (2020). Vitamins and Minerals: Types, sources and their functions. In R. E. Aluko (Ed.), *Functional Foods and Nutraceuticals* (pp. 149–172). Springer.

Bakane, P. H., Borkar, P. A., Gajabe, M., & Khakare, M. (2015). Physical properties of custard apple fruit (*Annona squamosa* L.). *International Journal of Agricultural Science Research*, *5*(4), 343–352.

Bermejo, A., Figadere, B., Zafra-Polo, M. C., Barrachina, I., Estornell, E., & Cortes, D. (2005). Acetogenins from *Annonaceae*: recent progress in isolation, synthesis and mechanisms of action. *Natural Product Reports*, *22*(2), 269–303.

Bhardwaj, R., Pareek, S., Sagar, N. A., & Vyas, N. (2020). Bioactive compounds of annona. In H. N. Murthy, & V. A. Bapat (Eds.), *Bioactive Compounds in Underutilized Fruits and Nuts* (pp 37–62). Springer.

Bhatnagar, P., Singh, J., Jain, M. C., & Singh B. (2012). Evaluation of landraces of custard apple (*Annona squamosa* L.). *Plant Archives*, *12*(2), 1045–1048.

Chandel, S. S., Dikshit, S. N., & Sharma, H. G. (2018). Collection and evaluation of custard apple (*Annona squamosa* L.) genotypes in Chhattisgarh plains. *Journal of Phamacognosy and Phytochemistry*, *2*, 149–152.

Chopra, R. N., Chopra, I. C., & Varma, B. S. (1998). *Supplement to Glossary of Indian Medicinal Plants*. National Institute of Science Communication, Council of Scientific Industrial Research.

Correa-Gordillo, J., Ortiz, J., Sanchez-Mejía, M., & Pachón H. (2012) Soursop (*Annona muricata* L.) antioxidant activity: A literature review. *Boletín Latinoamericano y del Caribe de Plantas Medicinales*, *11*, 111–126.

Ghawade. P. M., Supe, V. S., Pimpalpalle, L. V., & Tayade, S. A. (2018). Morphological characterization of custard apple genotypes. *Journal of Pharmacognosy and Phytochemistry*, *7*(1), 1029–1032.

Gleye, C., Raynaud, S., Hocquemiller, R., Laurens, A., Fourneau, C., Serani, L., Laprevote, O., Robolt, F., Leboeuf, M., Fournet, A., De Arias, A. R., Figadere, B., & Cave, A. (1998). Muricadienin, muridienins and chatenaytrienins, the early precursors of annonaceous acetogenins. *Phytochemistry*, *47*, 749–754

Guillopé R, Escobar-Khondiker M, Guérineau V, Laprévote O, Höglinger GU, & Champy P. (2011). Kaurenoic acid from pulp of *Annonacherimola* in regard to Annonaceae-induced Parkinsonism. *Phytotherpy Research*, *25*, 1861–1864.

Handique, K., Hazarika, D. N., Langthasa, S., Khanikar, H. B., & Deori, G. D. (2022). Morphological characterization of custard apple (*Annona reticulata*) grown in Brahmaputra valley of Assam. *The Pharma Innovation Journal*, *11*(5), 684–688.

Hemalatha, S., Amudha, P., Pushpa Bharathi, N., & Vanitha, V. (2017). Determination of bioactive phytocomponents from hydroethanolic extract of *Annona squamosa* (L.) Leaf by GC-MS. *International Journal of Pharmaceutical Sciences and Research*, *8*(6), 2539–2544.

Hemlatha, K., & Satyanarayana, D. (2015). Anti-inflammatory activity of *Annona squamosa* Linn. *Biomedical and Pharmacology Journal*, *2*(1), 17–20.

Himesh, S., Singhai, A. K., & Sarvesh, S. (2012). Quantification of ascorbic acid in leaves of *Annona squamosa*. *International Journal of Pharmacy and Pharaceutical Sciences*, *4*, 1–6.

Hosseinabadi, T. (2021). The Medicinal Importance of *Annona squamosa* fruits. *Journal of Exploratory Research in Pharmacology*, *6*(1), 1–2.

Jamkhande, P. G., & Wattamwar, A. S. (2015). *Annona reticulata* Linn. (Bullock's heart): Plant profile, phytochemistry and pharmacological properties. *Journal of Traditional and Complimentary Medicine*, *5*, 144–152

Kalidindi, N., Thimmaiah, N. V., Jagadeesh, N. V., Nandeep, R., Swetha, S., & Kalidindi, B. (2015). Antifungal and antioxidant activities of organic and aqueous extracts of *Annona squamosa* Linn. leaves. *Journal of Food and Drug Analysis*, *23*(4), 795–802.

Kalyani, R. L., Vijaykumar, P. P. N., Pammi, S. V. N., Rajkumar, M., Swamy, P. V., & Murthy, K. V. R. (2019). Biosynthesis of silver nanoparticles using *Annona squamosa* leaf extract with synergistic antibacterial activity. *Indian Journal of Pharmaceutical Sciences*, *81*(6), 1036–1044.

Katole, R. M., Gautam, J. M., & Mokat, D. (2018). Phytochemical study of *Annona squamosa* L. and *Annona reticulata* L. *International Journal of Research*, *5*, 156–175.

Kaur, R., Kaur, K., Kaur, P., & Singh, I. (2015). Sitaphal: Unexplored therapeutic potential. *Asian Journal of Research in Chemistry and Pharmaceutical Sciences*, *3*, 129–141.

Kumar, D., Yadav, D., Patidar, R., Kumar, P. N. S., Lobo, R., Setty, M. M., Prabu, S. L., Shirwalkar, A., & Shirwalkar, A (2018). Morpho anatomical and physicochemical studies of Annona reticulata, a common substitute for Annona squamosa. *Nigerian Journal of Natural Products and Medicine*, *12*, 17–23.

Kumar, J. A., Rekha, T., Devi, S. S., Kannan, M., Jaswanth, A., & Gopal, V. (2010). Insecticidal activity of ethanolic extract of leaves of *Annona squamosa*. *Journal of Chemical and Pharmaceutical Research*, *2*(5), 177–180.

Kumar, M., Changan, S., Tomar, M., Prajapati, U., Saurabh, V., Hasan, M., & Mekhemar, M. (2021). Custard apple (*Annona squamosa* L.) leaves: Nutritional composition, phytochemical profile, and health-promoting biological activities. *Biomolecules*, *11*(5), 614.

Ma, C., Chen, Y., Chen, J., Li, X., & Chen, Y. (2017). A review on *Annona squamosa* L.: Phytochemicals and biological activities. *The American Journal of Chinese Medicine*, *45*(05), 933–964.

Madona, L., Afrizal, & Efdi, M. (2021). *Anona* genus: Phytochemical, anticancer and activity structure relationship. *Internatinal Journal of Current Microbiology and Applied Sciences*, *10*(12), 91–110.

Meira, C. S., Guimarães, E. T., Macedo, T. S., da Silva, T. B., Menezes, L. R., Costa, E. V., & Soares, M. B. (2015). Chemical composition of essential oils from *Annona vepretorum* Mart. and *Annona squamosa* L.(Annonaceae) leaves and their antimalarial and trypanocidal activities. *Journal of Essential Oil Research*, *27*(2), 160–168.

Narwade, D. A., & Aher, A. N. (2019). A Review on: Phytochemical and Pharmacological Study of *Annona squamosa*. *Pharma Tutor*, *7*(1), 5–10.

Nguyen, M. T., Nguyen, V. T., Le, V. M., Trieu, L. H., Lam, T. D., Bui, L. M., & Danh, V. T. (2020). Assessment of preliminary phytochemical screening, polyphenol content, flavonoid content, and antioxidant activity of custard apple leaves (*Annona squamosa* Linn.). *IOP Conference Series: Materials Science and Engineering*, *736*(6), 062012.

Nugraha, A. S., Damayanti, Y. D., Wangchuk, P., & Keller, P. A. (2019). Anti-infective and anticancer properties of the *Annona* species: Their ethnomedicinal uses, alkaloid diversity, and pharmacological activities; *Molecules*, *24*(23), 4419.

Okhale, S. E., Akpan, E., Fatokun, O. T., Esievo, K. B., & Kunle, O. F. (2016). *Annona senegalensis* Persoon (Annonaceae): A review of its ethnomedicinal uses, biological activities and phytocompounds. *Journal of Pharmacognosy and Phytochemistry*, *5*, 211–219.

Pandey, N., & Barve, D. (2011). Phytochemical and pharmacological review on *Annona squamosa* Linn. *International Journal of Research in Pharmaceutical and Biomedical Sciences*, *2*(4), 1404–1412.

Pinto, A. C. Q., Cordeiro, M. C. R., & Andrade, S. R. M. (2005). *A monograph on Annona species*. International Centre for Underutilised Crops, University of Southampton.

Satyanarayana, T., Gangarao, B., Surendra, G., & Rajesh K. (2013). Pharmacognostical and Phytochemical studies of *Annona reticulata* Linn *International Journal of Research in Pharmacy and Chemistry*, *3*(2), 477–482.

Saha, R. (2011). Pharmacognosy and pharmacology of *Annona squamosa*: A review. *International Journal of Pharmacy & Life Science*, *2*(10), 1183–1189.

Sandeep, and Mittal, A. (2017). *Annona squamosa*: Seetaphal: A review. *IP International Journal of Comprehensive and Advanced Pharmacology*, *2*(3), 72–81.

Shukry, W. M., Galilah, D. A., Elrazek, A. A., & Shapana, H. A. (2019). Mineral composition, nutritional properties, vitamins, and bioactive compounds in *Annona squamosa* L. grown at different sites of Egypt. *Series Botany and. Environmental Science*, *1*, 7–22.

Soni, H., Malik, J., Yadav, A. P., & Yadav, B. (2018). Characterization of rutin isolated by leaves *Annona squamosa* by modern analytical techniques. *European Journal of Biomedical and Pharmaceutical Sciences*, *5*(6), 484–489.

Suchitra, M. R., & Shrinivasan, P. (2015). Sitaphal: Reemergence. *Research Journal of Pharmaceutical Biological and Chemical Sciences*, *6*(3), 1560–1565.

Thakur, D. R., & Singh, R. N. (1967). Pomological description and classification of some *Annonas*. *Indian Journal of Horticulture*, *24*(1), 11–19.

Tylor, J., Arnold, W., Thomas, J., & Mansor, S. M. (2011). Antimalarial alkaloids isolated from *Annona squamosa*. *Phytopharmacology*, *1*(3), 49–53.

Varadharaj, V., Janarthanan, U. D., Krishnamurthy, V., & Synnah, J. (2014). Assessment of phytonutrients in the ethanolic leaf extract of *Annona squamosa* (L.). *World Journal of Pharmacy and Pharmaceutical Sciences*, *3*, 725–732.

Verma, R. S., Joshi, N., Padalia, R. C., Singh, V. R., Goswami, P., & Chauhan, A. (2016). Characterization of the leaf essential oil composition of *Annona squamosa* L. from foothills of north India. *Medicinal and Aromatic Plants*, *5*(270), 2167–2412.

Vijayameena C, Sabhashini G, Loganayagi M, & Ramesh B. (2013). Phytochemical screening and assessment of antibacterial activity for the bioactive compounds in *Annona muricata. International Journal of Current Microbiology and Applied Sciences*, *2*, 1–8.

Win, M. O., & Myat, M. K. (2017). Pharmacological activities of *Annona squamosa*: Updated review. *International Journal of Pharmacy and Chemistry*. *3*(6), 86–93.

Wongs-Aree, C., & Noichinda, S. (2011). Sugar apple (*Annona squamosa* L.) and Atemoya (*A. cherimola* Mill.× *A. squamosa* L.). In E. M. Yahia (Ed.), *Postharvest Biology and Technology of Tropical and Subtropical Fruits* (pp. 399–427). Woodhead Publishing.

Yadav, Y. K. (2019). Evaluation of antidiabetic and antioxidant potential of custard apple (*Annona squamosa*) Leaf extracts: A compositional study. *International Journal of Chemical Studies*, *7*, 889–895.

4 Indian Belladonna (*Atropa belladonna*)

Ramandeep Kaur, Sukhcharn Singh, and Raees ul Haq

4.1 INTRODUCTION

The perennial herb *Atropa belladonna* is a member of the *Solanaceae* family. A perennial herbaceous plant, it has a mysterious past filled with beauty, life, and death. The word *Atropa* is derived from "Atropos," one of the three fates that define one's fate in Greek mythology, and *Belladonna*, the Italian word for a "beautiful woman." In ancient Rome, women used to use an extract of this plant on their cheeks to give their skin a pinkish-red glow and on their eyes to make them look lovely (Lackcovie, 2017). Later, during the Middle Ages, the plant was traditionally used as a herb in Europe to treat a variety of diseases. The plant has a range of common names, including poison black cherry, belladonna, naughty man's cherries, deadly nightshade, devil's cherries, dwale, devil's herb, dwayberry plant, bunika and velebilje (Lee, 2007). Taxonomically speaking, belladonna belongs to the *Solanaceae* family, which also includes potatoes, tomatoes, eggplants, and other plants. *Atropa belladonna* is the formal name for belladonna. Four species of the genus Atropa are known for their distinctive five-lobed blooms and alternating leaves.

As many as 20 different tropane alkaloids, including scopolamine, atropine, and hyoscyamine, are found in the leaves, berries, and also the roots of *Atropa belladonna*, and they serve as the plant's chemical defense in times of stress (Wink & Roberts, 1998). The primary alkaloids of numerous solanaceous plants, including those native to the United States such as *Scopolia carniolica, Datura* spp., and *Hyoscyamine* spp. (often identified as daturine), are S-(-)-scopolamine (hyoscine) and S-(-)-hyoscyamine (Gupta, 2012; Burrows & Tyrl, 2013). S-(-)-phenylalanine is used in the biosynthesis of several alkaloids. Hyoscyamine is converted into atropine, a racemic combination made up of l-hyoscyamine and d-hyoscyamine in a ratio of 1:1, after consumption of *Atropa belladonna* (Ulbricht et al., 2004). Hyoscyamine has a total alkaloid content of 0.2% to 2.0% in the plant's above-ground sections, 0.2% to 1.2% in the roots, and around 65% in its berries. In the leaves and 68.7% of the alkaloid complex overall, hyoscyamine can be detected (Ulbricht et al., 2004).

The alkaloids extracted from *Atropa belladonna* were used in non-FDA-approved over-the-counter (OTC) medications for human usage in the nineteenth century as analgesics for Parkinson's disease, motion sickness, and colic, and also as an anticholinergic in cough-cold medication items linked with excessive sweating, hyperkinesis, and bronchial asthma, as well as in plasters for treating psychiatric disorders, neuralgia, rheumatism and Parkinson's disease.

4.2 OCCURRENCE

Although the plant species are grown across the world, they are primarily native to Europe, North Africa, and Western Asia. The plant grows well in the northern hemisphere as a subshrub (Butcher, 1947). Belladonna is defined as having the following morphology: a height of 1.5 to 2 meters; leaves that are around 18 cm long; purple flowers with a bell-like structure; and berries which look like blueberries in size and form, contain numerous seeds, have a sweet, pleasant flavor, and dark-purple juice (Stepp, 2004). Black/dark-blue belladonna berries may be mistaken for other edible berries, putting youngsters at risk for poisoning due to their appealing hue and sweet, agreeable flavor (Kay, 2008).

 DOI: 10.1201/9781003292340-4

Furthermore, the soil's acidic, basic, or neutral character has no bearing on the plant's ability to grow in any type of soil, whether sandy, loamy, or stony. Often, however, it was discovered on soil with a pH close to 6 (acidic) (Lee, 2007; Mann, 1994). June, July, and August are the blooming months in Southern Britain.

4.3 MORPHOLOGY OF THE PLANT

A. belladonna is a branched herbaceous perennial that commonly grows to a height of around 5 feet, though it can occasionally reach as tall as 6 feet. It has a stout, purple stem that is not divided at the base and splits into three or, more rarely, two to four branches, each of which can branch freely. The plant has thick, fleshy and white branched roots that are at least 6 inches long, leaves range in length from 3 to 10 inches and possesses an ovate berry along with a five-cleft calyx that spreads around the bottom, full of dark inky juice, shining black color, sweet and consumed mainly by birds. The blossoms occur solitary in the calyx of the foliage, mysterious and grimy purple color, tinged with green, about an inch long, pendant, bell-shaped, crinkled; the corolla with five big lobes or teeth, reflexed slightly; blossoming time extends to early September from June.

The species favors well-drained, limed soil in full sun or moderate shade, although the crop does best in light, permeable, and calcareous soil. It is essential that the soil should stay hydrated. Belladonna is usually grown by seed spread in flats, which takes four to six weeks to germinate. When the seeds are about an inch high, they are transplanted, watered, and then kept in shade for a few days. Another method for spreading belladonna is by taking cuttings from the young branch tips. Growing belladonna on a hillside produces particularly good results in terms of the high proportion of alkaloids. The species' maximum growth ranges between 500 and 550 N. latitude and a height of 300 to 600 ft, yet it may be as low as sea level in places with calcareous soil, excellent drainage, and sufficient cover. The use of farmyard manure or a combination of nitrate of soda, kainit, and basic slag can all significantly boost the crop. The plants are pruned until they are 2.5 to 3.0 feet apart at the beginning of winter in order to prevent the overcrowding of the plants in the following year. In their first year plants typically reach a height of one and a half feet and bloom in September. At this point only the tips and leaves are only harvested. In the autumn of the fourth year, the roots may be harvested. As faded leaves and plant parts only produce limited amounts of alkaloids, the collected pieces should be dried in the sun as soon as possible (Kay, 2008).

4.4 CHEMICAL CONSTITUENTS/PHYTOCHEMICAL OVERVIEW

Following high-resolution gas liquid chromatography (GLC) and GLC-MS, Hartmann et al. (1986) discovered 13 alkaloids in total from the roots of *A. belladonna*; however, the plant's above-ground sections only produced 7 alkaloids (hygrines are completely absent). Norhyoscyamine is often found in the shoots of the plant species, although scopolamine has only been shown to be generated in roots. The study also discovered hyoscyamine N-oxide in numerous plant sections. Tritium (^{3}H) and carbon-14 (^{14} C) labeled ornithine is a go through technique used to determine the metabolic pathways. Feeding a specific labelled ornithine to the plant, a natural plant component -N-methylornithrine, which is regarded as a tropane alkaloid precursor, was identified from *A. belladonna* (Hedges & Herbert, 1981).

From the leaf extract of *A. belladonna*, Arraez-Roman et al. (2008) developed a quick and simple CE-electrospray interface (ESI)-TOF-MS procedure to isolate a few significant compounds, including apoatropine, belladonnine, tropine, norhyoscyamine, hyoscyamine, and 6-hydroxyhyoscyamine and scopolamine. Using sodium acetate-2-14C, Baralle and Gros (1969) described the existence of cuscohygrine alongside other alkaloids. Cuscohygrine was described as a typical component alkaloid of *A. belladonna* (Van Haga & Reinouts, 1954).

Use of Sephadex gel filtration and column chromatography on silver nitrate saturated on a silica gel, and eight steroidal glycosides were extracted from the alcoholic extracts of *A. belladonna* seeds which were called atroposides A, B, C, D, E, F, G, and H (Shvets et al., 1996). It has been shown that the structures of atroposides A, C, E, and G are 3-O-alpha-D-galactopyramoside, 3-O-beta-D-glucopyranosyl (1-->4)-beta-D-galactopyranoside, 3-O-beta-D-glucopyranosyl (1-->2)-beta-D-glucopyranosyl (1-->4)-beta the compound beta-D-glucopyranosyl (1-->4)-beta-D-galactopyranoside.

The entire plant is loaded with alkaloids, primarily l-hyoscyamine and atropine, making it potentially dangerous, but also giving it medical properties. L-hyoscyamine is the most abundant alkaloid in the root, stem, leaves, and fruits, followed by atropine. These two alkaloids account for 99% of all the alkaloids found in the leaves. Atropine concentrations in the root range from 0.4 to 0.6%, whereas the fruit only has a 0.1% content. There are also lower levels of scopolamine, scopolamine, belladonnine, atropamine, N-methylpyrrolidine, and N-methylpyrroline in belladonna (Hedges & Herbert, 1981). However, depending upon a number of factors such as the harvest season, location, meteorological conditions, and soil type, the concentration of alkaloids might differ from one plant to another.

4.5 ALKALOID PRODUCTION AND ISOLATION

Atropine and scopolamine are found extensively in hairy roots (hairy roots were caused by inoculating the sterile plant stems of *A. belladonna* with *Agrobacterium rhizogenes*) (Kamada et al., 1986). The concentrations were enumerated by GLC, and results revealed that concentrations of both alkaloids was similar in the axenic cultures as well as that of normally grown plants. Atropine was proposed to be extracted from roots using an integrated process by Dimitrov et al. (2005), which included liquid membrane isolation (using di-isopropyl ether functions role of liquid membrane and sulfuric acid functions as a stripping agent) and extraction. Using water and 70% ethanol as solvents, Srivastava and Chadha (1963) investigated the effects of commonly used two surface-active agents (Tween 80 and 20-0.2%) on the extraction of the belladonna plant, resulting in a relatively quick and thorough extraction of the medication (Tween 80 gave better results).

4.6 BIOCHEMISTRY

Complete amino acid arrangement of ferredoxin from *A. belladonna* were generated by the research of Mino (Mino et al., 2005). By automatically degrading the complete S-carboxymethylcysteinyl proteins, as well as the peptides produced by enzymatic digestion, belladonna and Hyoscyamus niger were both successfully tested. The amino acid sequences of the two ferredoxins differed by 1 to 8 amino acids from those of other plants that contain tropane alkaloids (*Datura arborea Datura stramonium, Datura metel*, and *Scopolia japonica*), and by 9 to 23 amino acids from those of other ferredoxins of solanaceous family, indicating close proximity of the tropane-alkaloid-containing plants.

According to Salonen and Simola (1989), glutamate but not NH4 + or proline or proline plus ornithine can boost growth and NRA (nitrate reductase activity) in NO3 containing media. According to Bensaddek et al. (2001), the ratio of scopolamins to hyoscyamine is strongly influenced (dose-dependent) by the nitrate and ammonium levels in the growth medium. During organogenesis, Phillipson and Handa (1976) discovered significant variations in the species' N-oxide content, with ripe fruit having the greatest levels. Sporer et al. (1992) studied the seasonal fluctuations in alkaloid content as well as alkaloid patterns in belladonna plants and hyoscyamine was predominantly examined in the study. Two peaks were found to be significant: the maximum level of alkaloid yields were detected in the early morning and the early night while it was the highest in mature seeds when measured in the afternoon.

Falk and Doran (1996) evaluated the impact of inoculum morphology on the development of hairy roots in *A. belladonna*, along with the production of tropane alkaloids, and it was observed that hyoscyamine content accumulated at higher quantities in fully grown tissues of roots in comparison to the tips; the concentration was nevertheless unrelated to inoculum morphology. To demonstrate the impact of medium conditioning and the mass transfer of oxygen gas on the growth of roots, Kanokwaree and Doran (1997) hypothesized that a surge in the amount of hairy root tips in shaking glass flasks would result in a 40% reduction in growth rates in belladonna. An increase of around 15 times in the root biomass was reported after culturing for a whole month. The concentrations of different alkaloids include 6-hydroxyhyoscyamine (0.1%), hyoscyamine (0.7%), littorine (0.02%), and scopolamine (0.02%). Lee et al. (1998) examined growth and concentration of tropane in transformed root cultures of *A. belladonna* cultured in media containing 0.1% of chitin, 0.1% of chitosan, 5 mM Glutathione, and 0.1% of yeast extract. The chemicals added to the media (individually to 18-day-old cultures) had virtually no effect on alkaloid production, but the addition of 5 mM H_2O_2 caused a momentary release of alkaloids (tropane) from transformed roots with a peak concentration of 0.35 mg per flask.

The excretion of tropane alkaloids into the media was similarly enhanced by Cu2 and Cd2 (5 mM), although this resulted in the lysis of converted roots. Baricevic et al. (1999) studied the impact of water stress and nitrogen fertilization on the concentration of hyoscyamine scopolamine in nightshade roots. The experimental findings showed that the highest concentration of alkaloids was reached with 95% depletion of the soil's available water and 1.60 g/pot of nitrogen.

Ali (2000) examined how putrescine affected the germination and seedling development of *A. belladonna* under the impact of NaCl and discovered that presoaking seeds in 102 mM putrescine could reduce the negative effects of NaCl on the species' early growth while also boosting alkaloids and endogenous putrescine. Gibberellic acid, heavy water, and tiron (a water-soluble radical scavenger) were the factors, among others, in relation to the alkaloid content, growth, and morphology of *A. belladonna* aside from these stages of development (Elzenga & Bruyn, 1956; Brumett & Sciuchetti, 1960; Uphaus et al., 1965; Akita & Ohta, 2000).

Belladonna alkaloids, a mydriatic used in ophthalmology, primarily function by inhibiting the activity of the central nervous system. Additionally, they paralyse the pharyngeal nerves by acting on the reflex action of swallowing. As a result of its paralyzing properties, it is frequently used to treat conditions, including nocturnal urine incontinence, dropsy, excessive hydrochloric acid output in the stomach, and duodenal ulcer. Belladonna has additional applications as an anticholinergic, an antiasthmatic, and an antispasmodic (in gallstones and renal colic). It is used in pulmonology to treat whooping cough and bronchial spasm issues. Parkinson's disease has traditionally been treated with belladonna extracts and a combination of atropine and levadopa.

As a result of its antisecretory properties, it is utilized as a neuroregulator for intestines in cases of ulcerative colitis, irritable colon, etc., as well as in stomach ulceration. It is also used as an antidote for mushroom poisoning (loaded with muscarinic alkaloids) and to prevent colic. It is often associated with certain laxatives. Belladonna's alkaloids can permeate the skin and enter the body through it; thus, it should only be administered topically as an ointment or oil to ease pain. Similar to atropine, belladonna's scopolamine causes a broad and rapid paralysis of the central nervous system in response to extremely tiny dosages, but, unlike atropine, its side effects, such as decreased salivation and perspiration and dilated pupils, are short-lived (Cabezon, 1997). Its fruits are utilized as aphrodisiacs and euphorics in Moroccan culture and by students to improve memory (Esteve and Sierra, 1971).

4.7 BIOACTIVE COMPOUNDS AND SYNTHESIS

To regulate their interactions with pollinators, herbivores, and other biotic and abiotic stressors, plants develop a chemically diversified variety of specific flavorings, perfumes, and medicinal compounds. One class of these specialized metabolites, known as tropane alkaloids, is produced by phylogenetically separate plant families; this includes the drugs cocaine and hyoscyamine and scopolamine. The forerunner to semi-synthetic tropane medicines, scopolamine is created in the roots of *Solanaceae* plants. To fill in the gaps in the biosynthetic pathway from putrescine to scopolamine, a set of 11 tissue-specific transcriptomes were produced from the delicate perennial Solanaceous plant *Atropa belladonna*, commonly known as deadly nightshade. These include the manufacture of tropinone, the formation of phenyllactic acid from phenylalanine, and the conjugation and activation of phenyllactic acid in the presence of tropine to generate littorine, a precursor to scopolamine through the late route. The first compound in this route, tropinone, contains the distinctive 8-azabicyclo[3.2.1] octane tropane core (Figure 4.1).

This pharmacore was created by a traditional biomimetic chemical synthesis, but the mechanism for tropinone biosynthesis in all species that make these molecules has remained a mystery. The major metabolite phenylalanine is converted into the specific acyl group for both hyoscyamine and scopolamine, tropic acid, via the intermediary, phenyllactic acid. Amino acids can be devoted to specific metabolism by aromatic aminotransferases. The ester of tropine and phenyllactic acid produces littorine, which is a precursor to hyoscyamine and scopolamine. Since the discovery of

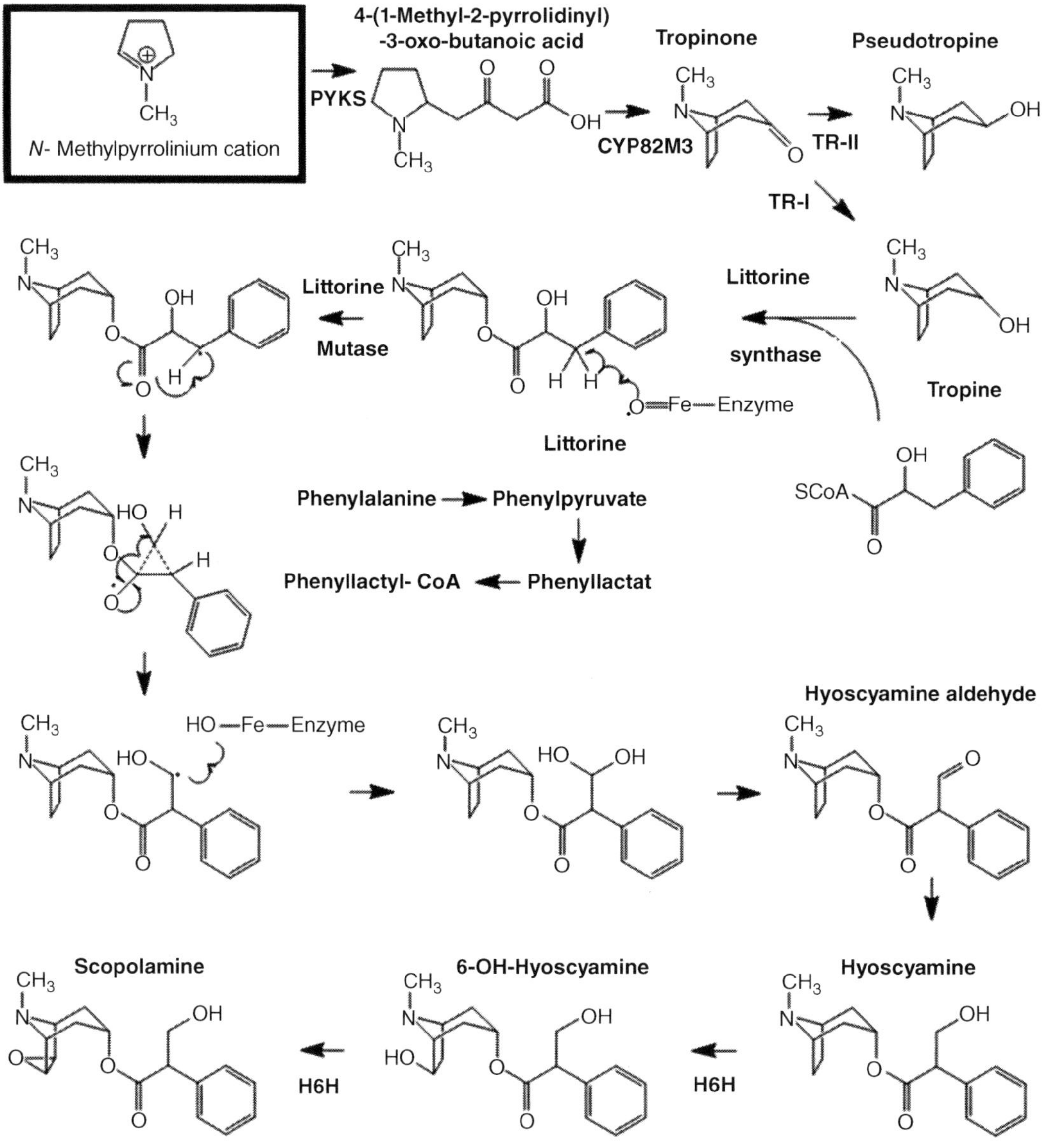

FIGURE 4.1 Synthesis of different bioactive compounds.

littorine, the mechanism for phenyllactic acid activation and esterification in tropane alkaloid biosynthesis has remained a mystery.

4.8 EXTRACTION AND ISOLATION TECHNIQUES

The extraction of different bioactive compounds from the plant involves conventional as well as innovative extraction procedures with the former involving solvent extraction and the latter employing ultrasound, microwaves, supercritical fluids, pressurized liquid extraction and enzyme-assisted extraction, as shown in Figure 4.2.

4.9 TROPANE

Tropane is a bicyclic amine with two carbon atoms and a pyrrolidine and piperidine ring that share a nitrogen atom. It is the structural component shared by all tropane alkaloids. Tropane alkaloids can be found in any part of the plant with roots possessing up to 1.3%, leaves containing 1.2%, stalks with 0.65%, blossoms around 0.6%, berries up to 0.7% and seeds comprising 0.4% tropane alkaloids. When a plant is budding and flowering, its leaves contain the greatest amount of alkaloids, and

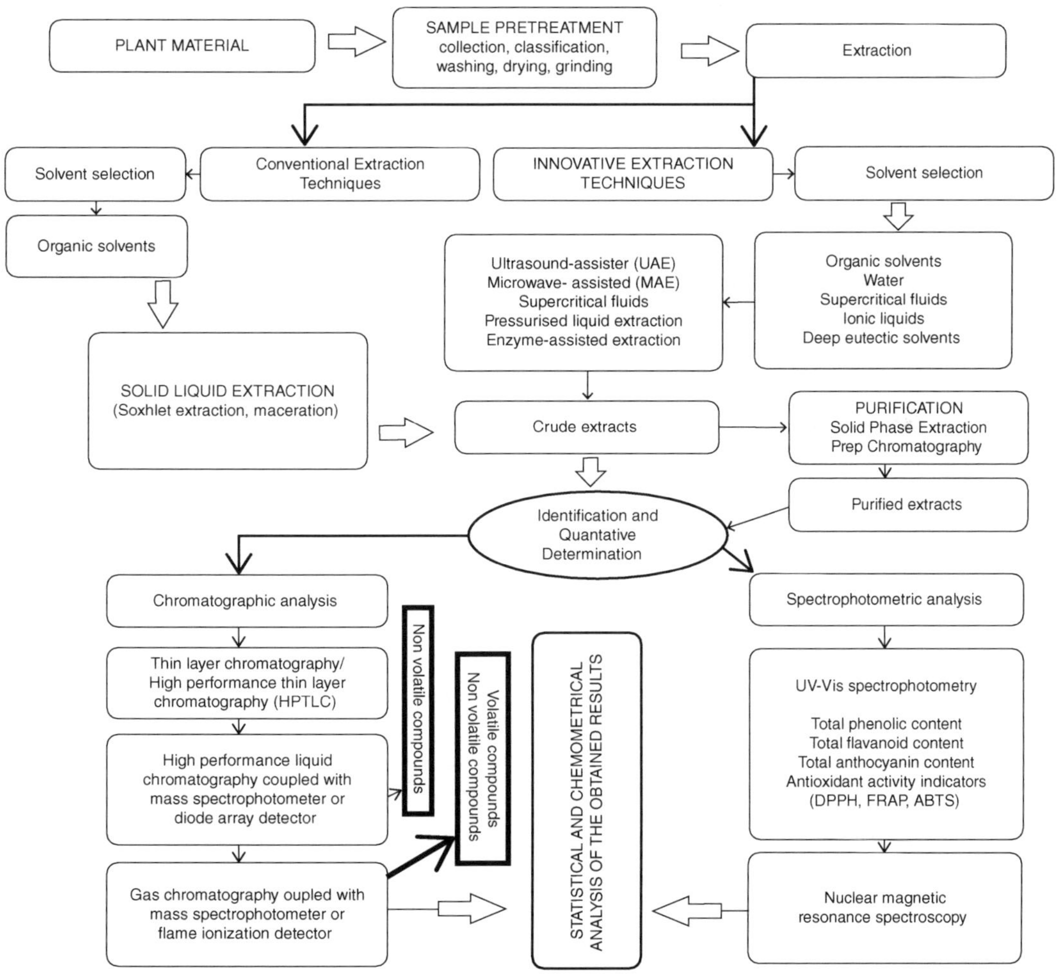

FIGURE 4.2 Extraction and isolation of bioactive compounds from *Atropa*.

the plant's roots are most poisonous at the end of its vegetation period (Rätsch, 2005). According to Hazlinsky (1956), bees transform belladonna nectar into honey which also contains tropane alkaloids. Traditionally, they are extracted from the cultured plant leaves. Utilizing cell suspension cultures could be an alternative to extracting from plant tissue (Fuyita & Tabata, 1987). However, it has been demonstrated that root differentiation was correlated with tropane alkaloid biosynthesis (Endo & Yamada, 1985). Therefore, the furry root arrangement offers a fascinating way of dealing with creating these auxiliary metabolites (Hamill et al., 1987; Verpoorte et al., 1999). Hairy root cultures have shown that they can quickly produce biomass and that they contain a lot of tropane alkaloids (Jouhikainen et al., 1999).

4.9.1 Synthesis of Tropane

Using ptO2 in the presence of 2 N HCl, tropinone was reduced to produce tropane (Scheme 10.4). Due to its simplicity and biomimetic approach, Robinson's tropinone synthesis was regarded as a classic in total synthesis (Birch, 1993). Although tropinone is a bicyclic molecule, the reactants used to make it are fairly straightforward: succinaldehyde, methylamine, and CH3)2CO dicarboxylic corrosive (or even CH3)2CO). Because biosynthesis makes use of the same building blocks, the synthesis is an excellent illustration of a biogenetic or biomimetic reaction. It likewise exhibits a couple response in a one-pot blend.

4.10 HYOSCYAMINE AND SCOPOLAMINE

Scopolamine, along with hyoscyamine, its biochemical precursor, are lethal alkaloids which originating from the roots of *A. belladonna*. The enantiomer of alkaloid atropine is hyoscyamine, with both alkaloids possessing hallucinogenic effects and being extremely poisonous. "Insane root", another name for the mandrake, is an anticholinergic and used for different medical uses, such as treating disorders related to the gastrointestinal system, when taken in low doses. Since the Bible, and perhaps even earlier, they have been used medicinally and in religious ceremonies in plant extracts. Scopolamine is sometimes illegally used to disorientate people or render them unconscious, making them vulnerable to robbery or rape as well as murder. Despite its negative effects, it has also been tried as a "truth drug."

4.10.1 Synthesis of Hyoscyamine and Scopolamine

Ornithine decarboxylase decarboxylates l-ornithine to putrescine to begin the biosynthesis of scopolamine. Putrescine N-methyltransferase converts putrescine to N-methylputrescine by a process of methylation (Ziegler & Facchini, 2008). The deamination of putrescine to 4-methyl aminobutanal, which then undergoes a spontaneous ring formation to N-methylpyrrolium cation, is catalyzed by putrescine oxidase, which specifically recognizes methylated putrescine. Hygrine is produced when the pyrrolium cation condenses with acetoacetic acid in the subsequent step. This reaction was not catalyzed by any enzymatic activity that could be demonstrated. Hygrine moves around even more to make tropinone. Next, tropinone reductase I turns tropinone into tropine, which then condenses into littorine with phenylalanine-derived phenylacetate. A Cyp80F1-type cytochrome P450 (Li et al., 2006) converts littorine to hyoscyamine aldehyde through oxidation and rearrangement.

4.10.2 Extraction

Using an ultrasound device, the comminuted particles from *Atropa* species were extracted for a period of three times in a span of 30 minutes, using a solution consisting of chloroform (15 ml), methanol (15 ml), and ammonia (25%) in a volumetric ratio of 15:15:1. After being kept at room temperature for one hour, it was passed through a filter paper, washed a few times with one milliliter of chloroform, and then evaporated until the loss of moisture or when a fairly dry sample is obtained. After thoroughly mixing the chloroform and 1 N sulfuric acids, the residue was dissolved. The chloroform fraction was separated, which was followed by alkalization of the aqueous solution on ice with ammonium hydroxide at a pH of 10. Once with 2 ml of chloroform and twice with 1 ml of chloroform, alkaloids were extracted. The residue is washed with 1–2 ml of chloroform following the mixing of anhydrous Na_2SO_4. At 40 °C, the solvent was evaporated to dryness under vacuum, and the residue, or total extract, was dissolved in an inappropriate volume of methanol (1–2 ml) (Kamada et al., 1986).

4.11 APPLICATION OF ACTIVE INGREDIENTS

Tropane derivatives are one of the pharmaceuticals with the greatest economic significance. More than 20 active pharmaceutical ingredients (APIs) are being produced by various pharmaceutical industries with tropane moiety consisting of basics of their structures. These substances are used as anaesthetics, bronchodilators, antispasmodics, antiemetics, and mydriatics. Antimuscarinic (antagonist) hyoscyamine belongs to a muscarinic acetylcholine receptor antagonist present at parasympathetic sites in the stomach secretions, heart muscle, salivary glands, sweat glands, in the gastrointestinal tract smooth muscle, the sinoatrial node, and the central nervous system and prevents the action of acetylcholine. It dries secretions, lowers blood pressure, and enhances both heart rate and cardiac output. According to Kapoor and Raju (2013), it may conflict with serotonin. The anticholinergic power of atropine of similar doses is comparable to 98% hyoscyamine. Scopolamine, another major drug derived from belladonna, has 92% of atropine's antimuscarinic potency. Various gastrointestinal conditions, such as irritable bowel syndrome, peptic ulcers, spasms, cystitis, colic, diverticulitis, and pancreatitis, can be treated with hyoscyamine. It has also been used in palliative care to control respiratory symptoms (involving mucus secretions) and alleviate some heart problems, Parkinson's disease symptoms, and "hypermucus secretions."

The derivatives of tropane are regarded as being among the economically important plant-based pharmaceuticals. Different pharmaceutical industries are involved in manufacturing of over 20 known active pharmaceutical ingredients which consist of tropane in their basic structures. The APIs found applications in antispasmodics, mydriatics, anaesthetics, antiemetics, and bronchodilators.

4.12 CONCLUSION

A range of different medicinal plants were used in the 20th century that were ignored as a cure for different diseases due to insufficient information. However, the medicinal properties of these plants may prove of value in protecting humanity.

The active ingredients in belladonna have a competitive effect and are antagonistic towards all muscarinic acetylcholine receptors. Provisional evidence suggests belladonna is effective in the treatment of several clinical conditions, viz. in the treatment of urethral stent pain, airway obstruction in sleeping babies, acute encephalitis, myocardial ischemia injury, the symptoms of menopause, irritable bowel syndrome, and severe headaches. Given these potential benefits, an understanding of the medicinal properties of belladonna could make it a potential therapeutic option for various diseases.

REFERENCES

Akita, M., & Ohta, Y. (2000). The effect of Tiron, a water soluble radical scavenger, on growth, morphology and alkaloid content of adventitious roots in *Atropa belladonna*. *Plant Cell Reports*, *19*(7), 705–709.

Ali, R. M. (2000). Role of putrescine in salt tolerance of *Atropa belladonna* plant. *Plant Science*, *152*(2), 173–179.

Arraez-Roman, D., Zurek, G., Babmann, C., Antonio, S. C., & Alberto, F. G. (2008). Characterization of *Atropa belladonna* L. compounds by capillary electrophoresis-electrospray ionization-time of flight-mass spectrometry and capillary electrophoresis-electrospray ionization-ion trap-mass spectrometry. *Electrophoresis*, *29*(10), 2112–2116.

Baralle, F. E., & Gros, E. G. (1969). Biosynthesis of cuscohygrine in *Atropa belladonna* from sodium acetate. *Phytochemistry*; *8*(5), 853–855.

Baricevic, D., Umek, A., Kreft, S., Maticic, B., & Zupancic, A. (1999). Effect of water stress and nitrogen fertilization on the content of hyoscyamine and scopolamine in the roots of deadly nightshade (*Atropa belladonna*). *Environmental and Experimental Botany*, *42*(1), 17–24.

Bensaddek, L., Gillet, F, Edmundo, J., Saucedo, N., & Fliniaux, M. (2001). The effect of nitrate and ammonium concentrations on growth and alkaloid accumulation of *Atropa belladonna* hairy roots. *Journal of Biotechnology*, *85*(1), 35–40.

Birch, A. J. (1993). Investigating a scientific legend: The tropinone synthesis of Sir Robert Robinson F.R.S. *Notes and Records of the Royal Society of London*, *47*(2), 226–277.

Brumett, R. E., & Sciuchetti, L. A. (1960). Changes induced by gibberellic acid on growth and alkaloid patterns in *Datura stramonium* L and in *Atropa belladonna* L. *Journal of the American Pharmacists Association*, *49*(5), 274–277.

Burrows, G. E., & Tyrl, R. J. (2013). *Toxic Plants of North America*. John Wiley & Sons.

Butcher, R. W.. (1947). *Atropa belladonna* L. *Journal of Ecology*, *34*(2), 345–353.

Cabezon, M. C. (1997). *Diccionario de plantas medicinales. Según la medicina Tradicional Marroquí*. Editorial Noesis.

Dimitrov, K., Metcheva, D., & Boyadzhiev, L. (2005). Integrated processes of extraction and liquid membrane isolation of atropine from *Atropa belladonna* roots. *Separation and Purification Technology*, *46*(1–2), 41–45.

Elzenga, G., & Bruyn, J. W. (1956). Interrelation of alkaloid content and stage of development of 1- and 2-year-old *Atropa belladonna* L. *Euphytica*, *5*(3), 259–266.

Endo, T., & Yamada, Y. (1985). Alkaloid production in cultured roots of three species of Duboisia. *Phytochemistry*, *24*(6), 1233–1236.

Esteve, C. F., & Sierra, R. F. C. (1971) *Algunas consideraciones acerca de las condiciones acerca de las condiciones ecológico- edáficas en el desarrollo de la "Atropa baetica". Simposio Internacional de Farmacobotánica*.

Falk, L. R., & Doran, P. M. (1996). Influence of inoculum morphology on growth of *Atropa belladonna* hairy roots and production of tropane alkaloids. *Biotechnology Letters*, *18*(9), 1099–1104.

Fuyita, Y., & Tabata, M. (1987). Secondary metabolites from plant cells-pharmaceutical applications and progress in commercial production. *Plant Tissue and Cell Culture*, 169–185.

Gupta, R.C., 2012. *Veterinary Toxicology: Basic and Clinical Principles*, revised. ed. Elsevier.

Hamill, J. D., Parr, A. J., Rhodes, M. J. C., Robins, R. J., & Walton, N. J. (1987). New routes to plant secondary products. *Biotechnology*, *5*, 800–804.

Hartmann, T., Witte, L., Oprach, F., & Toppel, G. (1986). Reinvestigation of the alkaloid composition of *Atropa belladonna* plants, root cultures, and cell suspension cultures. *Planta Medica*, *52*(5), 390–395.

Hazlinsky, B. (1956). Poisonous honey from deadly nightshade. *Zeitschrift für Bienenforschung*, *3*(3), 93–96.

Hedges, S. H., & Herbert, R. B. (1981). δ-N-Methylornithine: A natural constituent of *Atropa belladonna*. *Phytochemistry*, *20*(8), 2064–2065.

Jouhikainen, K., Lindgren, L., Jokelainen, T., Hiltunen, R., Teeri, T. H., & Oksman-Caldentey, K. M. (1999). Enhancement of scopolamine production in *Hyoscyamus muticus* L. hairy root cultures by genetic engineering. *Planta*, *208*, 545–551.

Kamada, H., Okamura, N., Satake, M., Harada, H., & Shimomura, K. (1986). Alkaloid production by hairy root cultures in *Atropa belladonna*. *Plant Cell Reports*, *5*(4), 239–242.

Kanokwaree, K., & Doran, P. M. (1997). Effect of inoculum size on growth of *Atropa belladonna* hairy roots in shake flasks. *Journal of Fermentation and Bioengineering*, *84*(4), 378–381.

Kapoor, A. K., & Raju, S. M. (2013). *Illustrated Medical Pharmacology*. JP Medical Ltd.

Kay, Q. O. N. (2008). Edible fruits in a cool climate: The evolution and ecology of endozoochory in the European flora. In C. Marshall and J. Grace (Eds.), *Fruit and Seed Production: Aspects of Development, Environmental Physiology and Ecology* (p. 240). Cambridge University Press.

Lacković, Z. (2017). Bunanje: XX century abuse of *Atropa belladonna* halucinogenic berries in continental Croatia. *Psychiatria Danubina*, *29*(3), 379–382.

Lee, K., Yamakawa, T., Kodama, T., & Shimomura, K. (1998). Effects of chemicals on alkaloid production by transformed roots of belladonna. *Phytochemistry*, *49*(8), 2343–2347.

Lee, M. R. (2007). *Solanaceae* IV: *Atropa belladonna*, deadly nightshade. *The Journal of the Royal College of Physicians of Edinburgh*, *37*, 77–84.

Li, R., Reed, D. W., Liu, E., Nowak, J., Pelcher, L. E., Page, J. E., & Covello, P. S. (2006). Functional genomic analysis of alkaloid biosynthesis in *Hyoscyamus niger* reveals a cytochrome P450 involved in Littorine rearrangement. *Chemistry & Biology*, *13*(5), 513–520.

Mann, J. (1994). *Murder, magic and medicine*. Oxford University Press.

Mino, Y., Yukita, M., Hiratsuka, N., & Wariishi, H. (2005). Amino Acid Sequences of Ferredoxins from *Atropa belladonna* and *Hyoscyamus niger*: Their Similarities to Those in Other Tropane-Alkaloid-Containing Plants. *Biological and Pharmaceutical Bulletin*, *28*(8), 1535.

Phillipson, J. D., & Handa, S. S. (1976). Hyoscyamine N-oxide in *Atropa belladonna*. *Phytochemistry*, *15*(5), 605–608.

Rätsch, C. (2005). *The encyclopedia of psychoactive plants: Ethnopharmacology and its applications*. Simon and Schuster.

Salonen, M., & Simola, L. K. (1989). Effect of nitrate, ammonium and some amino acids on growth and nitrate reductase activity in suspension cultures of *Atropa belladonna*. *Plant and Cell Physiology*, *30*(8), 1177–1181.

Shvets, S. A., Latsterdis, N. V., & Kintia, P. K. (1996). A chemical study on the steroidal glycosides from *Atropa belladonna* L. seeds. *Advances in Experimental Medicine and Biology*, *404*, 475–483.

Sporer, F., Sauerwein, M., & Wink, M. (1992). Diurnal and developmental variation of alkaloid accumulation in *Atropa belladonna*. *ISHS Acta Horticulturae 331: WOCMAP I – Medicinal and Aromatic Plants Conference: part 3 of 4*.

Srivastava, G. P., & Chadha, T. N. (1963). Effect of surface-active agents on the extraction of belladonna herb. *Journal of Pharmaceutical Sciences*, *52*(3), 299–301.

Stepp, J. R. (2004). The role of weeds as sources of pharmaceuticals. *Journal of Ethnopharmacology*, *92*(2–3), 163–166.

Ulbricht, C., Basch, E., Hammerness, P., Vora, M., Wylie, J., & Woods, J. (2004). An evidence-based systematic review of belladonna by the natural standard research collaboration. *Journal of Herbal Pharmacotherapy*, *4*(4), 61–90.

Uphaus, R. A., Crane, F. A., Blake, M. I., & Katz, J. J. (1965). Effects of heavy water on *Atropa belladonna*. *Journal of Pharmaceutical Sciences*, *54*(2), 202–205.

Van Haga, & Reinouts, P. (1954). Cuscohygrine, a normal constituent alkaloid of *Atropa belladonna*. *Nature*, *174*(4435), 833–834.

Verpoorte, R., van der Heijden, R., ten Hoopen, H. J. G., & Memelink, J. (1999). Metabolic engineering of plant secondary metabolite pathways for the production of fine chemicals. *Biotechnology Letters*, *21*, 467–479.

Wink, M., & Roberts, M. F. (1998). Compartmentation of alkaloid synthesis, transport, and storage. In: M. F. Roberts and M. Wink (Eds), *Alkaloids: Biochemistry, Ecology, and Medicinal Applications* (pp. 239–262). Springer.

Ziegler, J., & Facchini, P. J. (2008). Alkaloid biosynthesis: Metabolism and trafficking. *Annual Review of Plant Biology*, *59*(1), 735–769.

5 Redwood Pigweed (*Amaranthus retroflexus*)

Divyang Solanki, Qiu Yi Tan, Anima Manhswita, Srishti, Shreedevi Indapwar, Ritika Rathi, Gareema Pandey, Sangeeta Prakash, and Raees ul Haq

5.1 INTRODUCTION

Amaranthus retroflexus, commonly known in English as amaranth, careless weed and redroot pigweed, in French as *amarante recourbée* and in Spanish as *amaranto*, is found throughout the temperate regions of the world. It is regarded as being native to northern parts of America but now exists almost everywhere across the globe (Frankton and Mulligan, 1987). Amaranth is categorized as an annual growing herbaceous plant which can achieve a height of around two meters, and is used extensively in culinary dishes as well for medicinal purposes. In its natural setting, the plant grows as a weed in different horticultural crops grown in temperate regions; in Kashmir, the plant is extensively found in apple orchards as well as in field crops. *Amaranthus retroflexus* is regarded a as common dicotyledonous weed of family belonging to the *Amaranthaceae* family, which grows rapidly in large numbers, thereby significantly affecting the produce of agriculture crops (Bensch et al., 2003; Horak and Loughin, 2000; Alebrahim et al., 2012; Ma et al., 2015; Bajwa et al., 2019). In addition to soils with a high concentration of nitrogen, the longer daytime duration of summer favors plant growth, leaving a marked effect on production of corn, potato, and other row crops (Kenžević et al., 1995; Costea et al., 2003; Azadbakht et al., 2017).

Amaranthus retroflexus has a particular mark of phenotypic plasticity, enabling the plant to utilize the environmental information for deciding or moulding its shape, which can be seen in the form of variable morphological traits the plant undergoes to acclimatize to different habitats (Mandák et al., 2011; Gambino and Vilela, 2011). Amaranth has a good defence mechanism against reactive oxygen species (ROS) in the form of antioxidant enzymes that protects it during the stress induced by different factors, thereby enhancing the capability of the plant to different climates (Chaves and Oliveira, 2004; Coelho et al., 2017).

The plants of *Amaranthus* spp. are accumulators of oxalate and nitrate substances risking the ruminants feeding on these plants with nitrate poisoning and toxic nephropathy. Amaranth has also been associated with tubular nephrosis in kidneys; in general, however, the plant is not associated with such risks as it is only lightly grazed (Plumlee, 2004).

Amaranth has been regarded both as a medicinal plant and as a pseudo-cereal which has been consistently used as a source of nutrition in developing countries. In addition to basic nutrition, the plant provides biologically active protein hydrolysates and peptides. The bioactive peptides from redroot pigweed have inhibitory effects on pancreatic lipase and cholesterol esterace (Morales et al., 2020; Ajayi et al., 2021). The protein composition of amaranth is particularly rich, consisting of all the essential amino acids, with tryptophan and lysine dominating its composition (Szwejkowska and Bielski, 2012). Saturated and unsaturated fatty acids, tocopherol, and sterols form the lipid portion of the plant, with linoleic acid accounting for more than half the total (Kaźmierczak et al., 2011). Minute quantities of phytin, saponins, nitrates, hemagglutinins, polyphenols, protoalkaloids, and betacyans are present in different parts of the plant. Herbal medicines derived from amaranth have been claimed

 DOI: 10.1201/9781003292340-5

to treat pharyngitis, diarrhea, skin disorders, and ulcers (Esiyok et al., 2004). Extracts from the plant are known to exhibit potent antioxidant, anti-inflammatory, antimicrobial and hypoglycemic properties and can be used in the treatment of atherosclerosis, tuberculosis, cardiovascular, and nervous system-related issues (Szwejkowska and Bielski, 2012). Preparations from the plant seeds were found to inhibit the growth of ehrlich ascites carcinoma (EAC) cells, suggesting their apoptosis and therefore has the potential for using in cancer cure studies (Al-Mamun et al., 2016). As a whole, the plant has been a subject of interest due to pharmacological activity, diverse composition and the presence of bioactive compounds with biological significance (Karamac et al., 2019). In this chapter, amaranth has been evaluated on morphological, nutritional, phytochemical, and pharmacological perspectives.

5.2 CULTIVATION, OCCURRENCE, CLASSIFICATION, AND VARIETY IN SPECIES

Amaranthus retroflexus, also normally known as tumbleweed, is a common species of an annual growing herb; it flowers and fruits in the period from June to August, and its seeds ripen from August through October. It can grow to heights of between 50 and 120 cm (Batsatsashvili et al., 2017; Paniagua-Zambrana et al., 2020). It has grey pubescent stem, diamond-shaped leaves on the lower end of the plant, and narrower leaves on the upper part of the plant. Its flowers consist of 5 unisexual spatulate or linear petals with many slender long-pointed, green bracts. The flowers are in a compact inflorescence, with the terminal spike broadening towards the base. Its capsule is a one-seeded structure, measuring around 1 mm in length; colors may vary from black to brown. The capsules split circumferentially upon maturity, and approximately 230,000 to 500,000 seeds per plant can be produced (Stevens, 1957).

5.2.1 Systemic Classification

Amaranthus retroflexus is also commonly known by a wide range of names: wild beet amaranth, careless weed, common amaranth, red-root amaranth, pigweed, common tumbleweed, red-root pigweed, pigweed amaranth, rough pigweed. It was also previously known as *Amaranthus retroflexus var. salicifolius*.

According to the Natural Resources Conservation Service (1980) and CABI (2021), the classification and taxonomic tree of *Amaranthus retroflexus* is as shown in Table 5.1.

TABLE 5.1
Classification and the Taxonomic Tree of *Amaranthus retroflexus*

Classification	Name
Domain	*Eukaryota*
Kingdom	*Plantae* (Plants)
Subkingdom	*Tracheobionta* (Vascular plants)
Superdivision/Phylum	*Spermatophyta* (Seed plants)
Subphylum	*Angiospermae*
Division	*Magnoliophyta* (Flowering plants)
Class	*Magnoliopsida* (Dicotyledons)
Subclass	*Caryophyllidae*
Order	*Caryophyllales*
Family	*Amaranthaceae Martinov* (Amaranth family)
Genus	*Amaranthus* L. (Pigweed)
Species	*Amaranthus retroflexus* L. (Redroot amaranth)

5.2.2 Taxonomy

Amaranthus retroflexus had been reported to have a chromosome number of 34, making it a diploid (Grant, 1959; Murray, 1940). It can form a hybrid with similar species (i.e., *Amaranthus hybridus, Amaranthus powellii*, and *Amaranthus caudatus*). However, the F1 generation is particularly sterile (Murray, 1940). The inflorescences of hybrids will usually lead to the generation of odd shapes. There are many similarities between *Amaranthus retroflexus* and other species within the same family. *Amaranthus retroflexus* is commonly identified through its shorter and stouter growth nature. Table 5.2 offers a brief description of how other species of the genus can be differentiated from *Amaranthus retroflexus*.

5.2.3 Occurrence

Amaranthus retroflexus is native to North America and is found in all areas of the Caucasus. It can grow in various types and textures of soil flourishing in ground rich with nitrogen. It can grow in soil with pH ranging from 4.2 to 9.1 (Feltner, 1970), but it is less commonly seen in acidic soils such as those more prominently seen in the south-eastern USA. *Amaranthus palmeri*, however, is more abundant in those areas as it can withstand acidic soil. *Amaranthus retroflexus* is often found in cultivated fields, waste places, riverbanks, roadsides, and even in annual weed-dominated open habitats. However, according to Weaver and Mcwilliams (1980), it is rarely found in shaded or closed areas. *Amaranthus retroflexus* has a global distribution and is one of the most common weeds found in fields as well as horticultural row crops, particularly in the world's temperate regions. It serves as a host plant for numerous illnesses and pest insects. According to CABI (2021), it is a major host to *Autographa gamma* (the silver-Y moth) and the disease of *Choanephora cucurbibtarum* (choanephora fruit rot). It can also be a minor host for several organisms, such as *Aphis fabae* (black bean aphid), *Phytonemus pallidus* (strawberry mite), *Haematonectria haematococca* (potato dry rot), *Spodoptera eridania* (southern armyworm), and many more. Other examples of the plant serving as host to pests include *Berkeleyomyces basicola* (black root rot), *Pratylenchus penetrans* (nematode, northern root lesions) and Iris yellow spot virus (iris yellow spot).

5.3 NUTRITIONAL ATTRIBUTES OF THE PLANT

Given the impending issue of global warming, there is a need to find alternative crops which can fulfil the food demands of the world's growing population. According to Jacobsen et al. (2003),

TABLE 5.2
Differentiating Characteristics of Other Species of *Amaranthus* Compared to *Amaranthus retroflexus*

Plant Species	Differentiating Characteristics		
Amaranthus powellii	Inflorescences are thinner, longer branches	Perianth segments are straight and un-reflexed	-
Amaranthus hybridus			Acute perianth segments Shorter and more spiny bracteoles
Amaranthus palmeri, Amaranthus rudis, Amaranthus tuberculatus	Smooth and hairless stems and leaves	Separate plants for male and female flowers	North American species
Amaranthus spinosus, Amaranthus viridis	Smaller flowers	-	-
Amaranthus blitum	Indented leaf tips	-	-
Amaranthus graecizans, Amaranthus blitoides	Mostly axillary inflorescences	-	-

Amaranthus spp. is regarded as an underutilized species with the potential to be an alternative crop for future food security since it is highly nutritious and has an inherent ability to withstand severe climate conditions (i.e., high temperatures, drought, poor soil condition and absence of pests). The seeds and young leaves of the plant are edible with the former can be consumed raw, toasted, or even as sprouts while the seeds can be eaten as cereal. The young leaves of plants or sprouts may be consumed either raw or cooked. This edible plant, however, can result in serious indigestion in sheep, goats, pigs, cattle, and horses.

In a review by Alemayehu et al. (2015), it was observed that amaranth had an overall higher nutritional value than milk, soya bean, wheat, and maize. In a study conducted in Nigeria, the leafy parts of amaranth have served as an affordable source of protein and vitamins (Emokaro, 2007). When compared to commonly consumed vegetables of the Brazilian population, amaranth portrayed superior protein and lipids (Souza et al., 2022). Additionally, Schoenlechner et al. (2010) have reported that amaranth has a high content of lysine and amino acids which also contains sulphur, a chemical that is in insufficient quantities in other cereals (i.e., wheat, maize, and rice), and also limited in legumes. Zhu (2023) stated that proteins in amaranth have a level of antifungal and anticancer protection as well as the capability of modulating the gut microbiome. In the same study, the authors also reported that peptide-rich fractions obtained from the hydrolyzed proteins through enzyme activity can display anticoagulant, antihypertensive, anti-diabetic, anti-inflammatory, and antiviral properties. In addition, the amaranth peptide inhibits the acetylation of certain histone cores (i.e., H_3 and H_4, whereby these histones are part of DNA replication process) and subsequently creates a possible impact where it is anticarcinogenic due to the disturbance of the deacetylation and acetylation processes which cause the death of cells (Arêas et al., 2016). Furthermore, the inclusion of amaranth in human food consumption may reduce the risk of illnesses through multiple biological impacts such as antitumoral, antimicrobial, the lowering of cholesterol, and antioxidant effects.

The seeds or grains of amaranth consist of a wide range of nutrients, such as starch, dietary fibers, proteins, lipids (i.e., squalene), phenolics (i.e., betalains), phytosterols, vitamins and minerals (Balakrishnan and Schneider, 2022; Baraniak and Kania-Dobrowolska, 2022; Caselato-Sousa and Amaya-Farfán, 2012). Furthermore, amaranth grains have a relatively high protein content (in the range 13–18%) and consist of an amino acid profile that is almost perfect for human ingestion (Arêas et al., 2016). Amaranth grains are also reported to have high amounts of multiple minerals, such as magnesium, zinc, iron, calcium, and potassium (Berghofer and Schoenlechner, 2002; Gimplinger et al., 2008). According to Arêas et al. (2016), although amaranth has similar lipid profiles to other cereals, it differs because of the presence of squalene and unsaponifiable fraction. Amaranth grains are seen as useful dietary supplements for the treatment of malnutrition children due to the high levels of lysin, arginine, and histidine. Amaranth grains are also gluten-free (Zhu, 2023), which is beneficial as it can accommodate celiac disease and gluten-intolerant populations.

Souza et al. (2022) found *Amaranthus* spp. to have higher overall vitamin C levels than other conventionally consumed vegetables. Vitamin C not only increases the bioavailability of iron, but is also an antioxidant which helps with the synthesis of several amino acids and hormonal compounds. A deficiency of vitamin C could potentially lead to the development of scurvy (Carr and Natarajan, 2018). The chemoprotective effect from the presence of flavonoids, tannins, and saponins in plants could also potentially make a contribution as antioxidant and antitumor agents (Sreelatha et al., 2012). Among the other examples of antioxidant compounds present in amaranth are tocotrienols, tocopherols, flavonoids, and phenolic compounds (Arêas et al., 2016).

5.4 PHYTOCHEMICAL OVERVIEW OF THE PLANT

In recent times plant polyphenols have increasingly come to the attention of the scientific community, mostly because of their strong antioxidant capabilities and their ability to prevent several diseases and ailments. The *Amaranthus* genus is a globally widespread, phytonutrient-rich plant which has been acknowledged for its traditional use as a remedy for several illnesses and disorders

(Kongdang et al., 2021). Although the redroot pigweed (*Amaranthus retroflexus*) is often regarded as a weed plant in agricultural areas among the *Amaranthus* genus, its consumption as a nutrient source in various parts of the world is evident. In numerous places, *A. retroflexus* is eaten as food in its entirety. The first records note how this herb was utilized medicinally by Native Americans. In fact, several tribes believed that eating the entire plant, when combined with bread, honey, and a white dove's heart, could act as a love potion. Additionally, preparations from *A. retroflexus*'s aerial portions (parts of the plant that are present above ground) have been employed as a mystical herb in religious ceremonies (Fiorito et al., 2017). *A. retroflexus* leaves have been found to be a good source of iron, as well as vitamins A and C. The toasted or roasted seeds of *Amaranthus* are also consumed. They can also be ground into flour that is then used to make bread, eaten as a cereal option, or added to salads. In the southern Molise region of Italy, aerial sections devoid of flowers and leaves are eaten just like spinach.

According to a phytochemical screening by Carvalho et al. (2019), organic acids, amino acids, proteins, azulenes, carotenoids, steroids, coumarins, their derivatives, and saponins were found to be present in the ethanolic extracts of *A. retroflexus* leaves. Of these, saponins and coumarins are related to the allelopathic activity of *A. retroflexus* (Carvalho et al., 2019). In addition, fatty acids, particularly linoleic and linolenic acid, flavonoids (such as rutin and quercetin), alkaloids (such as amaranthine), sesquiterpenes, phenolic acids, and volatiles are among the examples of secondary metabolites identified to date (Fiorito et al., 2017; Goering et al., 1970; Kalinova and Dadakova, 2009).

According to studies determined using UHPLC-UV-MS assay, the leaves of *A. retroflexus* contain a significant amount of rutin (332.25 µg/g in the natural product *Amaranthi folium*). The plant stems also have a significant concentration of rutin. *A. retroflexus* contains protocatechuic acid (1–16 µg/g), chlorogenic acid (15.83 µg/g), caffeic acid (3.62 µg/g), p-coumaric acid (0.3–4.3 µg/g), and ferulic acid (6–19 µg/g). According to the assay, quercetin and its derivatives are not this species' main components (present in low concentrations). It has been noted that the leaves contain more flavonoid content than the seeds do (Saracin et al., 2023).

Regarding this plant's nutraceutical worth, particularly in terms of its presence in phytochemicals with positive effects on human health, little research has been documented. A study carried out by Fiorito et al. specifically checked for the presence of a few biologically active unprenylated and O-prenylated phenylpropanoids, including (1,4) boropinic acid, ferulic acid (2,1), 4′-geranyloxyferulic acid (3,5), umbelliferone (4,2), umbelliprenin (5,8), apigenin (6,3), auraptene (7,7) and 7-isopentenyloxycoumarin (7, 6) in various parts and extracts of *A. retroflexus*.

Of these, the detected analytes were umbelliferone, apigenin, 4′-geranyloxyferulic acid (GOFA), 7-isopentenyloxycoumarin, auraptene and umbelliprenin by extraction methods such as maceration and ultrasound. An overview of the detected phytochemicals from respective parts of *A. retroflexus* under by the most efficient extraction procedures is given in Table 5.3.

TABLE 5.3
Detected Secondary Metabolites (µg/g of Dry Plant) in *A. retroflexus* Extracts under the Best-applied Extraction Procedures

Sample	Part of the plant	Method	Solvent	Quantity
Umbelliferone	Roots	Maceration 96 h	EtOH	29.58 ± 0.02
Apigenin	Roots	Ultrasounds (10 min)	Acetone	46.80 ± 2.33
4′-geranyloxyferulic acid	Roots	Maceration 96 h	EtOH	4.06 ± 0.12
7-isopentenyloxycoumarin	Flowers	Maceration 96 h	Acetone	12.36 ± 0.27
Auraptene	Roots	Maceration 96 h	EtOH	8.21 ± 0.18
Umbelliprenin	Roots	Ultrasounds (10 min)	Hexane	32.15 ± 2.45

Adapted from (Fiorito et al., 2017).

Thus, with the exception of 7-isopentenyloxycoumarin (a phenyloxycoumarin), which was found to be most abundant in flowers (12.36 μg/m), the highest concentrations of quantified compounds were found in the roots of plants. These included umbelliferone (hydroxycoumarin), 4′-geranyloxy-ferulic acid (hydroxycinnamic acid), apigenin (flavone), umbelliprenin and auraptene (phenyl-oxy-coumarins). These secondary metabolites are currently given a large amount of attention due to their dietary benefits in terms of cancer prevention, chemoprevention, and their anti-inflammatory actions. Thus, the information provided here reinforces the possibility for the named edible herb to serve as a food that can improve human welfare (Fiorito et al., 2017).

As mentioned above, this well-known allelopathic weed in the globe, redroot pigweed (*A. retro-flexus*), contains a variety of organic components that are also involved in its allelopathic interactions. Amaranth extract can inhibit the viability and the spread of NB4 cells and can also have a beneficial anti-bacterial effect on Candida fungi and Gram-positive bacteria, according to preliminary tests of the effect of redroot pigweed extract on leukemia and various human pathogenic microorganisms. The identification of amaranth bioactive components, primarily terpenoid compounds (51.71%) as the primary group, and carvacrol (11.33%) as an important compound, was made possible by the phytochemical analysis of redroot pigweed extract (Bakhshayeshan-Agdam et al., 2022). This research offers a promising picture of the medicinal significance of this plant, which contains a variety of chemical substances with allelopathic and therapeutic activities.

The food science community has recently become interested in *A. retroflexus* since it has been demonstrated to be a reliable nutritious alternative to cereals with significant potential nutritional value. Among its benefits are that it is gluten-free and that it has high levels of high-quality protein and unsaturated fatty acids. However, there is little exclusive research about this plant's nutraceutical worth, particularly when it comes to the phytochemical content which has positive impacts on human health. Therefore, study into this subject continues to be a challenging but worthwhile discipline.

5.5 EXTRACTION AND ISOLATION TECHNIQUE

The bioactive compounds found in *A. retroflexus* have been reported for their antioxidant and antimicrobial (Marinas et al., 2014), anti-cancer (Dorostgo et al., 2020), biopesticidal (Azimi et al., 2020), and nutraceutical properties (Fiorito et al., 2017). It has therefore attracted the interest of various healthcare organizations who are keen to exploit it to secure the nutritional and health benefits for mankind. The isolation and extraction of bioactive compounds from *A. retroflexus* is a crucial step in their characterization and utilization. However, the appropriate timing of the collection of plant part and type of the extraction method used affects the availability and yield of the bioactive compound.

The most popular and commonly used traditional extraction techniques include maceration, serial extraction, Soxhlet extraction, and solvent extraction. Of these, Soxhlet extraction is the most advantageous as this technique requires a smaller amount of solvent and minimal extraction time.

The prolonged period of extraction, lower efficiency, high solvent requirement, and non-environmentally-friendly nature, are among the major shortcomings of these conventional techniques. To achieve the environmentally-friendly extraction of bioactive substances, novel techniques such as supercritical fluid, microwave-assisted, accelerated solvent, and ultrasound-assisted extraction (either singly or in combination) have been explored for the extraction of bioactive compounds from *A. retroflexus*. These are characterized by higher yield efficiency, less extraction time, automation, and the non-toxic and minimal consumption of extraction solvents.

Techniques makes use of different solvents, such as ethanol, water, methanol, hexane, and acetone, which have the capacity to dissolve the desired compound. The ratio and polarity of the solvent, sample composition, retention time, and temperature all combine to determine the efficiency of the technique (Garcia-Salas et al., 2010). Different technologies employed for the extraction of bioactive compounds are outlined in Table 5.4.

TABLE 5.4
Different Approaches Employed for the Extraction of Bioactive Compounds from *A. retroflexus*

Plant part	Country	Solvent-Extraction method	Bioactive compounds	Application	References
Roots	Italy	Acetone- Ultrasounds for 10 min	Apigenin	Nutraceutical value	Fiorito et al., 2017
		Hexane- Ultrasounds for 10 min	Umbelliprenin		
Leaves	Romania	Ethanol- Hydro-alcoholic extraction	Leave extract	Acute toxicity, Cytotoxicity, Photo toxicity	Dinu et al., 2017
	India	70% ethanol - Soxhlet's apparatus	Leave extract	DPPH (1,1-Diphenyl-2-Picryl-hydrazyl) free radical scavenging activity	Pandit and Setty, 2015
Pollen	Iran	Acetone-Acetonic extraction	Pollen extract	Antibacterial activity	Bashiri et al., 2022
		Chloroform-methanol (2 : 1)-Folch			
Seed	Iran	Methanol: water (80:20)-Soxhlet's apparatus	Seed extract	Anti-cancer property	Dorostgo et al., 2020
	Lithuania	Supercritical carbon dioxide extraction	Oil and Squalene	Antioxidant	Kraujalis and Venskutonis 2013

In order to analyse the plant extracts, a range of techniques can be used: High-Performance Liquid Chromatography (HPLC) (Fiorito et al., 2017), Mass Spectroscopy (MS), HPLC-Ultraviolet-Visible (HPLC-UV-VIS) chromatograms (Khanam and Oba, 2013), Canonical Discriminant Analyses (Hamidzadeh et al., 2021), and the calorimetric method (de Souza et al., 2022). At present, a series of Ultra HPLC-UV-MS analyses also has gained attention due to their high precision, sensitivity, and the minimal requirements of sample and time (Saracin et al., 2023).

5.5.1 Conventional Technologies

5.5.1.1 Maceration

In the maceration process, a closed system is used in which the powdered form of the sample is soaked in an appropriate solvent for 96 h intervals at room temperature under constant shaking conditions (Olejar et al., 2015); this is then followed by the filtration. Fiorito et al. (2017) have employed this technique for the extraction of nutraceutical bioactive compounds such as umbelliferone, auraptene, and 4-geranyloxyferulic acid (GOFA) from root extract using ethanol as solvent and isopentenyloxycoumarin from the flower of *A. retroflexus* using acetone as solvent.

5.5.1.2 Turbo Extraction Method

This extraction method relies on two or three repetitions of continuous swirling of the sample in the solvent. These shearing forces result in a decrease in particle size, resulting in the effective dissolution of compounds (Sonaglio et al., 2003). Regarding the usage of different solvents, turbo extraction methods have been categorized into three types: aqueous, hydroalcoholic, and methanolic. De Souza et al. (2022) examined the ability of these three turbo extraction methods for the extraction of phenolic content of different *Amaranthus* species. Of these, the hydroalcoholic turbo-extraction was found most effective.

5.5.1.3 Soxhlet Extraction

This technique enables the continuous extraction of substances which are not readily soluble. The powdered plant sample is added to the solvent and kept in a thimble. After that it is inserted into the Soxhlet extractor under the reflux condenser. The heated solvent produces the vapors, which are allowed to condense and drop back to the sample. This circulation of solvent through the extractor allows the continuous extraction of the desired compounds in aqueous form (Azwanida, 2015). In general, organic solvents such as chloroform, methanol, ethanol, n-hexane, propanol, acetone, and ethyl acetate are used because of their ease of mixing, which thereby improves the yield of the bioactive compound extraction process. However, the prolonged time period and the thermal degradation of a bioactive compound due to the risk of organic vapor exposure makes it imperative to find alternative techniques for the production of high-quality extracts.

5.5.2 Environment-friendly Novel Technologies

5.5.2.1 Supercritical Fluid Extraction

Carbon dioxide, which is non-toxic, non-polar, stable, and cheap, is used in superficial fluid extraction. Different combinations of pressure and temperature, usually above the critical point of carbon dioxide, makes this method more convenient for extraction of thermolabile compounds (Morgan, 2013). Krulj et al. (2016) employed three different extraction techniques viz. Soxhlet, accelerated solvent extraction, and supercritical fluid for the squalene and oil extraction from *Amaranthus* sp. 300 bar pressure, 40 °C temperature and 0.194 kg h^{-1} mass flow conditions of supercritical fluid extraction. This led to the extraction of 61.1 and 3.8 g kg^{-1} of oil and squalene, respectively, in 3 h. However, a combination of 300 bar, 40 °C, 2 h and 200 bar, 130 °C, 30 min. was found best for the extraction of oil (64%) and squalene (60%), respectively, from the seeds of *Amaranthus* sp. (Wejnerowska et al., 2013). It has been reported that the utilization of polar co-solvent, along with carbon dioxide, enhanced the efficiency of this technique for the extraction of polar compounds.

5.5.2.2 Accelerated Solvent Extraction

This technique is principally used to extract phenolic chemicals that are thermotolerant. The solute-solvent desorption is facilitated by a pressure range of 3.3–20.3 MPa and a temperature range of 40–200 °C (Zhang et al., 2023). In comparison to supercritical and Soxhlet extraction methods, Krulj et al. (2016) found that accelerated solvent extraction, which was performed at 10MPa, 90 °C, in 20 min. extraction time using n-hexane solvent, produced the best yields of squalene (4.7 g kg^{-1}) and oil (78.1 g kg^{-1}) from *Amaranthus* grain.

5.5.2.3 Ultrasound Assisted Extraction

This method makes use of the induced mechanical power to create tiny bubbles near to the sample, which help in the diffusion of cell content. To create the cavitation, a frequency range of 20–2000 kHz is used. In comparison to other approaches, this method is simple, cheap, easy to handle, and highly efficient, and requires minimal extraction time with the reduced consumption of solvents (Shirzad et al., 2017). Tabio-Garca et al. (2021) optimized the extraction of polyphenols and betalains from *A. hypochondriacus* using HCl as the solvent at 41.80 °C and 188.84 mW/ml. The response surface methodology was used by Roriz et al. (2021) to extract betacyanins and tocopherols from *A. caudatus* flowers.

5.6 APPLICATION OF ACTIVE INGREDIENTS

Redroot amaranth (*Amaranthus retroflexus*) contains several bioactive components that contribute to its medicinal properties. The following sections outline a number of the key bioactive components found in redroot amaranth.

5.6.1 Flavonoids

Quercetin, kaempferol, rutin, and luteolin are specific flavonoids found in Redroot amaranth, each with its own unique properties and potential health benefits (Yang et al., 2008). All of these flavonoids have the primary responsibility for their antioxidation, anti-inflammation, and anti-cancerous properties, and they have several cardiovascular health benefits, as detailed in Table 5.5 (Kumar et al., 2021).

5.6.2 Phenolic Compounds

Numerous phenolic substances, including tannins and phenolic acids like gallic acid, caffeic acid, and protocatechuic acid, are found in Redroot amaranth (Patterson, 1981). These substances have both anti-inflammatory and antioxidant properties (Dhakad et al., 2018), which add to the plant's total therapeutic value.

Gallic acid: Gallic acid is a phenolic acid known for its antioxidant properties. It is also known to reduce cholesterol levels through the improvement of blood vessel function.

Caffeic acid: This has potential protective effects against oxidative stress-related conditions, including cardiovascular diseases and neurodegenerative disorders (Odukoya et al., 2006). Furthermore, it plays a potential role in skin health, including its ability to protect against UV-induced damage and reduce inflammation in the skin (Zhang et al., 2023).

Protocatechuic acid: This is known for its potential neuroprotective effects and its ability to alleviate symptoms in neurodegenerative diseases. (Additionally, it has been investigated for its hepatoprotective effects, supporting liver health (Zhang et al., 2023).)

TABLE 5.5
Functions of Flavonoids and Fatty Acids

Compound	Activity
Flavonoids	Scavenge free radicals
	Inhibits inflammatory mediators
	Induces apoptosis
	Reducing blood
	Improving blood flow
	Enhancing blood vessel function
	Inhibit the growth of cancer cells
Linoleic acid	Skin health (Hansen et al., 1958)
	- Maintains the integrity of the skin barrier
	- Moisture retention
	- Promotes skin elasticity
	Cardiovascular health (Kris-Etherton et al., 2004)
	- Reduces total cholesterol level
	- Improves lipid profile
	Immune function (Field & Schley, 2004) Modulates immune responses
Oleic acid	Heart health (Martirosyan et al., 2007)
	- Reduces LDL cholesterol level
	Anti-inflammatory (Bassaganya-Riera et al., 2002)
	- Modulates neutrophils production
	- Wound healing
	Antioxidant (Gresta et al., 2020) Capture free radicals

5.6.3 Squalene

This is a naturally occurring hydrocarbon molecule, found in abundance in Redroot amaranth. The capacity of squalene to slow the development of cancer cells and trigger apoptosis (cell death) is recognized to have potential anticancer effects (Dastan et al., 2014). Additionally, it possesses antioxidant qualities by neutralizing free radicals that can aid in preventing oxidative damage (Sarker et al., 2022). Squalene's antioxidant activity also makes it potentially useful in protecting the skin against oxidative damage caused by UV radiation and environmental pollutants (Page, 2008). In addition to all these properties, it is a precursor to cholesterol synthesis in the body.

5.6.4 Essential Fatty Acids

Redroot amaranth's seeds contain two important fatty acids: linoleic and oleic. These are involved in several physiological activities in the body. They also improve cardiovascular health and control inflammation.

Linoleic acid: This is an essential omega-6 polyunsaturated fatty acid that plays a crucial role in maintaining healthy cell membranes and is a precursor to other important molecules in the body, such as prostaglandins and leukotrienes (Sarker et al., 2022). Some potential benefits of linoleic acid are seen in Table 5.5.

Oleic acid: Oleic acid is a monounsaturated fatty acid, belonging to the omega-9 family. Some of the potential benefits of oleic acid are shown in Table 5.5.

5.6.5 Vitamins and Minerals

Redroot amaranth is a good source of various vitamins and minerals, including vitamin C, vitamin A, calcium, iron, and magnesium (Jimoh et al., 2018). These nutrients are crucial for the maintenance of general health and have a significant impact on a number of biological processes.

Vitamin C: Redroot amaranth contains vitamin C (ascorbic acid), which plays a crucial role in the immune system, promotes collagen synthesis for healthy skin and connective tissues, aids in iron absorption, and acts as an antioxidant to protect cells from damage (Jimoh et al., 2018).

Vitamin A: Redroot amaranth is a source of vitamin A, which is essential for vision, growth and development, immune function, and the maintenance of healthy skin. It is also involved in various physiological processes, including cell differentiation, gene expression, and immune response (Korn, 2019).

Calcium: Calcium is an essential mineral for building and maintaining strong bones and teeth. It also plays a role in muscle contraction, nerve function, and blood clotting (Mirzaei et al., 2017).

Iron: Redroot amaranth contains iron, an essential mineral involved in the production of red blood cells and oxygen transport throughout the body (Bratovcic and Saric, 2019).

Magnesium: Magnesium is a mineral involved in numerous biochemical reactions in the body. It plays a role in energy production, protein synthesis, muscle and nerve function, and bone health. Magnesium is also important for maintaining a healthy heart rhythm and supporting the immune system.

5.6.6 Proteins and Peptides

Amaranth is known to contain a variety of storage proteins containing amino acids, such as globulins and albumins. It produces defence peptides as part of their immune response against pathogens and pests and have antimicrobial or insecticidal properties (Roskamp et al., 2013).

5.7 PHARMACOLOGICAL APPLICATIONS AND CLINICAL TRIALS

The weed plant of Redroot pigweed (*Amaranthus retroflexus*) is well known for its allelopathic effects throughout the world. Although it's a major concern, in recent years it has been found to have diverse therapeutic effects such as anti-cancerous, antibacterial, and antioxidant properties due to the presence of various biochemicals and polyphenols. The important properties and their role are discussed in the section below.

5.7.1 Anti-cancer Properties

A study on Redroot pigweed (Bakhshayeshan-Agdam et al., 2017), focusing on the chemical applications of the plant, showed that it can be used as a potential drug for the treatment of leukemia cancer patients. They also found a strong relationship between the allelopathic traits and the anti-cancerous potential of the plant. The MTT (3-(4,5-dimethylthiazol-2-yl)-2,5-diphenyltetrazolium bromide) assay of redroot pigweed ethanolic extract on human acute promyelocytic leukemia (APL) cancer cell lines (NB4 cells) resulted in successful inhibition of the cells. In a time- and dose-dependent manner, it hindered the proliferation and the viability of the NB4 cells. When the cells were treated with a concentration of 400 μg/ml of the extract, 70% viability decrease was observed in a period of 72 h. In a continuation of this study by the same author (Bakhshayeshan-Agdam et al., 2022), they used Hoechst staining and Annexin V-PI double staining to revel the apoptotic effect and apoptosis induction, respectively, of the redroot pigweed extract qualitatively on NB4 cells. However, they also saw that at higher concentrations of the pigweed extract (160 μg ml^{-1}) it also inhibits the proliferation and viability of the normal cells (HDFa cells) by less than 30%. In the same study they also mentioned about the excellent antibacterial effects of the extract on the Gram-positive bacteria and fungi *Staphylococcus Aureus*, *Staphylococcus epidermidis* and *Candida albicans* (in accordance with the other established literature). They also found that the components in the extracts of *A. retroflexus* were principally terpenoids, and carvacrol, which accounted for highest percentage (19.96%). According to the published research, these compounds are called allelochemicals, which are known for their therapeutic properties (Cheng and Cheng, 2015). These are used as anti-cancer agents, which is an overall advantage as weed plants can be used as anti-cancer remedies and in other drug production. Another species of *Amaranthus*, *A. mantegazzianus*, exhibit anti-tumor effects. In a study (Barrio and Añón, 2010), they found that protein isolated from the plant was responsible for the anti-tumor activity on four different tumor cells: MC3T3E1, UMR106, Caco-2, and TC7 cell lines. MPI was seen to inhibit cell adhesion and induce necrosis and apoptosis in the malignant cells of UMR 106.

5.7.2 Anti-bacterial and Anti-fungal Properties

A study was done to compare the effects of the silver nanoparticles with the alcoholic extract of *A. retroflexus* on some pathogenic bacterial species. The results showed that both silver nanoparticles as well as alcoholic extracts affected the bacteria growth. However, a synergistic property was established between them when they were both used in combination, and it resulted in enhanced antimicrobial activity. They also found and stated that *A. retroflexus* has a more significant inhibitory effect on the Gram-positive bacteria than on the Gram-negative ones (Rasi-Bonab et al., 2018). The inhibitory effects are seen on *S. aureus, Klebsiella pneumoniae*, *Bacillus subtilis*, *Candida famata* (fungal strain) and other fungi, either alone, or in combination with silver nanoparticles (Bahrami-Teimoori et al., 2017; Rasi-Bonab et al., 2018; Saritha et al., 2015). Antimicrobial activities of the different combinations of the A. retroflexus plant extracts are presented in Table 5.5. In another piece of research, Poiată et al. (2016), the researchers used various extracts of *A. retroflexus* to test the ability of inhibition of the microbial growth by the plant. They used the various combinations (Table 5.1) and monitored the antimicrobial activity by measuring the size of the zone of inhabitation that formed around the sample. They found that the chloroform sample from the seeds of the plant at the concentration of 16 mg/ml showed an excellent inhibitory activity against both Gram-positive and Gram-negative bacteria; it was the most active of all the combinations they used, showing the largest zones of inhibition. They also confirmed the antifungal activity of *A. retroflexus* against *C. albicans*, which is a type of fungi. They suggested that antimicrobial/antifungal compounds from the plant, especially peptides, can be used in discovering novel antimicrobial agents in future. Manso et al. (2022) showed that the extracts which were obtained from the leaves and the inflorescences of *A retroflexus* vegetative organs showed antibacterial activity against *Acinetobacter baumannii*. Gallic acid equivalents/ml were used in the study to express the overall amount of polyphenols in the extract. It is also noteworthy that the extract of *A. retroflexus* inflorescences had the highest polyphenol concentration. This extract was also the most effective against the studied microbial strains, indicating that antibacterial activity and polyphenol concentration are linked, as suggested by earlier investigations. They also showed that *A. retroflexus* is active against clinical strains of *Pseudomonas aeruginosa*. Against clinical strains of *P. aeruginosa*, the antibacterial activity was enhanced by mixing an alcoholic extract of *A. retroflexus* leaves with trimetoprim-sulphamethoxazole, piperacillin-tazobactam, ciprofloxacin, ticarcillin-clavulanic acid, piperacillin, ceftriaxone, and colistin. The antibacterial effects of ceftriaxone, ticarcillin-clavulanic acid, colistin, piperacillin, norfloxacin, and oxacillin were enhanced by an extract from the inflorescences of *A. retroflexus*. The extracts from the leaves and inflorescences of the vegetative organs of several species were also proven effective against *S. cerevisiae*, *Candida utilis*, *C. albicans*, and *C. famata* (Table 5.6).

5.7.3 Antioxidant Properties

The antioxidant activities of redroot pigweed leaves have also been studied by some researchers. In one study, the antioxidant activity of the methanolic extract of the plant along with its terpene

TABLE 5.6
Antimicrobial Properties of the *A. retroflexus* Plant Extracts

Sr. No	Sample (0.2 ml Volume)	*S. aureus* 25923	*S. lutea* ATCC 9341	*E. Coli* ATCC 25922	*P. aeruginosa* ATCC 27853	*C. albicans* ATCC 10231
1	Root aqueous extract	12 ± 0.1	16 ± 0.3	0	0	20 ± 0.5
2	Leaves and stems- aqueous extract	13 ± 0.2	13 ± 0.1	0	0	23 ± 0.4
3	Leaves and stems aqueous extract (80%)	13 ± 0.4	13 ± 0.5	0	0	12 ± 0.1
4	Root- ethanol extract (80%)	15 ± 0.1	15 ± 0.3	0	0	16 ± 0.3
5	Seeds- ethanol extract (80%)	12 ± 0.3	15 ± 0.2	0	0	12 ± 0.5
6	Seeds- chloroform extract	18 ± 0.4	20 ± 0.1	20 ± 0.3	19 ± 0.4	28 ± 0.4
7	Ampicillin (10 μg)	27 ± 0.6	40 ± 0.5	21 ± 0.5	19 ± 0.5	nt**
8	Chloramphenicol (30 μg)	25 ± 0.3	38 ± 0.3	24 ± 0.3	18 ± 0.2	nt
9.	Nystatin (100 μg)	nt	Nt	nt	nt	20 ± 0.1

** ATCC- American type culture collection, - not tested.

constituents was evaluated (Pacifico et al., 2008). The experiments showed that the methanolic extract of the plant had a strong effect on dose-response antioxidant activity, whereas the pure nerolidol derivatives had similar antioxidant capacities to a standard α-tocopherol. According to them, through the measurement of the inhibition of the formation of thiobarbituric acid reactive species (TBARS), a peculiar effect on lipid peroxidation can be detected, and this is important as it resembles oxidative stress as implicated in several pathological processes. Similarly, in another study they used 70% ethanolic extract *of Amaranthus retroflexus* leaves (called AREE) to find the power and DPPH (1,1-Diphenyl-2-Picryl-hydrazyl) free radical scavenging activity through the use of in-vitro models (Jyoti and Setty, 2015). They used this to evaluate the antioxidant activity of AREE, finding that it decreased significantly, and concluded that AREE expresses significant antioxidant activities which may be due to the presence of several polyphenols in the parts of *A. retroflexus*.

Similarly, in recent published research they used UHPLC–UV–MS analysis to study the polyphenolic compound, called rutin, which is known to be used in various pharmaceutical preparations. They found different parts of the redroot pigweed plant to be a great source of rutin with 332.25 μg/g concentration in natural products, including various other polyphenols (Saracin et al., 2023). However, the rutin extracted from the leaves should only be used in pharmaceutical preparations as leaves contain the highest concentration of the chemical. These phenolic compounds are known to have antioxidant activities as they eliminate the radicals, thereby slowing the process of lipid oxidation and hydroperoxide formation. They also showed the inhibitory activity of rutin against *P. aeruginosa*. They stated that the combination of rutin with quercetin showed the best antibacterial activity with synergic effects. Another paper showed the various antioxidant activities of the whole *Amaranthus* species from the *Amaranthus* family. They described the various assays used in the assessment of radial scavenging activities of the plant, such as 1-dipheny-2-picrylhydrazyl (DPPH) radial scavenging activity, hydroxyl radical scavenging activity, phosphomolybdenum-scavenging activity, and ferric-reducing antioxidant power (FRAP) (Adegbola et al., 2020).

Species of *Amaranthus* have also been shown to have antimalarial properties. This was demonstrated by Susantiningsih et al. (2012), in an animal model study. They gave an ethanolic formulation of *A. spinosus* to mice infected with *Plasmodium berghei*, and this resulted in increased body weight, increased hemoglobin content, and blood schizonticidal activity, thereby indicating an antimalarial effect like chloroquine.

In recent years, multiple clinical trials have been conducted on various species of *Amaranthus*. In a review conducted by Peter and Gandhi (2017), they summarize various such studies. For example, in moderately hypercholesterolemic patients, the effects of the defatted Amaranth snacks of *A. caudatus* on the plasma lipids of such patients, taking extruded (50 g) amaranth continuously over 60 days led to reduced levels of LDL as well as HDL. In a study, 90 postmenopausal women were given a dose of *A. tricolor* leaf powder (9 g) over a period of 3 months. They found that it helped in the prevention of post-menopause complications, by increasing the levels of serum retinol, serum ascorbic acid, hemoglobin, GST Px and SOD, and a decrease in MDA levels and a significant decrease in the fasting blood glucose levels of the subjects. In another study, they used a formulation of sun-dried *Amaranthus* leaves to show the significantly improved bioavailability of serum β-carotene, retinol, and hemoglobin levels in preschool children. All this indicates that it might be of potential in help curing Vitamin A deficiency and anemia among lactating mothers and children, mostly in developing countries. In another study, they showed that a single oral dose of amaranth extract was able to increase the levels of NO3− and $NO2^-$ in the body for at least 8 hours. The increased levels have seen to improve the overall performance of people involved in heavy physical activities or sports.

Overall, researchers show than *A. retroflexus* has been reported to have antipyretic, anti-ulcer, and anti-inflammatory activities and can be used in the treatment of diseases like rheumatoid arthritis, the treatment of asthma, and diseases of the urinary tract. In summary, all these properties of the plant make it important from a medical point of view.

5.8 CONCLUSION

Amaranthus retroflexus is also commonly known as wild beet amaranth, careless weed, common amaranth, red-root amaranth, pigweed, common tumbleweed, red-root pigweed, pigweed amaranth, or rough pigweed. It is a valuable plant with a variety of uses, including medicinal and non-medicinal ones. *A. retroflexus* is enriched with flavonoids, phenolic compounds, essential fatty acids, and vitamins (vitamins C and A). Bioactive compounds were extracted from the roots, leaves, seeds, and pollen of this plant. Green technologies were also employed to derive health-supporting compounds, providing an alternative to utilize the components in a variety of applications.

REFERENCES

Adegbola, P. I., Adetutu, A., & Olaniyi, T. D. (2020). Antioxidant activity of *Amaranthus* species from the Amaranthaceae family—A review. *South African Journal of Botany*, *133*, 111–117.

Ajayi, F. F., Mudgil, P., Gan, C. Y., & Maqsood, S. (2021). Identification and characterization of cholesterol esterase and lipase inhibitory peptides from amaranth protein hydrolysates. *Food Chemistry X*, *12*, 100165.

Alebrahim, M. T., Majd, R., Rashed Mohassel, M. H., Wilcockson, S., Baghestani, M. A., Ghorbani, R., & Kudsk, P. (2012). Evaluating the efficacy of pre-and post-emergence herbicides for controlling *Amaranthus retroflexus* L. and *Chenopodium album* L. in potato. *Plant Protection*, *42*, 345–350.

Alemayehu, F. R., Bendevis, M. A., & Jacobsen, S. E. (2015). The Potential for Utilizing the Seed Crop Amaranth (*Amaranthus* spp.) in East Africa as an Alternative Crop to Support Food Security and Climate Change Mitigation. *Journal of Agronomy and Crop Science*, *201*(5), 321–329.

Al-Mamun, M. A., Husna, J., Khatun, M., Hasan, R., Kamruzzaman, M., Hoque, K. M. F., Reza M. A., & Ferdousi Z. (2016). Assessment of antioxidant, anticancer and antimicrobial activity of two vegetable species of *Amaranthus* in Bangladesh. *Complementary and Alternative Medicine*, *16*, 157.

Arêas, J. A. G., Carlos-Menezes, A. C. C. C., & Soares, R. A. M. (2016). Amaranth. In B. Caballero, P. M. Finglas, & F. Toldrá (Eds.), *Encyclopedia of Food and Health* (pp. 135–140). Academic Press.

Azadbakht, A., Alebrahim, M., & Ghavidel, A. (2017). The effect of chemical and non-chemical control methods on weeds in potato (*Solanum tuberosum* L.) cultivation in Ardabil province. Iran. *Applied Ecology and Environmental Research*, *15*, 1359–1372.

Azimi, S., Rahmani, S., & Pazhouhandeh, M. (2020). Effect of protein extracts of *Amaranthus retroflexus* (*Amaranthaceae*) and Cuminum cyminum (*Apiaceae*) on digestive proteinases and biological characters of Helicoverpa (*Heliothis*) armigera (*Hübner*)(Lepidoptera: Noctuidae). *The Canadian Entomologist*, *152*(5), 646–662.

Azwanida, N. N. (2015). A review on the extraction methods use in medicinal plants, principle, strength and limitation. *Medicinal and Aromatic Plants*, *4*(196).

Bahrami-Teimoori, B., Nikparast, Y., Hojatianfar, M., Akhlaghi, M., Ghorbani, R., & Pourianfar, H. (2017). Characterisation and antifungal activity of silver nanoparticles biologically synthesised by *Amaranthus retroflexus* leaf extract. *Journal of Experimental Nanoscience*, *12*, 1–11.

Bajwa, A. A., Zulfiqar, U., Sadia, S., Bhowmik, P., & Chauhan, B. S. (2019). A global perspective on the biology, impact and management of *Chenopodium album* and *Chenopodium murale*: two troublesome agricultural and environmental weeds. *Environmental Science and Pollution Research*, *26*, 5357–5371.

Bakhshayeshan-Agdam, H., Salehi-Lisar, S. Y., Sedghi Samarkhazan, N., Mahdavi, M., Motafakkerazad, R., Khodaie, F., Zarrini, G., & Razeghi, J. (2022). In vitro study of redroot pigweed effects as weed plant species on leukemia. *Natural Product Research*, *36*(17), 4411–4414.

Bakhshayeshan-Agdam, H., Sedghi, N., Lisar, S. Y., and Mahdavi, M. (2017). Anti-cancer effect of redroot pigweed as weed plant species. *Proceedings of International Congress on Biomedicine*, Tehran, Iran.

Balakrishnan, G., & Schneider, R. G. (2022). The role of amaranth, quinoa, and millets for the development of healthy, sustainable food productsandmdash; a concise review. *Foods*, *11*(16), 2442.

Baraniak, J., & Kania-Dobrowolska, M. (2022). The dual nature of amaranth-functional food and potential medicine. *Foods*, *11*(4), 618.

Barrio, D. A., & Añón, M. C. (2010). Potential antitumor properties of a protein isolate obtained from the seeds of *Amaranthus* mantegazzianus. *European Journal of Nutrition*, *49*(2), 73–82.

Bashiri, Z., Yousefi, M., Royce, S. G., & Fereidouni, M. (2022). Antibacterial activity of aqueous and lipid extracts of five common allergenic pollens. *Journal of Medicinal Plants*, *21*(83), 1–8.

Bassaganya-Riera, J., Hontecillas, R., & Beitz, D. C. (2002). Colonic anti-inflammatory mechanisms of conjugated linoleic acid. *Clinical Nutrition*, *21*(6), 451–459.

Batsatsashvili, K., Mehdiyeva, N. P., Kikvidze, Z., Khutsishvili, M., Maisaia, I., Sikharulidze, S., Tchelidze, D., Alizade, V. M., Zambrana, N. Y. P., & Bussmann, R. W. (2017). *Amaranthus retroflexus* L. Amaranthaceae. In: Bussmann R. W. (Ed.), *Ethnobotany of the Caucasus* (pp. 107–112). Springer International Publishing.

Bensch, C. N., Horak, M. J., & Peterson, D. (2003). Interference of redroot pigweed (*Amaranthus retroflexus*), Palmer amaranth (*A. palmeri*), and common waterhemp (*A. rudis*) in soybean. *Weed Science*, *51*(1), 37–43.

Berghofer, E., & Schoenlechner, R. (2002). *Grain Amaranth* (pp. 219–260). Springer.

Bratovcic, A., & Saric, E. (2019). Determination of essential nutrients and cadmium in the white quinoa and amaranth seeds. *Croatian Journal of Food Science And Technology*, *11*(1), 135–139.

CABI. (2021). *Amaranthus Retroflexus (Redroot Pigweed)* (Vol. CABI Compendium). CABI International CABI.

Carr, A. C., & Natarajan, R. (2018). *Vitamin C: Current Concepts in Human Physiology*. MDPI – Multidisciplinary Digital Publishing Institute.

Carvalho, M. S. S., Andrade-Vieira, L. F., Santos, F. E. D., Correa, F. F., das Graças Cardoso, M., & Vilela, L. R. (2019). Allelopathic potential and phytochemical screening of ethanolic extracts from five species of *Amaranthus* spp. in the plant model *Lactuca sativa*. *Scientia Horticulturae*, *245*, 90–98.

Caselato-Sousa, V. M., & Amaya-Farfán, J. (2012). State of Knowledge on Amaranth Grain: A Comprehensive Review. *Journal of Food Science*, *77*(4), 93–104.

Chaves, M. M., & Oliveira, M. M. (2004). Mechanisms underlying plant resilience to water deficits: prospects for water-saving agriculture. *Journal of Experimental Botany*, *55*, 2365–2384.

Cheng, F., & Cheng, Z. (2015). Research progress on the use of plant allelopathy in agriculture and the physiological and ecological mechanisms of allelopathy. *Frontiers In Plant Science*, *6*, 1020.

Coelho, É. M. P., Barbosa, M. C., Mito, M. S., Mantovanelli, G. C., Oliveira, R. S. Jr., & Ishii-Iwamoto, E. L. (2017). The activity of the antioxidant defense system of the weed species *Senna obtusifolia* L. and its resistance to allelochemical stress. *Journal of Chemical Ecology*, *43*, 725–738.

Costea, M., Weaver, S. E., & Tardif, F. J. (2003). The biology of Canadian weeds. 130. *Amaranthus retroflexus* L., *A. powellii* S. Watson and *A. hybridus* L. *Canadian Journal of Plant Science*, *84*, 631–668.

Dastan, D., Salehi, P., Ghanati, F., Gohari, A. R., Maroofi, H., & Alnajar, N. (2014). Phytotoxicity and cytotoxicity of disesquiterpene and sesquiterpene coumarins from Ferula pseudalliacea. *Industrial Crops and Products*, *55*, 43–48.

de Souza, K. K. C., Salimena, J. P., Xavier, J. B., de Souza, D. C., Bertolucci, S. K. V., Bittencourt, W. J. M., & Resende, L. V. (2022). Nutritional and bioactive compounds of *Amaranthus* spp. in Brazil. *Research, Society and Development*, *11*(13).

Dhakad, A. K., Pandey, V. V., Beg, S., Rawat, J. M., & Singh, A. (2018). Biological, medicinal and toxicological significance of Eucalyptus leaf essential oil: a review: Biological, medicinal and toxicological significance of Eucalyptus leaf essential oil. *Journal of the Science of Food and Agriculture*, *98*(3), 833–848.

Dinu, M., Anghel, A. I., Olaru, O. T., Şeremet, O. C., Calalb, T., Cojocaru-Toma, M., & Ancuceanu, R. (2017). Toxicity investigation of an extract of *Amaranthus retroflexus* L. (Amaranthaceae) leaves. *Farmacia*, *65*(2), 289–294.

Dorostgo, Z., Movafagh, H., Sarravi, R., Araei, S. A., & Rezaei, F. (2020). The effect of hydroalcoholic extract of *Amaranthus retroflexus* L. seeds on cox2 and her2 genes on the induced colorectal cancer in an animal model of balbc mouse. *Journal of Complementary Medicine Research*, *11*(3), 141–141.

Emokaro, C. O. (2007). Profitability and production constraints in dry season amaranth production in Edo South, Nigeria. *Journal of Food, Agriculture and Environment*, *5*(3), 281–283

Esiyok, D., Ötles, S., & Akcicek, E. (2004). Herbs as a food source in Turkey. *Asian Pacific Journal of Cancer Prevention*, *5*, 334–339

Feltner, K. C. (1970). The ten worst weeds of field crops; pigweed. *Crops and Soils*, *23*(13–14).

Field, C. J., & Schley, P. D. (2004). Evidence for potential mechanisms for the effect of conjugated linoleic acid on tumor metabolism and immune function: lessons from n–3 fatty acids. *The American Journal of Clinical Nutrition*, *79*(6), 1190–1198.

Fiorito, S., Epifano, F., Palmisano, R., Genovese, S., & Taddeo, V. A. (2017). A re-investigation of the phytochemical composition of the edible herb *Amaranthus retroflexus* L. *Journal of Pharmaceutical and Biomedical Analysis*, *143*, 183–187.

Frankton, C., & Mulligan, G. A. (1987). *Weeds of Canada (Publication 948)*. NC Press.

Gambino, P., & Vilela, A. (2011). Morphological traits and allocation patterns related to stress-tolerance and seed yield in wild and domesticated evening primrose (*Oenothera* L. *Onagraceae*). *Industrial Crops and Products*, *34*, 1269–1276.

Garcia-Salas, P., Morales-Soto, A., Segura-Carretero, A., & Fernández-Gutiérrez, A. (2010). Phenolic-compound-extraction systems for fruit and vegetable samples. *Molecules*, *15*(12), 8813–8826.

Gimplinger, D. M., Schulteauf'm Erley, G., Dobos, G., & Kaul, H. P. (2008). Optimum crop densities for potential yield and harvestable yield of grain amaranth are conflicting. *European Journal of Agronomy*, *28*(2), 119–125.

Goering, K., Rao, P. S., Fritts, D., & Carroll, T. (1970). New Starches VI. The structure of the starch chunks from *Amaranthus retroflexus*. *Starch – Stärke*, *22*(7), 217–221.

Grant, W. F. (1959). Cytogenetic studies in *Amaranthus*: III. Chromosome numbers and phylogenetic aspects. *Canadian Journal of Genetics and Cytology*, *1*(4), 313–328.

Gresta, F., Meineri, G., Oteri, M., Santonoceto, C., Lo Presti, V., Costale, A., & Chiofalo, B. (2020). Productive and Qualitative Traits of *Amaranthus* Cruentus L.: An Unconventional Healthy Ingredient in Animal Feed. *Animals (Basel)*, *10*(8), 1428.

Hamidzadeh Moghadam, S., Alebrahim, M. T., Tobeh, A., Mohebodini, M., Werck-Reichhart, D., MacGregor, D. R., & Tseng, T. M. (2021). Redroot pigweed (*Amaranthus retroflexus* L.) and lamb's quarters (*Chenopodium album* L.) populations exhibit a high degree of morphological and biochemical diversity. *Frontiers in Plant Science*, *12*, 593037.

Hansen, A. E., Haggard, M. E., Boelsche, A. N., Adam, D. J., & Wiese, H. F. (1958). Essential fatty acids in infant nutrition. III. Clinical manifestations of linoleic acid deficiency. *The Journal of Nutrition*, *66*(4), 565–576.

Horak, M. J., & Loughin, T. M. (2000). Growth analysis of four *Amaranthus* species. *Weed Science*, *48*, 347–355.

Jacobsen, S. E., Mujica, A., & Ortiz, R. (2003). The Global Potential for Quinoa and Other Andean Crops. *Food Reviews International*, *19*(1–2), 139–148.

Jimoh, M. O., Afolayan, A. J., & Lewu, F. B. (2018). Suitability of *Amaranthus* species for alleviating human dietary deficiencies. *South African Journal of Botany*, *115*, 65–73.

Jyoti, P., & Setty, S. R. (2015). Antioxidant activities of *Amaranthus retroflexus* leaves. *International Journal of Pharmaceutics and Drug Analysis*, *3*(7), 223–225.

Kalinova, J., & Dadakova, E. (2009). Rutin and total quercetin content in amaranth (*Amaranthus* spp.). *Plant Foods for Human Nutrition*, *64*, 68–74.

Karamac, M., Gai, F., Longato, E., Meineri, G., Janiak, M. A., Amarowicz, R., & Peiretti, P. G. (2019). Antioxidant activity and phenolic composition of amaranth (*Amaranthus caudatus*) during plant growth. *Antioxidants*, *8*, 173.

Kaźmierczak, A., Bolesławska, I., & Przysławski J. (2011). Amaranth—its use in the prevention and treatment of certain civilization-related diseases. *Nowiny Lekarskie*, *80*, 192–198.

Kenžević, S. Ž., Weise, S. F., & Swanton, C. J. (1995). Comparison of empirical models depicting density of (*Amaranthus retroflexus* L.) and relative leaf area as predictors of yield loss in maize (*Zea mays* L.). *Weed Research*, *35*, 207–214.

Khanam, U. K. S., & Oba, S. (2013). Bioactive substances in leaves of two amaranth species, *Amaranthus* tricolor and A. hypochondriacus. *Canadian Journal of Plant Science*, *93*(1), 47–58.

Kongdang, P., Dukaew, N., Pruksakorn, D., & Koonrungsesomboon, N. (2021). Biochemistry of *Amaranthus* polyphenols and their potential benefits on gut ecosystem: A comprehensive review of the literature. *Journal of Ethnopharmacology*, *281*, 114547.

Korn, L. (2019). Plants for the prevention and treatment of Type 2 Diabetes: Salish Country and plants from around the world. *Fourth World Journal – Center for World Indigenous Studies*, *17*(2), 44–61.

Kraujalis, P., & Venskutonis, P. R. (2013). Optimisation of supercritical carbon dioxide extraction of amaranth seeds by response surface methodology and characterization of extracts isolated from different plant cultivars. *The Journal of Supercritical Fluids*, *73*, 80–86.

Kris-Etherton, P. M., Hecker, K. D., & Binkoski, A. E. (2004). Polyunsaturated Fatty Acids and Cardiovascular Health. *Nutrition Reviews*, *62*(11), 414–426.

Krulj, J., Brlek, T., Pezo, L., Brkljača, J., Popović, S., Zeković, Z., & Bodroža Solarov, M. (2016). Extraction methods of *Amaranthus* sp. grain oil isolation. *Journal of the Science of Food and Agriculture*, *96*(10), 3552–3558.

Kumar, V., Mishra, D., Joshi, M. C., Mishra, P., & Tanwar, M. (2021). Herbs and spices- New processing technologies. *Syzygium aromaticum*: Medicinal Properties and Phytochemical Screening. In: Rabia, S.A. (Ed.) *Herbs and Spices-New Processing Technologies*. IntechOpen.

Ma, X., Wu, H., Jiang, W., Ma, Y., & Ma, Y. (2015). Interference between redroot pigweed (*Amaranthus retroflexus* L.) and Cotton (*Gossypium hirsutum* L.): growth analysis. *PLoS One 10*, e0130475.

Mandák, B., Zákravský, P., Dostál, P., & Placková, I. (2011). Population genetic structure of the noxious weed *Amaranthus retroflexus* in Central Europe. *Flora*, *206*, 697–703.

Manso, T., Lores, M., & de Miguel, T. (2022). Antimicrobial Activity of Polyphenols and Natural Polyphenolic Extracts on Clinical Isolates. *Antibiotics*, *11*(1), 46.

Marinaş, I. C., Chifiriuc, C., Oprea, E., & Lazăr, V. (2014). Antimicrobial and antioxidant activities of alcoholic extracts obtained from vegetative organs of *A. retroflexus*. *Roumanian Archives of Microbiology and Immunology*, *73*(1–2), 35–42.

Martirosyan, D. M., Miroshnichenko, L. A., Kulakova, S. N., Pogojeva, A. V., & Zoloedov, V. I. (2007). Amaranth oil application for coronary heart disease and hypertension. *Lipids in Health and Disease*, *6*(1), 1–1.

Mirzaei, M., Rastgoo, M., Ghalibaf, K. H. M., & Zand, E. (2017). Evaluation of the Inhibitory Effects of Different Salts on 2,4-D Efficacy in Control of Kochia (*Kochiascoparia* L.) and Redroot pigweed (*Amaranthus retroflexus* L.). *Majallah-i hifāzat-i giyāhān*, *31*(2), 267–276.

Morales, D.,, Miguel, M., & Garcés-Rimón M. (2020). Pseudocereals: A novel source of biologically active peptides. *Critical Reviews in Food Science and Nutrition*, *61*, 1537–1544.

Morgan, E. D. (2013). Natural products: supercritical fluid chromatography. In: *Reference Module in Chemistry, Molecular Sciences and Chemical Engineering*. Elsevier.

Murray, M. J. (1940). The genetics of sex determination in the family *Amaranthaceae*. *Genetics*, *25*(4), 409–431.

Natural Resources Conservation Service. (1980). *Amaranthus retroflexus L.* United States Department of Agriculture (USDA). https://plants.usda.gov/home/plantProfile?symbol=AMRE

Odukoya, O. A., Inya-Agha, S. I., Segun, F. I., Sofidiya, M. O., & Ilori, O. O. (2006). Antioxidant activity of selected Nigerian green leafy vegetables. *Planta Medica*, *72*(11), 169.

Olejar, K. J., Fedrizzi, B., & Kilmartin, P. A. (2015). Influence of harvesting technique and maceration process on aroma and phenolic attributes of Sauvignon blanc wine. *Food Chemistry*, *183*, 181–189.

Pacifico, S., D'Abrosca, B., Golino, A., Mastellone, C., Piccolella, S., Fiorentino, A., & Monaco, P. (2008). Antioxidant evaluation of polyhydroxylated nerolidols from redroot pigweed (*Amaranthus retroflexus*) leaves. *LWT - Food Science and Technology*, *41*(9), 1665–1671.

Page, L. (2008). *Healthy Healing's Detoxification: Programs to Cleanse, Purify and Renew*. Healthy Healing Publication.

Pandit, J., & Setty. S. R. (2015). Antioxidant activities of *Amaranthus retroflexus* leaves. *International Journal of Pharmaceutics and Drug Analysis*, *3*(7), 223–225.

Paniagua-Zambrana, N. Y., Bussmann, R. W., Echeverría, J., & Romero, C. (2020). *Amaranthus caudatus* L. *Amaranthus deflexus* L. *Amaranthus hybridus* L. *Amaranthus retroflexus* L. *Amaranthus spinosus* L. *Amaranthaceae*. In N. Y. Paniagua-Zambrana, & R. W. Bussmann (Eds.), *Ethnobotany of the Andes* (pp. 199–207). Springer International Publishing.

Patterson, D. T. (1981). Effects of allelopathic chemicals on growth and physiological responses of soybean (*Glycine max*). *Weed Science*, *29*(1), 53–59.

Peter, K., & Gandhi, P. (2017). Rediscovering the therapeutic potential of *Amaranthus* species: A review. *Egyptian Journal of Basic and Applied Sciences*, *4*(3), 196–205.

Plumlee, K. H. (2004). Plants. In K. H. Plumlee (Ed.) *Clinical Veterinary Toxicology* (pp. 339–442). Mosby.

Poiată, A., Apetrei, C., & Ivanescu, B. (2016). Antimicrobial effects of different extracts from *Amaranthus retroflexus* L. *Analele Stiintifice ale Universitatii" Alexandru Ioan Cuza" din Iasi Sec. II a. Genetica si Biologie Moleculara*, *17*(2), 75–80.

Rasi-Bonab, F., Afaghi-Gharamaleki, A., Feizi, H., Alipoor-Dolatabad, S., Alizadeh, S., & Sadat-Amini, M. (2018). Evaluation of silver nanoparticles and alcoholic extract of *Amaranthus retroflexus* on pathogenic bacteria. *Health Biotechnology and Biopharma*, *2*, 12–23.

Roriz, C. L., Xavier, V., Heleno, S. A., Pinela, J., Dias, M. I., Calhelha, R. C., & Barros, L. (2021). Chemical and bioactive features of *Amaranthus caudatus* L. flowers and optimized ultrasound-assisted extraction of betalains. *Foods*, *10*(4), 779.

Roskamp, J. M., Chahal, G. S., & Johnson, W. G. (2013). The effect of cations and ammonium sulfate on the efficacy of dicamba and 2,4-D. *Weed Technology*, *27*(1), 72–77.

Saracin, A. P., Andrei, B. I. Ț. Ă., Saracin, I. A., Constantinescu, E., & Tanasie, Ș. E. (2023). Determination by UHPLC–UV–MS of polyphenol content of *Amaranthus retroflexus*. *Notulae Botanicae Horti Agrobotanici Cluj-Napoca*, *51*(1), 13102.

Saritha, K., Rajesh, A., Manjulatha, K., Setty, O. H., & Yenugu, S. (2015). Mechanism of antibacterial action of the alcoholic extracts of *Hemidesmus indicus* (L.) R. Br. ex Schult, *Leucas aspera* (Wild.), *Plumbago zeylanica* L., and *Tridax procumbens* (L.) R. Br. ex Schult. *Frontiers in Microbiology*, *6*, 577.

Sarker, U., Lin, Y.-P., Oba, S., Yoshioka, Y., & Hoshikawa, K. (2022). Prospects and potentials of underutilized leafy Amaranths as vegetable use for health-promotion. *Plant Physiology and Biochemistry*, *182*, 104–123.

Schoenlechner, R., Mandala, I., Kiskini, A., Kostaropoulos, A., & Berghofer, E. (2010). Effect of water, albumen and fat on the quality of gluten-free bread containing amaranth. *International Journal of Food Science and Technology*, *45*(4), 661–669.

Shirzad, H., Niknam, V., Taheri, M., & Ebrahimzadeh, H. (2017). Ultrasound-assisted extraction process of phenolic antioxidants from Olive leaves: a nutraceutical study using RSM and LC–ESI–DAD–MS. *Journal of Food Science and Technology*, *54*, 2361–2371.

Sonaglio, D., Ortega, G. G., Petrovick, P. R., & Bassani, V. L. (2003). Desenvolvimento tecnológico e produção de fitoterápicos. *Farmacognosia: da planta ao medicamento*, *5*, 289–332.

Souza, K. K. C. D., Salimena, J. P., Xavier, J. B., Souza, D. C. D., Bertolucci, S. K. V., Bittencourt, W. J. M., & Resende, L. V. (2022). Nutritional and bioactive compounds of *Amaranthus* spp. in Brazil. *Research, Society and Development*, *11*(13), e331111335618.

Sreelatha, S., Dinesh, E., & Uma, C. (2012). Antioxidant properties of Rajgira (*Amaranthus* paniculatus) leaves and potential synergy in chemoprevention. *Asian Pacific Journal of Cancer Prevention 13*(6), 2775–2780.

Stevens, O. A. (1957). Weights of seeds and numbers per plant. *Weeds*, *5*(1), 46–55.

Susantiningsih, T., Ridwan, R., Prijanti, A. R., Sadikin, M., & Freisleben, H.-J. (2012). Schizonticidal effect of a combination of *Amaranthus spinosus* L. and *Andrographis paniculata* Burm. f./Nees extracts in *Plasmodium berghei*-infected mice. *Medical Journal of Indonesia*, *21*(2), 66–70.

Szwejkowska, B., & Bielski S. (2012). Wartość prozdrowotna nasion szarłatu (*Amaranthus cruentus* L.) *Postępy Fitoterapii*, 4, 240–243.

Tabio-García, D., Paraguay-Delgado, F., Sánchez-Madrigal, M. Á., Quintero-Ramos, A., Espinoza-Hicks, J. C., Meléndez-Pizarro, C. O., & Espitia-Rangel, E. (2021). Optimisation of the ultrasound-assisted extraction of betalains and polyphenols from *Amaranthus* hypochondriacus var. Nutrisol. *Ultrasonics Sonochemistry*, *77*, 105680.

Weaver, S. E., & Mcwilliams, E. L. (1980). The biology of anadian weeds.: 44. *Amaranthus retroflexus* L., *A. powellii* S. Wats. and *A. hybridus* L. *Canadian Journal of Plant Science*, *60*(4), 1215–1234.

Wejnerowska, G., Heinrich, P., & Gaca, J. (2013). Separation of squalene and oil from *Amaranthus* seeds by supercritical carbon dioxide. *Separation and Purification Technology*, *110*, 39–43.

Yang, R.-Y., Lin, S., & Kuo, G. (2008). Content and distribution of flavonoids among 91 edible plant species. *Asia Pacific Journal of Clinical Nutrition*, *17*, 275–279.

Zhang, L.-L., Chen, Y., Li, Z.-J., Fan, G., & Li, X. (2023). Production, function, and applications of the sesquiterpenes valencene and nootkatone: a comprehensive review. *Journal of Agricultural and Food Chemistry*, *71*(1), 121–142.

Zhu, F. (2023). Amaranth proteins and peptides: Biological properties and food uses. *Food Research International*, *164*, 112405.

6 Corn/Wild Mint (*Mentha arvensis*)

Kanchan Sharma and Raees ul Haq

6.1 INTRODUCTION

Mint is a herb renowned for its distinctive flavor and taste. Its leaves are widely used to add flavor in different recipes and also for their health benefits. There exists numerous kinds of mint species in the plant kingdom identified under the genus *Mentha*. It includes 30 species prominently found in the temperate regions of Africa, Australia, North America, Asia and Europe (Dorman et al., 2003; de Sousa Barros et al., 2015). In the Indian Himalayan region, the plant is widely distributed in plains, valleys, and mountains. Mint plant is a perennial herb that exhibits great morphological variation which is evident by the attribution of higher number of different taxonomy names by the taxonomists in the last two centuries (Anca-Raluca et al., 2011). The genus *Mentha* is divided into various species corresponding to the number of chromosomes and morphological characteristics (Kapp, 2015; Tucker & Naczi, 2006). The species arvensis under the genus *Mentha* is capable of producing hybrids in the same way as other species of the same genus such as *M. aquatic* L., *M. longifolia* L., etc. (Kapp, 2015; Tucker & Naczi, 2006). It is commonly called corn mint, wild mint, menthol mint, Japanese mint, or field mint.

In earlier times, different herbs were used as a medicine for the treatment of diseases; as time progressed, it was their different isolates and essential oils, rather than the herbs themselves, which were used. The role of essential oils has been extensively used in form of aroma therapy for the treatment of disorders. These low-boiling essentials oils are concentrated throughout the different parts of plant (rhizome to leaf) and the diversity of their chemical composition impart it with many biological functions which can have hepatoprotective, antioxidant, or antiseptic properties, among many others (Sokovic et al., 2007; Sokovic et al., 2010; Dorman et al., 2003; Prabuseenivasan et al., 2006; Miladi et al., 2013). Considerable interest in aromatic plants has resulted in the generation of numerous studies that have been performed and are currently spreading worldwide due to the biological activity shown by different chemicals.

The mint plant is one of the eminent aromatic plants of the *Lamiaceae* family, which is known for different odoriferous members (Bakkali et al., 2008; Hedge, 1992). Essential oils obtained from this family have been reported to possess antimicrobial, antiseptic, and anti-inflammatory activities (Mimica-Dukic et al., 2004; Bozin et al., 2006; Kumar et al., 2013; Perveen et al., 2012; Perveen et al., 2014). The World Health Organization reports that a significantly large population of the globe are dependent on the use of herbal medications for the treatment of common health issues. Compounds such as ethers, phenols, alcohols, aldehydes, and ketones impart biological activity and fragrance to the essential oils. The detailed composition of the essential oil has been documented by different studies (Sokovic et al., 2010; Pino et al., 1998; Pino et al., 1999; Pino et al., 1999a; Verma et al., 2010; Hussain et al., 2010; Padalia et al., 2013) and the chief compounds include monoterpene hydrocarbons, oxygenated monoterpenes, diterpenoids, sesquiterpenoids, and sesquiterpene hydrocarbons. Oil taken from *Mentha arvensis* is the second-largest essential oil production in the world (Sanganeria, 2005), comprising phenylpropanoids and terpenoids with monoterpenes accounting

 DOI: 10.1201/9781003292340-6

for more than 90% of the composition (Sangwan et al., 2001; Srivastava et al., 2002). The presence of menthol, carvone, linalool, menthone, menthofuran, linalyl acetate, isomenthone, piperitenone oxide, α-terpineol, menthol acetate, and p-menthone has been recorded in the essential oil of *Mentha arvensis* (Pino et al., 1996a, 1996b; Singh et al., 2005; Gassenmeier, 2006; Vivek et al., 2009; Souza et al., 2014; Verma et al., 2010; Hussain et al., 2010; Padalia et al., 2013). In conventional or traditional medicine, the plant has been used as a nasal decongestant, for skin and gastric illnesses, and as a carminative agent to relieve flatulence (Maciel do Nascimento et al., 2009). The plant has proven its potential in a range of different scientific observations. This chapter draws together the evidences which make the plant an interesting topic to researchers.

6.2 SYSTEMATIC CLASSIFICATION

Kingdom	Plantae
Division	Spermatophyta
Subdivision	Angiospermae
Class	Dicotyledones
Order	Lamiales
Family	Lamiaceae
Genus	*Mentha*
Species	*arvensis*

Mentha arvensis is further classified into various subspecies (Quattrocchi, 1947), as outlined below:

i. *Mentha arvensis* subsp. *austriaca*,
ii. *Mentha arvensis* subsp. *agrestis*,
iii. *Mentha arvensis* subsp. *arvensis*,
iv. *Mentha arvensis* subsp. *palustris*,
v. *Mentha arvensis* subsp. *lapponica*,
vi. *Mentha arvensis* subsp. *parietariifolia*,
vii. *Mentha arvensis* var. *piperascenes* (Chinese or Japanese or Korean mint),
viii. *Mentha arvensis* subsp. *haplocalyx*.

6.3 ORIGIN

This plant is an angiosperm under the mint family *Lamiaceae*, originating in the temperate regions of Africa, Australia, North America, western and central Asia, eastern Himalaya and the Europe (USDA, 2023; Dorman et al., 2003; de Souza et al., 2016). It is an herbaceous and perennial herb, a rootstock creeper with many runners, which can grow up to 10 to 60 centimeters in length (Figure 6.1). The leaves, paired in opposites, are broad and hairy and possessed of coarsely serrated margins. The color of its flowers ranges from pale purple to pink; they grow at the base of the leaves on a stem, which is erect or can be semi-sprawling. It grows well in moist places and is, therefore, found in abundance along streams. It is prone to attacks from rust fungus and mildew, the latter of which usually occurs in the foggy and humid weather. Of these two, the former one is more common and its tendency to spread and damage the whole farm of mint once it spreads make it the more dangerous to the herb. The only solution to prevent it from damaging the rest of the farm is to immediately remove the infected plants to reduce the probability of contaminating the remainder of the plants (Sievers & Lowman, 1933).

FIGURE 6.1 *Mentha arvensis* in its natural habitat.

6.4 TRADITIONAL USES

The plant's leaves have been used widely as a flavoring agent in various food preparations, such as chutney, tea, bread, soups, and salads, and also in the preparation of some curds (Anwar et al., 2019; da Cruz-Almeida et al., 2018). The folk medical use of the leaves of *Mentha arvensis* includes its capability to treat cold and/or digestion by adding its leaves into tea (Fagan, 2019). It is also known for its potential benefits in treating colic and flatulence. In some countries, such as Spain, it is common for people to consume mint tea after dinner, a drink which is called poleo, as this is believed to have digestive and tonic properties since its constituent oils help in the breakdown of fats, making food easier to digest (Jäger et al., 2007; Khadraoui et al., 2015; Fatiha et al., 2016; Abdelli et al., 2016). There is evidence of its use since Greek, Roman, and Egyptian civilizations both for cooking purposes and for the preparation of medicines. The Romans, for example, are known to have used its leaves to add flavor to wines and sauces. Since ancient times, it has been used by the Indian people to tone the stomach, relieve muscle strains, ease gastric acidity, and stimulate mind and body (Hossain et al., 2014).

6.5 COMPOSITION

It is the iconic source of natural menthol and is, thus, also known as menthol mint (McKay & Blumberg, 2006; Hajlaoui et al., 2009). The chemical analysis of *Mentha arvensis* has confirmed the presence of many phytochemicals and essential oils. The phytochemicals present belong to different classes: glycosides or acylated flavonoids, sterols and steroidal glycosides, lignans, hydrocarbons, fatty acids, proteins, free sugars, cinnamic acids, organic acids, alkaloids, tocopherols, etc. However, the primary components of these species include essential oils, flavonoids, and phenolics. The prominent flavonoids in this species include eriocitrin, diosmin, eriodictyol, isorhoifolin, narirutin, naringenin-7- O-glucoside and apigenin (Triantaphyllou et al., 2001; Ali et al., 2002; Fialova et al., 2008; Cirlini et al., 2016; Al-Sabahi et al., 2016; Miguel et al., 2016; Mamadalieva et al., 2020). The principal component present in its essential oils is monoterpenes. The major acids discovered in its water extract are caffeic acid, along with its derivatives: rosmarinic, chlorogenic and salvianolic acids. Additionally, linalool, menthone, carvone and pulegone are all major constituents of its essential oils (Brahmi et al., 2015). Table 6.1 presents the overall composition of this plant *Mentha arvensis*.

There is enough scientific evidence to support the fact that the variability of essential oil composition of this herb, both quantitatively and qualitatively, depends upon several factors, including its origin, year, plant age, harvesting time, weather conditions, fertilization, geographical location, and vegetative development (Kokkini, 1991; Mamadalieva et al., 2020) (Figure 6.2).

The typical essential oil composition of *Mentha arvensis* principally consists of menthol, at about 32%, linalool, amounting to 18%, pulegene 12% and 11% of trans-geraniol according to the study carried out by Al-Sabahi and his co-workers (Al-Sabahi et al., 2016). The concentrations of menthol, menthone and menthyl acetate in *Mentha* as reported by the Pino and others are 52%, 26%, and

TABLE 6.1
General Composition of *Mentha arvensis*

S. No.	Major Component	Constituents	Reference
1	Essential oils	Menthol, menthone, linalool, pulegene, and trans-geraniol methyl acetate, and isomenthone, menthyl acetate, limonene, isopulegol, 1,8-cineol, 3-octanone	de Sousa Guedes et al., 2016; Al-Sabahi et al., 2016, Pino et al., 1996a; Hussain et al., 2010; Chagas et al., 2020 Bokić et al., 2020
2	Polyphenols	Caffeic acid, chlorogenic acid, rosmarinic acid, salvianolic acid, isosalvianolic acid, gallic acid, ferulic acid, syringic acid, coumaric, vanillic, lithospermic acid	Kapp et al., 2013; Lv et al., 2012; Kapp, 2015; Riachi and De-Maria et al., 2015; Pereira et al., 2016; Dorman et al., 2003; Areias et al., 2001; Perez et al., 2014; Brahmi et al., 2015; Yamamura et al., 1998; Zheng et al., 2002; Fiamegos et al., 2004; Proestos et al., 2005; Kivilompolo & Hyotylainen, 2007; Chen et al., 2006; Dorman et al., 2003;
3	Flavonoids	Flavones, flavanones, luteolin, eriocitrin, eriodictyol, luteolin, apigenin, catechin, rutin, diosmein, diosmin, naringenin, luteolin, naringenin-7- O-glucoside, narirutin, isorhoifolin, and apigenin,	Dorman et al., 2003; Triantaphyllou et al., 2001; Kivilompolo & Hyotylainen, 2007; Sankara & Nair, 1972; Tomas-Barberan et al., 1988; Yamamura et al., 1998; Zheng et al., 2002; Fiamegos et al., 2004; Proestos et al., 2005; Chen et al., 2006; Bimakr et al., 2010; Guedon & Pasquier, 1994; Fecka and Turek, 2007; Farnad et al., 2014; Voirin et al., 1994
4	Lignans	Spicatolignan A and B, medioresinol, medioresinol sulfate,	Zheng et al., 2007; Bimakr et al., 2010; Fecka & Turek, 2007; Hadjmohammadi et al., 2013;
5	Fatty acids	Linoleic acid, linolenic acid, palmitic acid, diacylglycerol, triacylglycerol	Perez et al., 2014; Kunnumakkara et al., 2009; Kizil et al., 2010; Choudhury et al., 2006; Maffei & Scannerini, 1992;
4	Pigments	carotenoids, chlorophylls, neoxanthin, violaxanthin	Raju et al., 2007; Dambrauskiene et al., 2008; Curutchet et al., 2014
5	Vitamins	tocopherols, ascorbic acid	Riachi & De-Maria, 2015; Perez et al., 2014; Dambrauskiene et al., 2008
6	Other compounds	triterpenoids, steroids, ursolic acid, lutein, sugars, saponins, alkaloids, quinines, anthraquinones	Ertas et al., 2015; She et al., 2010; Maffei & Scannerini, 1992; Padmini et al., 2008

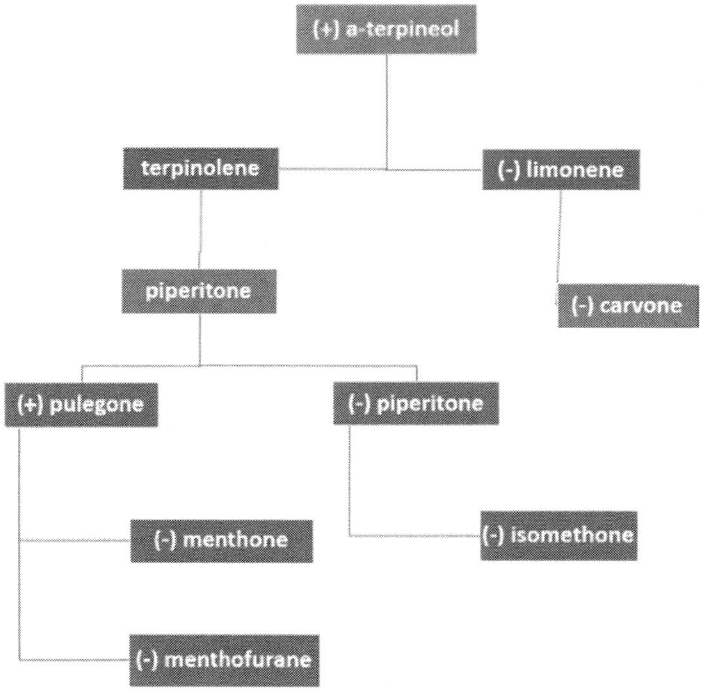

FIGURE 6.2 Biosynthesis of aromatic monoterpenes reported in *Mentha arvensis* (Bertoli et al., 2011).

11%, respectively (Pino et al., 1996b). By contrast, the levels of menthol and isomenthone, as measured by Hussain et al. (2010), are around 79–81% and 6%, respectively. In their study of the plant, Bokhari and others observed the values of 56%, 11%, 8%, and 5% for menthol, menthone, methyl acetate and isomenthone, respectively, (Bokhari et al., 2016). The concentration of secondary metabolites, as reported by different studies, are: menthol about 57%, isomenthone at 21%, menthyl acetate along with limonene at around 4% and 4% isoisopulegol (de Sousa Guedes et al., 2016); 16% 13-octanone and 23% 8-cineol (Bokić et al., 2020); 86% menthol, 4% menthone and isomenthone, and pulegone 1% (Chagas et al., 2020). Different phenolics, flavonoids, and fatty acids, viz. luteolin-7-O-glucoside, rosmarinic acid, caffeic acid, eriocitrin, naringenin-7-O-glucoside, apigenin, luteolin and isorhoifolin have been reported (Dorman et al., 2003; Al-Sabahi et al., 2016).

6.6 IDENTIFICATION AND CHARACTERIZATION

Prior to the extraction of essential oil from *M. arvensis*, the plant which is above the ground level is cut at the budding stage before being air-dried at room temperature. The essential oils are then obtained via hydro distillation, primarily from leaves and inflorescences. The composition of essential oils obtained in this manner can be studied by either chromatographic or mass spectrometric methods (Kurilov et al., 2009).

The broth dilution assay of the essential oils of *Mentha arvensis* was carried out to test its antibacterial activities using 0.2% sterile peptone water on different bacterial strains, which included *Escherichia coli, Enterococcus faecalis, Candida albicans, Salmonella typhimurium, Pseudomonas aeriginosa* and *Staphylococcus aureus*. To prepare the assay, the microbial suspension was prepared and poured on Muller-Hinton agar as solid media, which is then filtered and incubated for 26 h at 38 °C with cephalexin along with ciprofloxacin used as positive controls and the results inferred that the essential oils effectively acted upon all bacterial strains used in the study. The antimicrobial

TABLE 6.2
Antibacterial and Antifungal Activity of *M. arvensis*

S. No.	Microorganism Used in the Test	Amount of Essential Oil (μL)	% inhibition
1	*Pseudomonas aeriginosa*	0.2	39.0
2	*Staphylococcus aureus*	0.2	1.5
3	*Salmonella typhimurium*	0.2	30.0
4	*Escherichia coli*	0.2	6.0
5	*Candida albicans*	0.2	70.0
6	*Enterococcus faecalis*	0.2	51.0

activity of menthol is listed in Table 6.2 (Mickienė et al., 2011; Horváth & Koščová, 2017; Khan et al., 2019) (Table 6.2).

The antimicrobial properties of leaves of *Mentha arvensis* includes its antibacterial, antiviral, and antifungal properties. The mechanism of their action is not completely known; however, research showed that its pharmacological properties vary with the kind and quality of its components (Anca-Raluca et al., 2011). Its biological activity is also affected by the location and presence of functional groups within molecules. There are various mechanisms by which the essential oil (Figure 6.3) inhibits microorganisms such as hydrophobicity, lipophilicity, the destabilization of microbial cell membranes, proton motive force depletion, the impairment of bacterial enzyme systems, and damage to the proteins forming cell membranes and cell contents. This leads to lipid portioning, causing damage to cell membranes and mitochondria (Božović et al., 2015; Lopez-Romero et al., 2015). The damage caused to the cells and its components make the cells prone to lysis and ultimately causes leakage of the cell contents. Most of the constituents of essential oils primarily exhibit antimicrobial properties against bacterial cells. Other mechanisms by which they exhibit antimicrobial actions include depolarization of the cell membranes or the increased permeability of cell walls, making the cells prone to lysis (Saad et al., 2013; Chouhan et al., 2017) (Figure 6.4).

The antifungal properties of *Mentha arvensis* was tested using agar media on *Candida albicans* with samples incubated at a temperature of 30°C for a period of 7 days within the culture media, using fungicide fluconazole as positive control. The test results revealed a positive correlation between antifungal activity and essential oil concentration as the former increases with the increase in the latter one. The essential oil of *Mentha arvensis* is also effective against the growth of other fungal strains, as reported by Mickiene et al. (2011) and Khan et al. (2019) (Table 6.3).

Colorimetric analysis was conducted using gallic acid as a standard to determine its total phenolic content, which is directly related to its antioxidant activity. Free radicals make the body more prone to diseases such as cancer and heart problems. These can be prevented by phenolic compounds as the constituent components of these compounds protect the body against harmful metabolites produced during fat metabolism (Khan et al., 2019; Saboon et al., 2017; Koleva et al., 2002). The free radical scavenging activity of the essential oils of *Mentha arvensis* were determined with the help of a DPPH (2,2-diphenyl-1-picrylhydrazyl) assay. It was revealed from the test results that the free radical scavenging activity of plants depends not only on its phenolic content but also on other secondary metabolites present, such as vitamins, flavonoids, and carotenoids (Alizadeh et al., 2013; Mancini et al., 2015).

6.7 PHARMACOLOGICAL BENEFITS

The pharmacological properties of *Mentha arvensis* are the result of the presence of the secondary metabolites such as essential oils and phenolic compounds. The former one is mainly dominant in

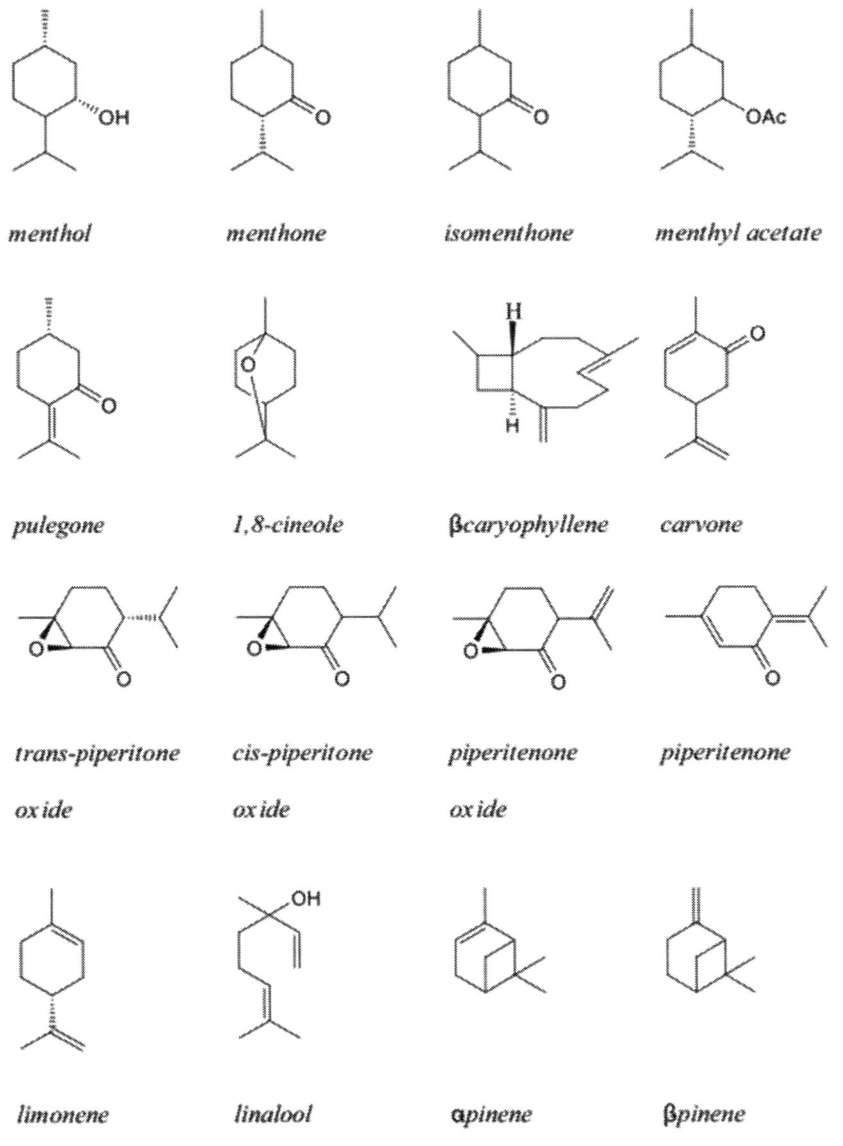

FIGURE 6.3 Various components of essential oils present in *Mentha arvensis*.

monoterpenes and sesquiterpenes. It is commercially used as a flavoring agent in many food preparations due to its antioxidant, antimicrobial, and sensory characteristics (Nguyen & Mittal, 2007; Espina et al., 2014; Kapp, 2015; Salehi et al., 2018). The health benefits posed by the mint species are mainly due to its constituents' essential oils. The plants have anti-inflammatory, carminative, anti-infectious, astringent, antimicrobial, and stomachic properties. In addition, it imparts good antiseptic, decongest, analgesic, and antispasmodic properties. Several research studies have documented its fungicidal effects. It has physiological effects, especially on the nervous and gastrointestinal systems of the human body (Anwar et al., 2019). Moreover, some studies have also reported its effects on mental abilities, especially on cognition, mood, and memory (Abou-Jawdah et al., 2002; Babar et al., 2015; Kennedy et al., 2018; Liang et al., 2020). Documented and recognized biological

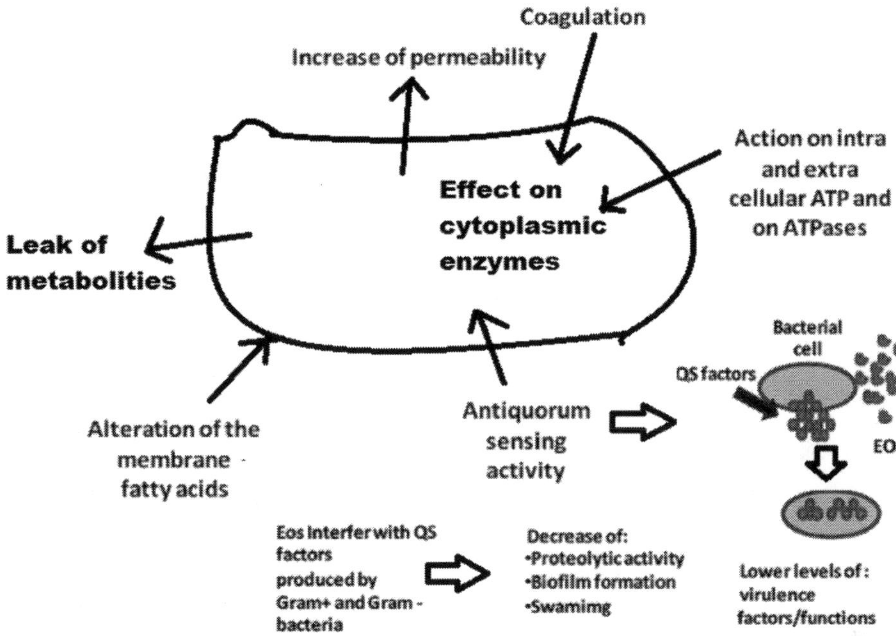

FIGURE 6.4 Mechanism of action of essential oil components on target bacterial cells.

TABLE 6.3
Antifungal Activity of *M. arvensis* on Inhibition of *mycelia* Growth

Amount of essential oil (µl ml^{-1})	Growth of the mycelia (mm)	% inhibition
0	38	0
0.7	14.67	10
1.30	7.67	36.5
2.7	3.67	72.30
5.5	0.00	88.60
10.6	0.00	90.25
23	0.00	94.6
32	0.00	97.1
42	0.00	97.75
55	0.00	99
100	0.00	100

Source: Khan et al., 2019; Mickiene et al., 2011.

activities of *Mentha arvensis* includes antibacterial, antifungal (Table 6.4), antiviral, antioxidant, and cytotoxic properties (Dorman et al., 2003; Bokhari et al., 2016; Chagas et al., 2020; Coutinho et al., 2009; de Sousa Guedes et al., 2016; Hussain et al., 2010). Its leaves are also used to treat different types of diseases such as neuralgia, toothache, laryngitis, rheumatic pain, and bronchitis (Al-Sabahi et al., 2016). Its essential oils and extracts are useful in various herbal cosmetics, foods, and skin and hair care products (Mamadalieva, 2020).

TABLE 6.4
***In vitro* Antimicrobial Activity of Crude and Isolated *Mentha arvensis* Extracts**

S. No.	Tested Sample	Microorganism	Main Findings		Reference
1	EO		IZ (mm)	MIC (µl/ml)	Bokhari et al., 2016
		Bacillus subtilis,	20.7	25.0	
		S. aureus,	22.3	12.5	
		E. coli,	22.3	12.5	
2	EtOH extract	*Acinetobacter baumannii* ATCC 10545	Inhibition zone of 34.5 mm was detected when the concentration of extract for dosage was 100 µg/ml.		Zhang et al., 2015
3	EtOH extract	*S. aureus* 358 SA358	Resilient activity of the extract against multiresistant- and methicilin-resistant *S. aureus*.		Coutinho et al., 2009
4	EO		IZ (mm)		Wannissorn et al., 2005
		Clostridium. perfringens	90		
		C. jejuni	90		
		Salmonella typhimurium	47.5		
5	EO	Twelve isolates of *Aeromonas* spp.	Superior action on MIC and MBC were observed ranging from 1.25 to 16.67 µl/ml.		Chagas et al., 2020
6	EOs, menthol, menthone, piperiten-one oxide, and carvone	*S. aureus* ATCC 25923, *B. subtilis* ATCC 10707, *Escherichia coli* ATCC 25922, *Aspergillus flavus* ATCC 32612, *Alternaria solani* ATCC 11078, *Fusarium solani* ATCC 36031, *Rhizopus solani*, *Alternaria alternata* ATCC 20084, *Aspergillus niger* ATCC 10575, *Rhizopus* spp.	Among the tested samples *M. arvensis* showed maximum antimicrobial activity with higher IZ of 14–33 and 16–30 (mm) and lowest MIC values of 20–330 and 56–139 (µl/ml) against the bacterial and fungal strains, respectively. Isolated menthol antimicrobial activity was higher with IZ of 19–33 (mm) and MIC values of 30–107 (µl/ml) higher than the stronger drugs (amoxycillin and fluconazole).		Hussain et al., 2010

IZ – Zone of Inhibition; MIC – Minimum Inhibitory Concentration; EO – Essential oil; EtOH – Ethyl alcohol.

The essential oils present in it are widely used in aromatherapy against various disorders. These compounds actually have a low boiling point and are primarily produced by odoriferous glands in a plant; they possess great solubility in organic solvents. They typically become accumulated in different parts of plants, such as the stem, leaves, fruit, seeds, barks, and rhizome. In industry, they are also used in the preparation of scents, cosmetics, cleansers, solvents, and detergents (Miladi et al., 2013; Bakkali et al., 2008; Hedge, 1992).

It also exhibits cytotoxic and anti-proliferative properties against different cancer cells, namely breast cancer, colon cancer, lung cancer, cervical cancer, and colon cancer (Sharma et al., 2014; Yi & Ywetzstein, 2011; Shirazi et al., 2004; Mazzio & Soliman, 2009).

6.8 CONCLUSION

The plant of *Mentha arvensis*, especially its leaves, is rich in many biologically active components possessing multiple health benefits, including antibacterial, antifungal, antispasmodic, antiseptic, anti-infectious, digestive, vasoconstrictor, digestive, decongestant, nervine stimulant, and anti-ulcerative properties. This is evident from its folk use and has also been observed in scientific studies. Additionally, it has a good flavor, making it very useful for culinary purposes. Regarding its multidimensional properties, there is a greater need to explore other aspects of its constituent components. Many studies on different species of the *Mentha* family have been conducted, but each species, such as *M. arvensis*, need to be explored in great detail to ascertain their potent beneficial and toxic traits.

REFERENCES

Abdelli, M., Moghrani, H., Aboun, A., & Maachi, R. (2016). Algerian *Mentha pulegium* L. leaves essential oil: Chemical composition, antimicrobial, insecticidal and antioxidant activities. *Industrial Crops and Products*, *94*, 197–205.

Abou-Jawdah, Y., Sobh, H., & Salameh, A. (2002). Antimycotic activities of selected plant flora, growing wild in Lebanon, against phytopathogenic fungi. *Journal of Agricultural and Food Chemistry*, *50*, 3208–3213.

Ali, M. S., Saleem, M., Ahmad, W., Parvez, M., & Yarudani, R. (2002). A chlorinated monoterpene ketone, acylated glycosides of betasitosterol and a flavonoid glycoside of *Mentha* longifolia. *Phytochemistry*, *59*, 889–895

Alizadeh, A., Alizadeh, O., Amari, G., & Zare, M. (2013). Essential oil composition, total phenolic content, antioxidant activity and antifungal properties of Iranian *Thymus daenensis* subsp. *daenensis* Celak. as in influenced by ontogenetical variation. *Journal of Essential Oil Bearing Plants*, *16*, 59–70.

Al-Sabahi, J. N., Hanif, M. A., Al-Maskari, A. Y., Al Busaidi, M. S. M., Al Maskari, M. Y., & Al-Haddabi, M. H. (2016). Chemical report on wild growing *Mentha arvensis* and *Citrullus colocynthus* from Oman. *Journal of Essential Oil Bearing Plants*, *19*(3), 719–726

Anca-Raluca, A., Boz, I., Padurariu, C., Atofani, D., Coisin, M., & Zamfirachie, M. M. (2011). Comparative biochemical and physiological research on taxa of *Mentha* L. genus. *Journal of Plant Development*, *18*, 41–45.

Anwar, F., Abbas, A., Mehmood, T., Gilani, A.-H., & Rehman, N. (2019). *Mentha*: A genus rich in vital nutra-pharmaceuticals—A review. *Phytotherapy Research*, *33*, 2548–2570.

Areias, F. M., Valentao, P., Rade, P. B., Ferreres, F., & Seabra, R. M. (2001). Phenolic fingerprint of peppermint leaves. *Food Chemistry*, *73*, 307–311.

Babar, A., Ali, A.-W. N., Saiba, S., Aftab, A., Alam, K. S., & Firoz, A. (2015). Essential oils used in aromatherapy: A systemic review. *Asian Pacific Journal of Tropical Biomedicine*, *5*(8), 601–611.

Bakkali, F., Averbeck, S., Averbeck, D., & Idaomar, M. (2008). Biological effects of essential oils – a review. *Food and Chemical Toxicology*, *46*, 446–475.

Bertoli, A., Leonardi, M., Krzyzanowska, J., Oleszek, W., & Pistelli, L. (2011). *Mentha longifolia* in vitro cultures as safe source of flavouring ingredients. *Biochimica Polonica*, *58*(4), 581–587

Bimakr, M., Abdul Rahman, R., Saleena, T. F., Ganjloo, A., Md-Salleh, L., Selamat, J., Hamid, A., & Zaidul, I. S. M. (2010). Comparison of different extraction methods for the extraction of major bioactive flavonoid compounds from spearmint (*Mentha spicata* L) leaves. *Food and Bioproducts Processing*, *89*, 67–72.

Bokhari, N., Perveen, K., Al Khulaifi, M., Kumar, A., & Siddiqui, I. (2016). In vitro antibacterial activity and chemical composition of essential oil of *Mentha arvensis* Linn. leaves. *Journal of Essential Oil Bearing Plants*, *19*, 907–915.

Bokić, B. S., Rat, M. M., Kladar, N. V., Anakov, G. T., & Bozin, B. N.(2020). Chemical diversity of volatile compounds of mints from southern part of Pannonian plain and Balkan peninsula-new data. *Chemistry and Biodiversity*, *17*(8), e2000211.

Bozin, B., Mimica-Dukic, N., & Anackov, G. (2006). Characterization of the volatile composition of essential oils of some *Lamiaceae* species and the antimicrobial and antioxidant activities of the entire oils. *Journal of Agricultural and Food Chemistry*, *54*, 1822–1828.

Božović, M., Pirolli, A., & Ragno, R. (2015). *Mentha* suaveolens Ehrh. (*Lamiaceae*) essential oil and its main constituent piperitenone oxide: Biological activities and chemistry. *Molecules*, *20*, 8605–8633

Brahmi, F., Hauchard, D., Guendouze, N., Madani, K., Kiendrebeogo, M., Kamagaju, L., Stévigny, C., Chibane, M., & Duez, P. (2015). Phenolic composition, in vitro antioxidant effects and tyrosinase inhibitory activity of three Algerian *Mentha* species: *M. spicata* (L.) *M. pulegium* (L.) and *M. rotundifolia* (L.) Huds (Lamiaceae). *Industrial Crops and Products*, *74*, 722–730.

Chagas, E. C., Majolo, C., Monteiro, P. C., de Oliveira, M. R., Gama, P. E., Bizzo, H. R., & Chaves, F. C. M. (2020). Composition of essential oils of *Mentha* species and their antimicrobial activity against *Aeromonas* spp. *Journal of Essential Oil Research*, *32*, 209–215.

Chen, G., Gao, H., Zheng, J., Wu, B., Yang, X., & Wu, L. (2006). Study of chemical constituents in active parts of *Mentha spicata* III. *Journal of Chinese Material* and *Medica*, *31*(7), 560–562

Choudhury, R. P., Kumar, A., & Garg, A. N. (2006). Analysis of Indian mint (*Mentha spicata*) for essential, trace and toxic elements and its antioxidant behaviour. *Journal of Pharmaceutical and Biomedical Analysis*, *41*, 825–832.

Chouhan, S., Sharma, K., & Guleria, S. (2017). Antimicrobial activity of some essential oils-present status and future perspectives. *Medicines*, *4*(3), 58.

Cirlini, M., Mena, P., Tassotti, M., Herrlinger, K. A., Nieman, K. M., Dall'Asta, C., & Del Rio, D. (2016). Phenolic and volatile composition of a dry spearmint (*Mentha spicata* L.) extract. *Molecules*, *21*, 1007.

Coutinho, H. D., Costa, J. G., Lima, E. O., Falcão-Silva, V. S., & Siqueira-Júnior, J. P. (2009). Potentiating effect of *Mentha arvensis* and chlorpromazine in the resistance to aminoglycosides of methicillin-resistant *Staphylococcus aureus*. *Vivo*, *23*, 287–290.

Curutchet, A., Dellacassa, E., Ringuelet, J. A., Chaves, A. R., & Vina, S. Z. (2014). Nutritional and sensory quality during refrigerated storage of fresh-cut mints (*Mentha* × *piperita* and *M. spicata*). *Food Chemistry*, *143*, 231–238

da Cruz-Almeida, E. T., de Medeiros-Barbosa, I., Tavares, J. F., Barbosa-Filho, J. M., Magnani, M., & de Sousa, E. L. (2018). Inactivation of spoilage yeasts by *Mentha spicata* L. and *M.* × *villosa* Huds. Essential oils in cashew, guava, mango, and pineapple juices. *Frontiers in Microbiology*, *9*, 1111.

Dambrauskienė, E., Viškelis, P., & Karklelienė, R. (2008). Productivity and biochemical composition of *Mentha piperita* L. of different origin. *Biologija*, *54*(2), 105–110.

de Sousa Barros, A., de Morais, S. M., Ferreira, P. A. T., Vieira, I. G. P., Craveiro, A. A., dos Santos Fontenelle, R. O., de Menezes, J. E. S., da Silva, F. W. F., & de Sousa, H. A. (2015). Chemical composition and functional properties of essential oils from Mentha species. *Industrial Crops and Products*, *76*, 557–564.

de Sousa Guedes, J. P., da Costa Medeiros, J. A., de Souza e Silva, R. S., de Sousa, J. M. B., da Conceição, M. L., & de Souza, E. L. (2016). The efficacy of *Mentha arvensis* L. and *M. piperita* L. essential oils in reducing pathogenic bacteria and maintaining quality characteristics in cashew, guava, mango, and pineapple juices. *International Journal of Food Microbiology*, *238*, 183–192

de Souza, E. L., da Cruz Almeida, E. T., & de Sousa Guedes, J. P. (2016). The potential of the incorporation of essential oils and their individual constituents to improve microbial safety in juices: A review. *Comprehensive Reviews in Food Science and Food Safety*, *15*, 753–772.

Dorman, H. J. D., Kosar, M., Kahlos, K., Holm, Y., & Hiltunen, R. (2003). Antioxidant properties and composition of aqueous extracts from *Mentha* species, hybrids, varieties, and cultivars. *Journal of Agricultural and Food Chemistry*, *51*, 4563–4569.

Ertaş, A., Gören, A. C., Haşimi, N., Tolan, V., & Kolak, U. (2015). Evaluation of antioxidant, cholineserase inhibitory and antimicrobial properties of *Mentha longifolia* subsp. noeana and its secondary metabolites. *Records of Natural Products*, *9*(1), 105–115.

Espina, L., García-Gonzalo, D., & Pagán, R. (2014). Impact of essential oils on the taste acceptance of tomato juice, vegetable soup, or poultry burgers. *Journal of Food Science*, *79*, S1575–S1583.

Fagan, D. (2019). *Wildflowers of Oregon: A Field Guide to over 400 Wildflowers, Trees, and Shrubs of the Coast, Cascades, and High Desert*. Guilford, CT.

Farnad, N., Heidari, R., & Aslanipour, B. (2014). Phenolic composition and comparison of antioxidant activity of alcoholic extracts of peppermint (*Mentha piperita*). *Journal of Food Measurement and Characterization*, *8*, 113–121.

Fatiha, B., Adjaoud, A., Marongiu, B., Porcedda, S., Piras, A., Falconieri, D., & Chibane, M. (2016). Chemical composition and in vitro antimicrobial, insecticidal and antioxidant activities of the essential oils of *Mentha pulegium* L. and *Mentha rotundifolia* (L.) Huds growing in Algeria. *Industrial Crops and Products*, *88*, 96–105

Fecka, I., & Turek, S. (2007). Determination of water soluble polyphenolic compounds in commercial herbal teas from *Lamiaceae*: Peppermint, melissa, and sage. *Journal of Agricultural and Food Chemistry*, *55*, 10908–10917.

Fialova, S., Tekel'ova, D., Mrlianová, M., & Grancai, D. (2008). The Determination of Phenolics Compounds and Antioxidant Activity of Mints and Balms Cultivated in Slovakia. *Acta Facultatis Pharmaceuticae Universitatis Comenianae*, *55*, 96–102.

Fiamegos, Y. C., Nanos, C. G., Vervoort, J., & Stalikas, C. D. (2004). Analytical procedure for the in-vial derivatization-extraction of phenolic acids and flavonoids in methanolic and aqueous plant extracts followed by gas chromatography with mass-selective detection. *Journal of Chromatography*, *10*(41), 11–18.

Gassenmeier, K. (2006). Identification and quantification of L-menthyl lactate in essential oils from *Mentha arvensis* L. from India and model studies on the formation of L-menthyl lactate during essential oil production. *Flavour and Fragrance Journal*, *21*, 725–730.

Guedon, D. J., & Pasquier, B. P. (1994). Analysis and distribution of flavonoid glycosides and rosmarinic acid in 40 *Mentha piperita* clones. *Journal of Agricultural and Food Chemistry*, *42*, 679–684.

Hadjmohammadi, M., Karimiyan, H., & Sharifi, V. (2013). Hollow fibre-based liquid phase microextraction combined with high-performance liquid chromatography for the analysis of flavonoids in *Echinophora platyloba* DC and *Mentha piperita*. *Food Chemistry*, *141*, 731–735.

Hajlaoui, H., Trabelsi, N., Noumi, E., Snoussi, M., Fallah, H., Ksouri, R., & Bakhrouf, A. (2009). Biological activities of the essential oils and methanol extract of tow cultivated mint species (*Mentha longifolia* and *Mentha pulegium*) used in the Tunisian folkloric medicine. *World Journal of Microbiology and Biotechnology*, *25*, 2227–2238.

Hedge, C. (1992). A global survey of the biogeography of the Labiatae. In R. M. Harley, and T. Reynolds (Eds.) *Advances in Labiatae Science* (pp 7–17). Royal Botanic Gardens.

Horváth, P., & Koščová, J. (2017). In vitro antibacterial activity of *Mentha* essential oils against *Staphylococcus aureus*. *Folia Vet*, *61*, 71–77

Hossain, M. A., Al-Hdhrami, S. S., Weli, A. M., Al-Riyami, Q., & Al-Sabahi, J. N. (2014). Isolation, fractionation and identification of chemical constituents from the leaves crude extracts of *Mentha piperita* L grown in sultanate of Oman. *Asian Pacific Journal of Tropical Biomedicine*, *4*(1), S368–S372.

Hussain, A. I., Anwar, F., Nigam, P. S., Ashraf, M., & Gilani, A. H. (2010). Seasonal variation in content, chemical composition and antimicrobial and cytotoxic activities of essential oils from four *Mentha* species. *Journal of the Science of Food and Agriculture*, *90*, 1827–1836.

Jäger, A. K., Almqvist, J. P., Vangsøe, S. A. K., Stafford, G. I., Adsersen, A., & Van Staden, J. (2007). Compounds from *Mentha aquatica* with affinity to the GABA-benzodiazepine receptor. *South African Journal of Botany*, *73*, 518–521.

Kapp, K. (2015). *Polyphenolic and essential oil composition of Mentha and their antimicrobial effect* [Doctoral dissertation, University of Helsinki]. Finland.

Kapp, K., Hakala, E., Orav, A., Pohjala, L., Vuorela, P., Püssa, T., Vuorela, H., & Raal, A. (2013). Commercial peppermint (*Mentha* × *piperita* L.) teas: Antichlamydial effect and polyphenolic composition. *Food Research International*, *5*, 53–75.

Kennedy, D., Okello, E., Chazot, P., Howes, M. J., Ohiomokhare, S., Jackson, P., & Wightman, E. (2018). Volatile terpenes and brain function: Investigation of the cognitive and mood effects of *Mentha* × *piperita* L. essential oil with in vitro properties relevant to central nervous system function. *Nutrients*, *10*, 1029.

Khadraoui, A., Hachama, K., Khodja, M., Khelifa, A., Mehdaoui, R., Harti, H., & Chahboun, N. (2015). Extraction study and the antibacterial activity of phenol and flavonoid contents in *Mentha pulegium* L. from Algeria. *Journal of Materials and Environmental Science*, *6*, 2501–2508.

Khan, A. A., Amjad, M. S., & Saboon. (2019). GC-MS analysis and biological activities of *Thymus vulgaris* and *Mentha arvensis* essential oil. *Turkish Journal of Biochemistry*, *44*(3), 388–396.

Kivilompolo, M., & Hyotylainen, T. (2007). Comprehensive two-dimensional liquid chromatography in analysis of *Lamiaceae* herbs: Characterisation and quantification of antioxidant phenolic acids. *Journal of Chromatography*, *11*(45), 155–164.

Kizil, S., Hasimi, N., Tolan, V., Kilinç, E., Yuksel, U. (2010). Mineral content, essential oil components and biological activity of two *Mentha* species (*M. piperita L., M. spicata* L.). *Turkish Journal of Field Crops*, *15*(6), 148–153.

Kokkini, S. (1991). Chemical races within the genus *Mentha* L. In H. F. Liskens, and J. F. Jackson (Eds.), *Essential oils and waxes* (pp. 63–78). Springer.

Koleva, L. L., Van-Beek, T. A., Linssen, J. P., De-Groot, A., & Evstatieva, L. N. (2002). Screening of plant extracts for antioxidant activity: A comparative study on three testing methods. *Phytochemical Analysis*, *13*, 8–17.

Kumar, A., Varshney, V. K., Prasad, R., Rawat, M. S. M., & Stashenko, E. E. (2013). Antioxidant and antifungal activity of leaf essential oil of *Morina longifolia*. *Journal of Biologically Active Products from Nature*, *3*(3), 183–193.

Kunnumakkara, A. B., Chung, J. G., Koca, C., & Dey, S. (2009). Mint and its constituents. In B. B. Aggarwal, and A. B. Kunnumakkara (Eds.), *Molecular Targets and Therapeutic Uses of Spices: Modern Uses for Ancient Medicine*, (pp. 373–401). World Scientific.

Kurilov, D. V., Kirichenko, E. B., Bidyukova, G. F., Olekhnovich, L. S., & Ku, L. D. 2009. Composition of the essential oil of introduced Mint forms of *Mentha piperita* and *Mentha arvensis* species. *Doklady Biological Sciences*, *429*, 538–540.

Liang, X., Ding, D., Zhao, Q., Wu, W., Xiao, Z., Luo, J., & Hong, Z. (2020). Inability to smell Peppermint is related to cognitive decline: A prospective community-based study. *Neuroepidemiology*, *54*, 258–264.

Lopez-Romero, J. C., González-Ríos, H., Borges, A., & Simões, M. (2015). Antibacterial effects and mode of action of selected essential oils components against *Escherichia coli* and *Staphylococcus aureus*. *Evidence-Based Complementary and Alternative Medicine*, *7*(9), 54–65.

Lv, J., Huang, H., Yua L., Whent, M., Niu, Y., Shi, H., Wang, T. T. Y., Luthria, D., Charles, D., & Yu, L. C. (2012). Phenolic composition and nutraceutical properties of organic and conventional cinnamon and peppermint. *Food Chemistry*, *132*, 1442–1450.

Maciel do Nascimento, E. M., Rodrigues, F. F. G., Campos, A. R., & Martins da Costa, J. G. (2009). Phytochemical prospection, toxicity and antimicrobial activity of *Mentha arvensis* (Labiatae) from Northeast of Brazil. *Journal of Young Pharmacists*, *1*(3), 210–212.

Maffei, M., & Scannerini, S. (1992). Fatty acid variability in some *Mentha* species. *Biochemical Systematics and Ecology*, *20*(6), 573–582.

Mamadalieva, N. Z., Hussain, H., & Xiao, J. (2020). Recent advances in genus *Mentha*: Phytochemistry, antimicrobial effects, and food applications. *Food Frontiers*. *2*, 1–24.

Mancini, E., Senatore, F., Del-Monte, D., De-Martino, L., Grulova, D., & Scognamiglio, M. (2015). Studies on chemical composition, antimicrobial and antioxidant activities of five Thymus vulgaris L. essential oils. *Molecules*, *20*, 12016–12028.

Mazzio E. A., Soliman K. F. A. (2009). In vitro screening for the tumoricidal properties of international medicinal herbs. *Phytotherapy Research*, *23*, 385–398.

McKay, D. L., & Blumberg, J. B. (2006). A review of the bioactivity and potential health benefits of peppermint tea (*Mentha piperita* L.). *Phytotherapy Research*, *20*, 619–633.

Mickienė, R., Ragažinskienė, O., & Bakutis, B. (2011). Antimicrobial activity of *Mentha arvensis* L. and *Zingiber officinale* R. essential oils. *Biologija*, *57*, 92–97.

Miguel, M., Barros, L., Pereira, C., Calhelha, R. C., Garcia, P. A., Castro, M. Á., & Ferreira, I. C. F. R. (2016). Chemical characterization and bioactive properties of two aromatic plants: *Calendula officinalis* L. (flowers) and *Mentha cervina* L. (leaves). *Food* and *Function*, *7*, 2223–2232.

Miladi, H., Slama, R. B., Mili, D., Zouari, S., & Bakhrouf, A. (2013). Essential oil of *Thymus vulgaris* L. and *Rosmarinus officinalis* L.: Gas chromatography-mass spectrometry analysis, cytotoxicity and antioxidant properties and antibacterial activities against foodborne pathogens. *Natural Sciences*, *5*, 729–739.

Mimica-Dukic, N., Bozin, B., Sokovic, M., & Simin, N. J. (2004). Antimicrobial and antioxidant activities of *Melissa officinalis* L. (Lamiaceae) essential oil. *Journal of Agricultural and Food Chemistry*, *52*, 2485–2489.

Nguyen, P., & Mittal, G. (2007). Inactivation of naturally occurring microorganisms in tomato juice using pulsed electric field (PEF) with and without antimicrobials. *Chemical Engineering and Processing – Process Intensification*, *46*, 360–365.

Padalia, R. C. Verma, R. S. Chauhan, A., Sundaresan V., Chanotiya C. S. (2013). Essential oil composition of sixteen elite cultivars of *Mentha* from western Himalayan region India. *Maejo International Journal of Science and Technology*, *7*(1), 83–93.

Padmini, E., Prema, K., Geetha, B. V., & Rani, M. U. (2008). Comparative study on composition and antioxidant properties of mint and black tea extract. *International Journal of Food Science and Technology*, *43*, 1887–1895.

Pereira, E., Pimenta, A. I., Calhelha, R. C., Antonio, A. L., Verde, S. C., Barros, L., Santos-Buelga, C., & Ferreira, I. R. (2016). Effects of gamma irradiation on cytotoxicity and phenolic compounds of *Thymus vulgaris* L. and *Mentha x piperita* L. *LWT—Food Science and Technology*, *71*, 370–377.

Pérez, M. G. F., Rocha-Guzmán, N. E., Mercado-Silva, E., Loarca-Piña, G., Rosalía, R. (2014). Effect of chemical elicitors on peppermint (*Mentha piperita*) plants and their impact on the metabolite profile and antioxidant capacity of resulting infusions. *Food Chemistry*, *156*, 273–278.

Perveen, K., Bokhari, N. A., & Soliman, D. A. W. (2012). Antibacterial activity of *Phoenix dactylifera* L. leaf and pit extracts against selected Gram negative and Gram positive pathogenic bacteria. *Journal of Medicinal Plants Research*, *6*(2), 296–300.

Perveen, K., Bokhari, N. A., Siddiqui I., & Wahibi, M. S. (2014). *In vitro* antimicrobial activity of some medicinal plants found in Saudi Arabia. *Journal of Pure and Applied Microbiology*, *8*(3), 2129–2134.

Pino, J. A., Rosado, A., & Fuentes, V. (1996a). Chemical composition of the essential oil of *Mentha arvensis* L. var. *piperascens Malinv*. from Cuba. *Journal of Essential Oil Research*, *8*, 685–686.

Pino, J. A., Rosado, A., & Fuentes, V. (1999). Chemical composition of the leaf oil of *Mentha rotundifolia* (L.) Hudson from Cuba. *Journal of Essential Oil Research*, *11*(2), 241–242.

Pino, J. A., Rosado, A., & Fuentes, V. (1999a). Essential Oil of *Mentha citrata* Ehrh. Grown in Cuba. *Journal of Essential Oil Research*, *11*(4), 413–414.

Pino, J. A., Rosado, A., & Sanchez, E. (1998). Essential oil of *Mentha spicata* L. from Cuba. *Journal of Essential Oil Research*, *10*(6), 657–659.

Pino, J. A., Rosado, A., & Fuentes, V. (1996b). Chemical composition of the essential oil of *Mentha arvensis* L. var. *piperascens* Malinv. from Cuba. *Journal of Essential Oil Research*, *8*(6), 685–686.

Prabuseenivasan, S., Jayakumar M., & Ignacimuthu, S. (2006). *In vitro* antibacterial activity of some plant essential oils. *BMC Complementary Medicine and Therapies*, *6*, 39–46.

Proestos, C., Chorianopoulos, N., Nychas, G. J. E., & Komaitis, M. (2005). RP-HPLC analysis of the phenolic compounds of plant extracts: Investigation of their antioxidant capacity and antimicrobial activity. *Journal of Agricultural and Food Chemistry*, *53*, 1190–1195.

Quattrocchi, U. (1947). *CRC World Dictionary of Plant Names: Common names, Scientific Names, Eponyms, Synonyms, and Etymology. Vol. III M-Q*. CRC Press.

Raju, M., Varakumar, S., Lakshminarayana, R., Krishnakantha, T. P., & Baskaran, V. (2007). Carotenoid composition and vitamin A activity of medicinally important green leafy vegetables. *Food Chemistry*, *101*, 1598–1605.

Riachi, L. G., & De-Maria, C. A. B. (2015). Peppermint antioxidants revisited. *Food Chemistry*, *176*, 72–81.

Saad, N. Y., Muller, C. D., & Lobstein, A. (2013). Major bioactivities and mechanism of action of essential oils and their components. *Flavour and Fragrance Journal*, *28*, 269–279.

Saboon, A. M., Ahmad, M. S., & Mashwani, Z. R. (2017). Fermentation enhances redox protective activities of *Gymnosporia royleana* Wall. ex Lawson extracts. *Iranian Journal of Science and Technology, Transaction A: Science*, *43*, 1–7.

Salehi, B., Stojanović-Radić, Z., Matejić, J., Sharopov, F., Antolak, H., Kręgiel, D., & Sharifi-Rad, J. (2018). Plants of genus *Mentha*: From farm to food factory. *Plants*, *7*, 70.

Sanganeria, S. (2005). Vibrant India-Opportunities for the flavor and fragrance industry. *Perfumer and Flavorist*. *30*, 24–34.

Sangwan, N. S., Farooqi, A. H. A., Shabih, F., & Sangwan, R. S. (2001). Regulation of essential oil production in plants. *Plant Growth Regulation*, *34*, 3–21.

Sankara, S. S., & Nair, A. G. R. (1972). Flavonoids of the leaves of *Mentha spicata* and *Anzsochzlus carnosus*. *Phytochemistry*, *11*, 452–453.

Sharma, V., Shabir, H., Moni, G., & Kumar, S. A. (2014). In vitro cancer activity of extracts of *Mentha* spp. against human cancer cells. *Indian Journal of Biochemistry and Biophysics*, *51*(5), 416–419.

She, G. M., Xu, C., Liu, B., & Shi, R. B. (2010). Polyphenolic acids from mint (the aerial of *Mentha haplocalyx* Briq.) with DPPH radical scavenging activity. *Journal of Food Science*, *75*(4), 359–362.

Shirazi, F. H., Ahmadi, N., & Kamalinejad, M. (2004). Evaluation of northern Iran *Mentha pulegium* L. cytotoxicity. *Daru Journal of Pharmaceutical Sciences*, *2*(3), 106–110.

Sievers, A. F., & Lowman, M. S. (1933). *Commercial Possibilities of Japanese Mint in the United States as a Source of Natural Menthol*. US Dept. of Agriculture.

Singh, A. K., Raina, V. K., Naqvi, A. A., Patra, N. K., Kumar, B., Ram, P., & Khanuja, S. P. S. (2005). Essential oil composition and chemo arrays of menthol mint (*Mentha arvensis* L. f. *piperascens* Malinvaud ex. Holmes) cultivars. *Flavour and Fragrance Journal*, *20*, 302–305.

Sokovic, M., Glamoclija, J., Marin, P. D., Brkic, D., & Van Griensven, L. J. L. D. (2010). Antibacterial effects of the essential oils of commonly consumed medicinal herbs using an *in vitro* model. *Molecules*, *15*, 7532–7546.

Sokovic, M., Petar, D., Marin, B. B. D., & Van Griensven, L. J. L. D. (2007). Chemical composition and antibacterial activity of essential oils of ten aromatic plants against human pathogenic bacteria. *Food*, *1*(1), 1–7.

Souza, M. A. A., Araujo, O. J. L., Brito, D. M. C., Fernandes, M. S., Castro, R. N., & Souza, S. R. (2014). Chemical composition of the essential oil and nitrogen metabolism of menthol mint under different phosphorus levels. *American Journal of Plant Sciences*, *5*, 2312–2322.

Srivastava, R. K., Singh, A. K., Kalra, A. V., Tomar, K. S. R., Bansal, P., Patra, D. D., Chand, S., Naqvi, A. A., Sharma, S., & Kumar, S. (2002). Characteristics of *Mentha arvensis* cultivated on industrial scale in the Indo-Gangetic plains. *Industrial Crops and Products*, *15*, 189–198.

Tomas-Barberan, F. A., Husain, S. Z., & Gil, M. I. (1988). Distribution of methylated flavones in the Lamiaceae. *Biochemical Systematics Ecology*, *16*(1), 43–46.

Triantaphyllou, K., Blekas, G., & Boskou, D. (2001). Antioxidative properties of water extracts obtained from herbs of the species *Lamiaceae*. *International Journal of Food Sciences and Nutrition*, *52*, 313–317.

Tucker, A. O., & Naczi, R. F. C. (2006). *Mentha*: An overview of its classification and relationships. In B. M. Lawrence (Ed.), *Mint: The genus Mentha* (pp. 3–39). CRC Press, Taylor and Francis Group.

USDA-ARS, National Plant Germplasm System. (2023). Germplasm Resources Information Network-(GRIN). National Germplasm Resources Laboratory. https://npgsweb.ars-grin.gov/gringlobal/taxon/taxonomydetail?id=24069

Verma, R. S., Rahman, L., Verma, R. K., Chauhan, A., Yadav A. K., & Singh, A. (2010). Essential oil composition of menthol mint (*Mentha arvensis)* and peppermint (*Mentha piperita*) cultivars at different stages of plant growth from Kumaon region of Western Himalaya. *Open Access Journal of Medicinal and Aromatic Plants*, *1*(1), 13–18.

Vivek, S., Nisha, S., Harbans, S., Devendra, S. K., Vijaylata, P., Bikram, S., & Raghbir, G. C. (2009). Comparative account on GC-MS analysis of *Mentha arvensis* L. "corn mint" from three different locations of North India. *International Journal of Drug Development and Research*, *1*(1), 1–9.

Voirin, B., Saunois, A., & Bayet, C. (1994). Free flavonoid aglycones from *Mentha piperita*: Developmental, chemotaxonomical and physiological aspects. *Biochemical Systematics and Ecology*, *22*(1), 95–99.

Wannissorn, B., Jarikasem, S., Siriwangchai, T., & Thubthimthed, S. (2005). Antibacterial properties of essential oils from Thai medicinal plants. *Fitoterapia*, *76*, 233–236.

Yamamura, S., Ozawa, K., Ohtani, K., Kasai, R., & Yamasaki, K. (1998). Antihistaminic flavones and aliphatic glycosides from *Mentha spicata*. *Phytochemistry*, *48*(1), 131–136.

Yi, W., & Ywetzstein, H. (2011). Anti-tumorigenic activity of five culinary and medicinal herbs grown under greenhouse conditions and their combination effects. *Journal of Science and Food Agriculture*, *91*, 1849–1854.

Zhang, W.-J., Yang, K., You, C.-X., Wang, C.-F., Geng, Z.-F., Su, Y., & Deng, Z. W. (2015). Contact toxicity and repellency of the essential oil from *Mentha haplocalyx* Briq. against *Lasioderma serricorne*. *Chemistry and Biodiversity*, *12*, 832–839.

Zheng J., Zhao D. S., Wu B., Wu L. J. (2002). Study on chemical constituents in the herb of *Mentha spicata*. *China Journal of Chinese Materia Medica*, *27*(10), 749–751

Zheng, J., Chen, G. T., Gao, H. Y., Wu, B., & Wu, L. J. (2007). Two new lignans from *Mentha spicata* L. *Journal of Asian Natural Products Research*, *9*(5), 431–435.

7 Basil (*Ocimum basilicum*)

Gareema Pandey, Divyang Solanki, Anima Manhswita, Qiu Yi Tan, and Sangeeta Prakash

7.1 INTRODUCTION

Basil (*Ocimum basilicum* L.) is an annual herb in the mint family (*Lamiaceae*). It has been used for centuries and is a vital element in several cooking styles and traditions. The genus *Ocimum* comprises some 50 to 150 species and varieties which are confined to Asia's tropical climates, as well as the regions of Central and South Africa. The considerable variety seen among the component species is primarily responsible for the disagreement about the exact number of species within the genus. Variation can be seen among them in terms of the growth habits, morphology, leaves, flower color, chemical composition, and stems. Basil quickly cross-pollinates, and the ensuing diversity and variation have prompted some writers to rename portions of the species (Al-Maskari et al., 2012).

Basil has a branched, square stem, opposing leaves, bilabiate calyx and corolla, a verticillaster inflorescence, and black or brown seeds (also known as nutlets). In general, the *Ocimum* species includes both cultivated and wild herbs and shrubs (Marwat et al., 2011). There are other plants with the common name basil which are not part of the genus *Ocimum*, including 'wild basil' (*Clinopodium vulgare* L.), and 'basil thyme' (*Acinos arvensis* (Lam.) Dandy), which can cause confusion and misunderstanding (Al-Maskari et al., 2012).

There are several theories regarding the origin of the word basil. According to some, the word basil is an abbreviated version of the Greek term "*Basilikon phuton*," which means "royal herb." Others believe that the term *Ocimum basilicum* is derived from the "Greek words okimon," which means "smell," and "basilikon," which means "royal." As a plant, basil is related to a variety of beliefs and rituals. In Italy, it is a symbol of love; in France, it is known as a royal plant, and, during the Victorian period, it was employed as a symbol of good wishes. It was used by Jews to boost vigor during fasting, and an African tradition suggests that basil protects against Scorpions. However, some Europeans believe it to be a Satanic symbol (Marwat et al., 2011). The plant was allegedly associated with the growth of a scorpion in a man's brain after he inhaled it every day. The relationship with scorpions can still be evident in basil's association with the astrological sign of Scorpio today (Shahrajabian et al., 2020). Other legends claim that if basil was left to die in horse manure, it would create venomous animals. Depending on where you live, *O. basilicum* is given different names. In English, it is commonly referred to as common basil, sweet basil or basil, whereas it is known as "*babui tulsi* in India", notably in Hindi and Bengali. Basil is also known as "*basilica* (in French), *basilikum* or *basilienkraut* (in German), *basilico* (in Italian), *rehan* (in Arabic), and *albahaca* (in Arabic) (in Spanish)". In Arabic, it is called "*hebak* [or] *rihan*" (Al-Maskari et al., 2012).

Sweet basil (*O. basilicum*) is probably the most recognizable basil; nonetheless, it has a huge variety of cultivars that differ in terms of flavor, smell, and usage. There are currently around 160 identified cultivars in existence. Popular types include *'Cinnamon,' 'Dark Opal,'* and holy basil (*the species O. tenuiflorum L.*, previously known as *O. sanctum L.*). Cinnamon, licorice, lemon, and anise are just a few of the scents and flavors available. According to the species, these plants can be herbaceous or shrubby, with heights ranging widely, from 20 cm to 3 m (Al-Maskari et al., 2012). According to the species, their leaves could be glossy, smooth, curly, or hairy, and they can be either green or blue/purple in color. The flowers range in color from white to purple to lavender. Most

DOI: 10.1201/9781003292340-7

ordinary basil varieties are classified annuals; however, numerous perennial forms, such as *O. tenuiflorum*, occur in warm tropical climates. Basil's decorative relevance has expanded in recent years, as has the plant's economic worth internationally, due to its diverse range of shapes, colors, and sizes. Basil is also commonly planted as a decorative plant in public parks and residential gardens (Al-Maskari et al., 2012).

The leaves have a strong fragrant perfume that resembles cloves. Every good Hindu sleeps with a basil leaf on his breast. This is his ticket to Paradise. Sweet basil (*Ocimum basilicum*) drew a great deal of attention in ancient Indian medicine systems (Ayurvedic and Unani) for its efficacy in treating various ailments. Basil is also a well-known source of flavoring components and is utilized in traditional remedies as well as being a culinary plant. It is also a widely-used plant in Mediterranean and American cuisine (Marwat et al., 2011). Basil is additionally used in cosmetics such as soap, shampoos, lotions, oils, and scents. Its oil is used in aromatherapy and as a medication for stress, migraine, colds, and hay fever. Similarly, basil tea is beneficial for digestion, the elimination of gas, indigestion, constipation, diarrhea, and vomiting.

The plant is also used to cure mental weariness, nervous problems, and coughs (in combination with hyssop). The antispasmodic, aromatic, digestive, carminative, stomachic, tonic, and galactogogue characteristics of *O. basilicum*'s leaves and blooming parts are widely acknowledged in medicine. The leaves were used to treat loss of smell (anosmia), acne, snake bites, skin infections, and insect stings. The concentration of the essential oil of basil varies similarly between cultivars and species, which is assumed to be due to diverse geographic origins, ecological conditions, distinct chemotypes, genetic patterns, and variances in the nutritional state of plants. In Finland, 17 distinct collections of sweet basil were studied for their morphological characteristics and chemical composition. In terms of both these traits, there was considerable variety. When ten commercially available basil cultivars from Italy were tested, a comparable degree of heterogeneity was observed. Chemical investigations of the variations revealed particular relationships to their morphological characteristics (Al-Maskari et al., 2012). Two of the violet-leaf varieties included linalool chemotypes, while three of the large-leaved kinds contained methyl chavicol (p-methoxyallyl benzene or estragole) and linalool chemotypes. Most of the essential oil in basil is to be found in its leaves and flowers; there are minimal levels of essential oils in the branches and stems, but these amounts are not economically significant. Basil is very varied in terms of both its morphology and chemistry, and the differences appear to be greatly impacted by environmental conditions. Basil's origin, source, and growing circumstances all influence its plant applications, particularly its flavors, fragrances, and medicinal properties. The plant's adaptability is evident in its diverse range of applications, which will be described in greater depth later in the chapter (Al-Maskari et al., 2012).

7.2 ORIGIN OF BASIL

Basil has a lengthy and intriguing legend-filled past. It has been cultivated for 5,000 years and originated in Iran, India, and other tropical Asian locations. It is widely believed that "Alexander the Great (356–323 BCE) brought it to ancient Greece, that it found its way to England from India in the mid-1500s, and that it reached the United States in the early 1600s". It was planted in mediaeval gardens and is referenced in numerous "classic herbals, including Culpeper's, Gerard's, Parkinson's, and Dioscorides" (Li & Chang, 2016).

The use of sweet basil for healing was contested in "Roman times by Greek doctors Galen and Dioscorides, who believed it should not be used internally". Pliny, a Roman scholar and author, championed its usage, and it has subsequently used to decrease blood pressure, cleanse the blood, and lower blood sugar and cholesterol levels, along with as a general detoxifier. Gerard commended basil as a melancholy treatment, but he also reiterated Dioscorides' caution that too much basil "dulleth the sight and is of a hard digestion."

7.3 BASIL: CULTIVATION, OCCURRENCE, CATEGORIZATION, AND VARIETY IN SPECIES

Basil (*Ocimum basilicum* L.), a source of essential oil, is a culinary herb and medicinal plant that thrives in tropical and subtropical regions. Its essential oil is utilized in oral health and dental goods, as well as in the fragrance and food industries. "*Ocimum* belongs to the genus *Ocimum*, the family Lamiacaea, the order Lamiales, the class Magnoliopsida, the phylum Magnoliopsida, and the kingdom Plantae" (USDA, 2020). In laboratory tests, it germinates at a rate of 95–98%; in the field, by contrast, it germinates at a rate of 10–15%. The anatomy of a basil plant is made up of: (a) nodes, (b) the main stem, (c) future stem growth, (d) internodes, (e) leaves, and (f) the dominant growing tip. It may be utilized in a variety of ways, including dried leaves, flowers, essential oil, and as a decorative plant. Purple basils, sweet basils, cinnamon basil, lemon basils, fine leaf basil, bush basil, and anise basil are among the most common kinds of basil. Aphids, leafhoppers, and whiteflies are the most common pests of the plant.

The most common diseases of basil are *Fusarium, Botrytis*, and leaf spot. Previous research has revealed that it is susceptible to desiccation stress, which can result in a considerable decrease in essential oil content, dry and fresh matter yield, proline content, protein content, and glucose content. According to the study by Kalamartzis et al. (2020), basil cultivars should have improved water use efficiency in semi-arid environments, allowing plants to conserve more groundwater. Following their harvest, dry stem yields 1.2–2 (fresh weight 8–10) t/ha. Leaf spot (*Pseudomonas cichorii*), fusarium wilt (*Fusarium oxysporum* f. sp. *basilicum*), downy mildew (*Peronosporabelbahrii*), root rot (or damping off) (*Rhizoctonia solani*; Pythium spp.), and grey mould (*Botrytis cinerea*) are among the most frequent diseases to affect the plant (Shahrajabian et al., 2020).

7.3.1 Location

Although basil can be grown in a variety of environmental and climatic conditions, the best conditions are to be found in countries with warm climates. The primary ecological needs for the cultivation of basil are warmth, light, and moisture. Because this plant is sensitive to the cold, outdoor growing is limited to frost-free parts of the world. Pakistan, India, the Comoros Islands, Haiti, Madagascar, Réunion, Guatemala, Indonesia, Thailand, South Africa, Russia, Morocco, Egypt, France, Bulgaria, Israel, the United States (California, Arizona, and New Mexico), Hungary, Italy, Germany, Poland, Turkey, Greece, other Balkan countries, and Slovakia are among the locations where the plant is widely grown. Absolute output estimates for basil oil are difficult to come by, although the United States is likely to be the largest market for basil oil, followed by the European countries of Germany, France, the United Kingdom, and the Netherlands. Global figures on dried basil output are similarly difficult to pin down. A considerable amount of global output, namely that produced in the Mediterranean region, India, and California, is not exported; most basil in these countries is consumed locally. According to import figures, the United States is one of the world's largest users of dried basil. In the 1990s, the total volume of basil herb shipped into Europe was approximately 830–880 t/year (Al-Maskari et al., 2012).

7.3.2 Taxonomy

Most commercial basil varieties on the market are *O. basilicum*. Darrah (1980) classified *O. basilicum* cultivars into seven categories.

1. "tall and slender varieties, such as those seen in sweet basil".
2. "strong, large-leafed varieties, such as 'Lettuce Leaf,' commonly known as 'Italian' basil".
3. "dwarf basil varieties, which are short and have little leaves, such as 'Bush' basil".

4. "compact forms, *O. basilicum* var. *thyrsiflora*, generally known as 'Thai' basil".
5. "purpurescens, purple-coloured basil varieties with classic sweet basil flavour".
6. "purple varieties, such as 'Dark Opal,' a potential hybrid of *O. basilicum* and *O. forskolei* with lobed leaves, a sweet basil scent, and a clove-like aroma".
7. "Citriodorum basils, which include lemon-flavored varieties".

Basil taxonomy is complicated by the presence of many botanical varieties, chemotypes inside the species, and cultivar names which may differ little in their appearance (Simon et al., 1990). Each basil variety's aromatic character is controlled by genotype and impacted by the primary chemical components of essential oils (EOs), which are predominantly comprised of phenylpropanoids and monoterpenes (Marotti et al., 1996). Chemotypes are often characterized based on the presence of more than 20% of a molecule or component. The aroma chemicals discovered in basil chemotypes include linalool, eugenol, methylchavicol, methylcinnamate, and citral.

These chemotypes are named for their geographical conditions, examples including, for example, European, Reunion or Tropical, and Egyptian basil. Sweet basil, a European species, is said to have the best scent, with its main components being methylchavicol and linalool. Egyptian basil is similar to European basil, except that it contains more methylchavicol. The Reunion type is distinguished by a high methylchavicol content and is found in the Madagascar, Comoros Islands, Vietnam, and Thailand. Tropical basil, such as that found in India, Bulgaria, Pakistan, and Guatemala, is high in methylcinnamate, whereas basil from Java, Russia, and North Africa is high in eugenol (Simon et al., 1999).

7.4 NUTRITIONAL ATTRIBUTES OF THE PLANT

Throughout the preceding three decades or so, scientists have taken a considerable interest in plant research, notably in the industrialized countries of America and Europe. Today, about 60% of the global population uses herbs and natural ingredients for therapy, since they are recognized as a significant source of bioactive compounds. Plants contain a variety of bioactive compounds which have been demonstrated to have a variety of health-promoting properties, such as antioxidative, antibacterial, antiviral, antihypertensive, anti-inflammatory, and so on (Peschel et al., 2006). Polyphenols are used in the creation of nutritional supplements, nutraceuticals, functional food components, and cosmeceuticals as physiologically active molecules. So-called functional foods are those that, when ingested on a regular basis, generate a specific beneficial health benefit in addition to their fundamental nutritional attributes. These foods often include varying levels and types of bioactive chemicals. When a bioactive component is added to a food composition for a specific purpose, the resultant product may be classified as a functional food (Filip, 2017).

Among the health benefits of the new functional food are lower cholesterol levels, relief from lactose and gluten sensitivity, speedier recovery from diarrhea, and the suppression of cancer cell multiplication. Nutraceuticals, by contrast, are defined as dietary supplements that offer bioactive chemicals from food in a concentrated form, contained in a non-food matrix, and used to improve health at dosages greater than those acquired from typical diets. They are often taken in the form of pills, tablets, capsules, or powder. Nutraceuticals derived from vegetables and fruits, plants, fungus, microalgae, and algae have been proven to include a diverse variety of bioactive substances (constituents), such as carotenoids, polyphenols, anthocyanins, coumarins, etc. Polyphenols are secondary plant metabolites that have a range of forms and activities in plants and which are widely recognized in the human diet as antioxidants (Rice-Evans et al., 1997). Antioxidants are essential molecules that can protect the cell membrane from damage caused by free radical-induced oxidative stress (Rao et al., 2010). Natural antioxidants, such as the polyphenols found in medicinal and fragrant plants, are gaining popularity as a means of preventing oxidative damage (Wojdyło et al., 2007).

The human body has a number of mechanisms in place to safeguard cellular molecules from reactive oxygen species (ROS)-induced damage, including non-enzymatic and enzymatic antioxidant systems. Exogenous antioxidants, on the other hand, are continually necessary to maintain an acceptable quantity of antioxidants to balance the ROS in the human body. Many synthetic antioxidants, such as butylated hydroxyl toluene (BHT) and butylated hydroxyl anisole (BHA), are extremely efficient and can be used in industrial operations. Their usage has been restricted due to the fact that they are, in general, poisonous, mutagenic, and carcinogenic to human health (Arlett et al., 2012).

As a result, chemicals, particularly those derived from natural sources, can protect against ROS-mediated damage and may be useful in disease prevention and/or treatment. Because of their redox characteristics, phenolic compounds in herbs operate as antioxidants by acting as hydrogen donors, reducing agents, metal chelators, and free radical quenchers. Plants, plant combinations, or even extracted chemicals are frequently added to meals to produce beneficial characteristics. Basil is an aromatic and therapeutic herb which is widely available to the food industry, and it appears to offer promising potential (Filip, 2017). This plant is renowned for its use as a spice and ingredient in Southeast Asian and Italian cuisines. It is also commonly used in sauces and salads, the most well-known of which is the Italian pesto. Basil, whether dried or fresh, is also one of the most widely used culinary herbs in North America. It has also long been used in folk medicine to treat asthma, as a carminative, stomachic, antispasmodic and medicine for the inflammation of respiratory and urinary systems (Simon et al., 1999).

7.5 PHYTOCHEMICAL OVERVIEW OF THE PLANT

Sweet basil has a low fat and calorific value, but it is high in minerals and vitamin A. "2.5 g of basil leaves contains vitamin A (96.6 IU), less than 1 calorie, calcium (3.85 mg), potassium (11.55 mg), and lesser amounts of protein, vitamin C and other vitamins, minerals and fibre (five fresh leaves)" (CABI 2022). According to the US Department of Agriculture, the sweet basil leaf has a generally recognized as safe (GRAS) status in the range of 2–680 ppm and 0.01–50 ppm for its essential oil. The usage of extremely large amounts of oil is thought to pose a health concern owing to the presence of carcinogenic chemicals (Hosseini-Parvar et al., 2015).

O. tenuiflorum possesses essential oils that are principally concentrated in the green leaves and which thus have a distinct scent. This volatile oil, with a "leafy" aroma, is composed mostly of phenols, terpenes, and aldehydes. In addition to essential or fixed oils, the plant contains alkaloids, glycosides, tannins, and saponins. The plant's leaves also have high levels of ascorbic acid and carotene. The chemical qualities of basil leaves have been described based on multiple international investigations, and thus edaphic and regional variables are likely to impact distinct chemical constituents. The scent of different cultivars of *O. basilicum* differs according to the varying compositions of their essential oils (Muhammad et al., 2011). A large range of basil cultivars are found around the world, indicating a wide variation of chemical composition. Basil essential oil typically includes the following constituents: eucalyptol, α-terpineol, methyl eugenol, β-elemene, eugenol, linalool, α-bergamotene, germacrene D, cubenol, α-guaiene, camphor, τ-cadinol, α-caryophyllene, bornylacetate, and β-caryophyllene (Muhammad et al., 2011). It has been observed that key essential oil components include phenylpropanoids, terpenes, aldehydes, and alcohols; and essential oil composition is impacted by growing circumstances, location, cultivars, seasonal fluctuation, varied agronomic management, drying, and manufacturing processes and harvesting. The production and quality of essential oils produced by basil plants are also affected by field conditions. Basil's primary phenolic components are flavonol-glycosides and phenolic acids. Oleic acid, stearic acid, linoleic acid, palmitic acid, α-linolenic acid, myristic acid, lauric acid, arachidonic acid, and carpic acid are the main fatty acids to be found in the plant species (Castronuovo et al., 2019). Higher levels of temperature and light have an impact on antioxidant capacity. Vanillic, caffeic, quercetin, rosmarinic acids, rutin, chlorogenic, p-hydroxybenzoic acid and apigenin are the most prominent antioxidant chemicals found in basil (Shahrajabian et al., 2020).

7.6 BIOACTIVE COMPOUNDS IN BASIL—CHEMICAL CONSTITUENTS

Due to the presence of secondary metabolites such as tannins, essential oils, flavonoids, phenols, steroids, and anthocyanins, *Ocimum* species have long been known for their health benefits and have been used as a folk medicine (Filip, 2017). Recent methods and instruments utilized for the extraction of bioactive compounds are portrayed in Figure 7.1. Basil contains significant amounts of potassium, iron, and magnesium. Potassium and magnesium, two of the seven essential macrominerals, provide protection against several chronic diseases and also boost cardiovascular health and the transmission of nerve impulses. It has both diuretic and digestive-stimulating properties. Basil contains a variety of vital nutrients, including vitamin C, vitamin A, phosphorus, and calcium. It also contains high levels of carotenoids such as carotene, which are converted to vitamin A in the body.

Basil contains a lot of polyphenols, which are a diverse group of chemicals including phenolic acids, simple or complex flavonoids, and colorful anthocyanins. They inhibit peroxide oxidation in terms of pharmacological efficacy (Filip, 2017).

7.6.1 Essential Oil

Basil essential oil (EO) is extracted by steam distillation from the flowering tops and leaves and used to flavor foods, fragrances, pharmaceuticals, and dental and oral treatments. The global output of basil essential oils is around 100 t/year. The EO concentration of basil ranges from 0.5 to 1.0% for European chemotypes, while differences in EO content among countries may be related to the regions' varying climatic circumstances. According to Marotti et al. (1996), the EO concentration in the herbs of ten Italian basil cultivars ranged from 0.3 to 0.8%. Oil content ranged from tiny amounts to 2.65% in comprehensive research on 270 basil accessions in Germany. In contrast, the EO content of Fiji basil is just 0.2% (dominant compounds: methyleugenol 24.7%, methylcinnamate 23.6%,

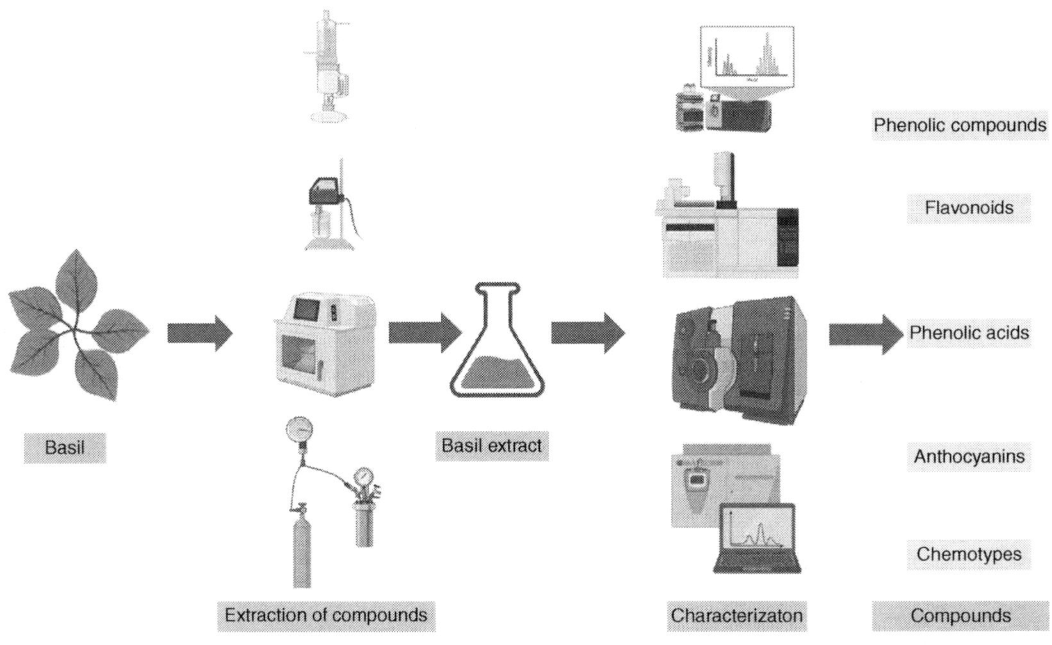

FIGURE 7.1 Techniques applied for the extraction of bioactive compounds.

and linalool 22.3%). In contrast, basil from Cuba has a much higher EO content of 1.9–2.5% (dominant compounds: 1,8-cineole 5.4%, linalool 5.0%, and methylchavicol 66.8%), whereas basil from Burkina Faso has a 0.7–1.8% oil content (dominant compounds: 1,8-cineole 60.2%, α-terpineol 6.5%, and β-pinene 5.7%) (Kéita et al., 2000). As can be seen from the above figures, the composition of basil essential oils varies greatly. Based on the primary components of the oil, essential oils extracted from *O. basilicum* have traditionally been divided into four different chemotypes with several subtypes: methylchavicol-rich, linalool-rich, methyleugenol-rich, and methylcinnamate-rich (Filip, 2017).

Different varieties of basil and producing locales provide oils with various tastes and values on the global market. The European basil EO, which principally contains linalool and methylchavicol, is often regarded as having the greatest quality and scent. It is an oil which is commonly used in the cosmetic and perfume industries. Basils with scented leaves (licorice, lemon, cinnamon) are used in honeys, jellies, baked goods, and vinegars. The plant's essential oil has a wide range of characteristics. It has been demonstrated to have antiviral, antioxidant, anti-inflammatory, antidiabetic, and HIV-1 inhibitory actions. Antibacterial activity against Gram-positive and Gram-negative bacteria, yeast, and moulds was also demonstrated. The antibacterial activity of essential oils of basil from Niazbo, France, and India, which are enriched in methylchavicol, methylcinnamate, and linalool, was tested with regard to 11 Gram-positive bacteria and 7 Gram-negative bacteria, respectively. Gram-positive bacteria were shown to be more resistant to these oils than Gram-negative bacteria. Basil oils were also shown to possess fungistatic properties. The essential oil demonstrated insecticidal efficacy, and the entire plant, as well as the essential oil, is a well-known insect repellent. According to Danesi et al. (2008), EOs from a purple basil cultivar prevent cardiomyocytes from oxidative stress better than green "Genovese" and "Lettuce Leaf" cultivars.

7.6.2 Phenolic Acids

Basil has a significant phenolic acid content, which contributes to its excellent antioxidant capacities. Rosmarinic, caffeic, chicoric, and caftaric acids are the most common phenolic acids found in basil. One of these, rosmarinic acid, is an important phytochemical due to its antioxidant and pharmacological characteristics. It was proven to be an extremely powerful active agent against human immunodeficiency virus type 1 (HIV-1). The phenolic acid concentration of basil varies according to the cultivar, as well as the extracts produced using various solvents and extraction procedures.

Total phenolic content and phenolic acid concentration in basil are substantially influenced by environmental circumstances during growth, such as temperature and photoperiod, the availability of soil nutrients, and seasonal, regional, and climatic variation. Furthermore, owing to age, storage circumstances, and/or the drying process, the same commercial basil products may experience some phenolic acid depletion. The primary function of phenolic acids is as antioxidants. The antioxidative activity of phenolic acids is related mostly to their redox characteristics, which can be useful in neutralizing free radicals, quenching singlet and triplet oxygen, and degrading peroxide. As a result, they have various benefits, such as being a preventative against cardiovascular disease and acting as antiplatelet and anti-inflammatory drugs (Filip, 2017).

7.6.3 Flavonoids and Anthocyanins

Flavonoids are plant-derived phenolic chemicals that act as antimicrobials, antioxidants, visual attractants, photoreceptors, light screening agents, and feeding repellents. Flavonoids has demonstrated several biological effects in the human body, such as antiviral, antiallergenic, vasodilating, and anti-inflammatory properties, and are essential components in the human diet. However, the antioxidant activity of flavonoids has sparked the greatest interest, owing to their capacity to decrease free radicals. Flavonoids extracted from basil are members of the flavones and flavonols subgroups, which exist as glycosylated derivatives. Grayer et al. (2002) discovered that the prominent components in *O. basilicum* were quercetin, luteolin, apigenin, and kaempferol, all of which were found as O-glycosides.

Anthocyanins are to be found in abundance in basils with intense purple pigmentation. Certain purple basil cultivars include high quantities of anthocyanins, which are water-soluble phenolic chemicals that are principally responsible for red, purple, and blue coloration in plants. They perform a variety of activities inside plants, ranging from defensive to protective duties to supporting growth, development, and reproduction. They have antioxidant properties as well as radical scavenging abilities, which have been connected to disease prevention in humans. The benefits of their use are related to neuroprotective effects, as observed, for example, in a lower risk of Parkinson's disease (Gao et al., 2012). In a study conducted by Simon et al. (1999), 14 distinct anthocyanins were found, the majority of which were cyanidin and peonidin glycosides. Total anthocyanins range from 0.07 to 0.19 mg/g fresh weight.

According to current research, plant polyphenolics may offer preventative properties against neurological, inflammatory disease, cardiovascular, several types of cancer, and diabetes. Basil is a medicinal herb used in traditional medicine, including Ayurveda, to treat stress and cardiovascular illness, including hypertension, and in Chinese medicine to treat cardiovascular disease, including hypertension. It is an effective anti-inflammatory drug which reduces inflammation by around 54%. Amrani et al. (2009) found that aqueous basil extracts had anti-platelet aggregation and antithrombotic properties in vivo. The new findings revealed a substantial drop in blood glucose levels as well as plasma total cholesterol, LDL-cholesterol, and triglyceride levels.

Basil has grown in popularity because of its numerous health advantages. Aromatic/medicinal plant extracts and essential oils (EOs) have been studied not just for illness prevention and treatment, but also as natural food preservatives. Due to the carcinogenicity of several of the synthetic antioxidants (BHT and BHA) used in food production, they were replaced by safe additives/extracts derived from natural sources with demonstrated antioxidant activity.

Natural antioxidants are increasingly being used to replace synthetic antioxidants in the food and pharmaceutical industries. As a result, there is an increasing interest in researching natural, non-toxic, and healthful herbal extracts with antioxidant characteristics that might be utilized as additives. Basil EO was used as an ingredient and antioxidant in beef burger products.

The advantages included the suppression of lipid oxidative degradation, which resulted in there being no changes in organoleptic qualities such as flavor, color, and texture. EO acts as antibacterial agents in these products. Despite all of these benefits, only a few studies have reported the use of basil in dishes. The plant's extracts and leaves are used to flavor and preserve olive oils (Perez et al., 2007). Because of their strong color and fragrant qualities, such items are widely valued by customers. Direct aromatization of olive oil with basil is a recent trend in the Mediterranean region, with both sensory and nutritional benefits (Veillet et al., 2010). Because of the health benefits of basil, it is commonly consumed with yoghurt in Turkey. To evaluate its functional and preservation capabilities, the leaves were inserted into Portugal's "Serra da Estrela" cheese, with the intention of developing unique culinary items using natural ingredients. Polysaccharides are abundant in basil seeds. This becomes gelatinous when soaked in water and is utilized in Asian cocktails and sweets.

7.7 PHARMACOLOGICAL APPLICATIONS AND CLINICAL TRAILS

For centuries, basil has been used as a traditional medicine. Various studies have highlighted the importance of the herb in modern medicine. From treating small ailments such as cold, cough, stomach ache, arthropod stings, constipation, and headaches, basil has the potential to protect cells from different phases of cancer. Studies have found that different parts of the basil plant have numerous different properties, such as antifungal, antioxidant, antiaging, antibacterial, antipyretic, antitumor, and many more (Figure 7.2).

FIGURE 7.2 Bio-functional properties of the basil pant.

7.7.1 Cosmetic and Anti-aging Activities

In many traditional as well as modern medicines, the leaves of basil and its extract are used in toothpastes; they have proven useful in toothache and oral healthcare and hygiene (Ahmed, 2016; Devi et al., 2014; Idu et al., 2009). Solanki et al. (2017) have shown that the ethanolic extract of basil leaves was faster and more effective in the healing of wounds when compared with the use of standard ointments in male albino rats. In another study, a mixture of the extract of *Trifolium pratense* with basil showed very good healing effects in terms of both the coverage and the fibroblast density recovery. The animal model study, by using a hydrogel made up of the same mixture, showed a 100% wound healing rate in 13 days (Antonescu et al., 2021).

The basil plant has also been used to making skin whitening creams and as a vehicle for transdermal drug delivery for more than two decades (Fang et al., 2004; Manosroi & Manosroi, 2003). Over the past few years, it has been used for other purposes, such as anti-aging and UV-protective creams (Jadoon et al., 2015; Rasul & Akhtar, 2011).

7.7.2 Anti-microbial Activities

Basil, and its extracts of polyphenols and polysaccharides, show a broad spectrum of anti-microbial activities. They are proven to be effective against *Bacillus subtilis, Staphylococcus aureus, Proteus vulgaris, Klebsiella pneumonia, Enterococcus faecalis*, and others. The Gram-positive bacteria show more susceptibility towards basil than the Gram-negative bacteria because of the structural variation in the cell walls of the two. It has shown to have an inhibitory effect on the bacteria that have similar clinical infections. They tend to work by decreasing the levels of inflammatory mediators present in the serum (Opalchenova & Obreshkova, 2003; Siddiqui et al., 2012; Vlase et al., 2014). Future studies are demanded to standardize the therapeutic dose, time of application and relative effects with other antibacterial agents.

Basil also possesses anti-fungal and anti-parasitic properties. In a study it was shown that the bioactive compounds of the basil plant inhibit the growth of *Aspergillus flavus* and *Aspergillus parasiticus*. These produce metabolic products which are carcinogenic, mutagenic, and teratogenic in nature. They are also efficient in inhibiting the toxicogenic stains of various fungi.

The essential oils extracted from basil inhibits larval-phase activities and are very effective against mosquitoes. It can also be used as an insecticide for the crops. A study showed that when basil extract was used on beans (*Phaseolus vulgaris*), it protected the crop against the common pest *Acanthoscelides obtectus*; as a result, the crop suffered less damage.

In a study it was found that basil essential oil was effective in the treatment of trichomoniasis, as it inhibited the growth of *T. vaginalis* trophozoites. In another study it was seen that it acts as an effective drug against *Plasmodium falciparum* (Hashemi et al., 2021; Deligianni et al., 2023).

Basil and its extracts, which include ursolic acid, linalool, and so on, have shown to be effective against several viruses such as RNA viruses, adenoviruses, hepatitis virus, herpes viruses, adenoviruses and enteroviruses. They can be used for both animals as well as humans (Dhama et al., 2021).

7.7.3 Anti-cancer/Antitumor Activity

Cancer has become the largest reason of death across the globe. Basil and its polysaccharides have been found to be very effective in combating cancer. The various mechanism of working includes preventing the destruction of basement membrane, initiating the apoptosis of tumor cells, etc. A study found that basil extract can inhibit SKOV3 cells, and it enhances the killing ability of NK cells, which indirectly increase the ability of the cells to stop the metastasis of tumors. The various types of cancer on which the basil extract work efficiently include colon cancer, breast cancer, cervix cancer, pancreas cancer, bladder cancer, and leukemia (Dhama et al., 2021; Sestili et al., 2018).

7.7.4 Antioxidant/Anti-Depressant/Neuro-related Activities

Studies have demonstrated the antioxidant activities of the basil plant. In a study, they found the way to enhance the status of antioxidants in rats which, in turn, worked as an antidepressant in that case. The antioxidant properties of basil polysaccharide also led to increased activity of SOD and glutathione-Px, whereas it decreased the content of malondialdehyde in the liver tissues of a mice (Muneefa et al., 2017). The antioxidant properties can also contribute to the chemoprotective and antidepressant properties of the plant. Similarly, they showed that the basil EO had anti-depressant properties and improved behavior changes. Another study showed that the leaf extract from basil had the ability to enhance the neuromuscular coordination, recognition ability, short-term memory, and explanatory behavior in male albino mice (Zahra et al., 2015). The antioxidant properties of the basil and its extracts have also shown it to have a hepatoprotective role. It functions in the treatment of hepatic fibrosis by preventing oxidative damage of the hepatic tissues (Alomar & Al-Attar, 2019).

7.7.5 Analgesic Properties

The methanolic extract of basil was demonstrated to have an analgesic effect on a Swiss mice, through the tail immersion method. It was shown that the concentration of 200mg/kg of basil extract had better activity when compared with the standard drug (Aspirin) (Golak et al., 2010). Similarly, another animal study done on mice showed the same effect. In this, the effect of basil was tested through the use of a hot plate and acetic acid induced writhing method, and the result found that the basil extract had a significant difference when compared to the other drugs administered (Umamageswar & Kudagi, 2015).

7.7.6 Immunity Modulatory/Enhancer Properties

All the immunity research carried out on the basil plant have been proven to have a positive effect by combating various immune-related disorders, which is mainly because of the presence of various alkaloids, flavonoids, phenolic and glycosidic compounds, tannins etc. It has been seen that the consumption of basil extracts results in the proliferation of lymphocytes, and modulates both humoral and cell-based immunity. Another study showed that the extracts can act effectively against immune-suppressive drugs such as cyclophosphamide; it also downregulates cytokines derived from regulatory T cells and TH1 and TH2. It enhances the DNA synthesis in a concentration-dependent manner of the peripheral blood mononuclear cells.

Some work carried out on animals like chickens and fishes has also proved to be beneficial. A study conducted on broiler chicken (Jahejo et al., 2019) showed that supplementing basil seed powder in their diets led to improved feed efficiency, increased nutrition retention, and increased white blood cells (WBC) count and hemoglobin. Similarly, in fishes it increased the red blood cells (RBC) and WBC count (Dhama et al., 2021).

7.7.7 Ulcer

Basil extracts have been effective in the treatment on gastrointestinal diseases in many medicine types such as Ayurveda. Clinical studies in recent years have proven the same. The studies done on rat models show that the extract from basil plant reduces the chances of ulcer, it is mainly because of the strengthening of gastric mucosa, as the production of acid and pepsin is decreased. It was found that the anti-secretory and lipoxygenase-inhibiting histamine antagonistic activities of the basil oil are responsible for anti-ulcer activities. The basil extracts are also effective against microorganisms which cause diarrheal infections and abdominal pain. Another study suggested that regular consumption of basil extracts protects from *H. pylori* infections (Dhama et al., 2021; Sestili et al., 2018).

7.7.8 Diabetes

Various studies of mice have shown that basil polysaccharides have an anti-hyperglycemic effect. The various methods by which the natural plant extract helps are by maintaining glucose homeostasis by digesting and absorbing carbohydrates, stimulating the production of glycogen from the liver, mimicking insulin in the body, and being insulin-sensitive so that there is no chance of toxicity. In a study done on diabetic rats, it was shown that it decreased the levels of blood glucose and the end products of advanced glycation (Widjaja et al., 2019). Similarly, in another study Kadan et al. (2016) conducted chemical and biophysical tests to prove the involvement of compounds form the basil in treating diabetes. They also conducted phytochemical analysis to analyse the various compounds of basil extract beneficial in the disease.

7.7.9 Hyperlipidemia and Cardiac Diseases

Hyperlipidemia is the accumulation of very high levels of fat in the blood; it, in turn, gives rise to cholesterol and triglycerides. The basil plant extracts have been found to have a hypolipidemic activity on humans. It is also beneficial in protecting against the cardiovascular diseases which arise as a result. A study showed that the basil extract reduced the levels of cholesterol and triglycerides in the liver and plasma of the mice that suffered from hyperlipidemia (Harnafi et al., 2008). Basil showed a positive effect in those subjects that suffered from blood pressure, and other cardiac diseases. In a study, Fathiazad et al. (2012) studied the role of basil on cardiac functions on artificially induced myocardial infraction. They observed that basil had a protective effect on the heart tissues against the infraction, which were linked to the antioxidant properties of the plant.

7.8 EXTRACTION AND ISOLATION TECHNIQUES FOR BIOACTIVE COMPOUNDS

7.8.1 Conventional Techniques

Synergistic effects of phenolic compounds give a natural antioxidative effect to the basil extracts and health-promoting activities, such as anti-inflammatory, antimicrobial, and antidiabetic properties (Romano et al., 2022). This herb is used medicinally to treat coughs, headaches, diarrhoea, warts, constipation, kidney failure, and worms (Pratama et al., 2020). Basil leaves are also a rich source of flavonoids, such as quercetin, anthocyanins, luteolin, and kaempferol (KA) (Altemimi et al., 2020). Various conventional methods employed for the extraction of bioactive compounds are presented in Table 7.1.

Approximately 70% of the drugs approved for infectious diseases and cancer in the past decade were derived from natural bioactive compounds (Newman & Cragg, 2016). The extraction of these bioactive compounds (BACs) depends upon the optimal conditions and technique used for isolation, purification, and identification. Due to the wide range of plant matrices and the physiochemical features of BACs (for example, many are thermally unstable), the extraction from plant components could prove difficult. As a result, the identification and optimizing of extraction parameters is a technological imperative to produce high-quality extracts.

Steam distillation, water distillation, combined water and steam distillation, maceration, cohobation, and enfleurage are all examples of conventional extraction (CE). Steam and hydrodistillation are the techniques most often used for extraction of BACs/EOs from plants (Soxhlet extraction). Steam distillation has been popularly utilized, particularly on a commercial scale (Tongnuanchan & Benjakul, 2014). The European Pharmacopoeia includes the technique of hydrodistillation with Clevenger-type apparatus. Density, selectivity, toxicity, reactivity, volatility, miscibility with aqueous media, viscosity, and purity are all factors that influence the solvent system (Prabu & Suriyaprakash, 2012). Extraction and isolation using conventional techniques is presented in Table 7.1. Traditionally, these bioactive chemicals are extracted using extraction processes which are principally focused on solvents such as hexane, methanol, ethanol, and so on, through either mixing or heating. Disadvantages of these conventional methods include 1) prolonged extraction time, 2) expensive solvents, 3) loss of solvent through evaporation, 4) lower selectivity, and 5) breakdown of thermolabile compounds.

These issues can be resolved using other extraction methods such as "green extraction technologies". In particular, supercritical fluids are used in the production of volatile oils and extracts, having proven advantages such as selectivity, rapidity, and simplicity (Figure 7.3). These extracts can be used for cosmetics, food, and pharmaceutical companies (Romano et al., 2022).

The techniques of Gas Chromatography-Mass Spectrometry (GC-MS) (Sagbo et al., 2020), High-Performance Liquid Chromatography (HPLC) (Doctor et al., 2020), and Thin-Layer Chromatography (TLC) (Legerská et al., 2020) can all be used to analyses plant extracts. In

TABLE 7.1
Conventional Methods for the Extraction of Bioactive Compounds

Variety of Basil	Country	Method of Extraction	Bioactive Compounds	Bioactivity	Reference
Purple basil (*Ocimum basilicum* L.)	Brazil	Maceration and addition of ethanol and citric acid.	Total anthocyanins, flavonoids, total phenolic content	DPPH radical activity	(Pedro et al., 2016)
Basil (*Ocimum basilicum* L.)	Poland	Acetone, water, acetic acid, Methanol, mixing and centrifugation	Total phenolic content	Reducing power and Free radical scavenging activity	(Złotek et al., 2016)
Thai basil (*Ocimum basilicum* var. *thyrsiflora*) leaves	Vietnam	75% methanol, shaking in water bath and centrifugation	Total Phenolic and Flavonoid Content	Alpha-amylase Inhibitory activity, DPPH Radical Scavenging activity, Amyloglucosidase Inhibition activity	(Ma & Le, 2021)
Ocimum basilicum var. purpurascens (Lameaceae)	Portugal	Maceration using ethanol/water	26 phenolic compounds, 13 non-anthocyanin molecules, with rosmarinic acid, 13 anthocyanins, with cyanidin-3-(6,6′-di-p-coumaroyl)-sophoroside-5-glucoside being the main compound.	Antimicrobial activity, Hepatotoxicity	(Fernandes et al., 2019)
Fresh basil (*O. basilicum* L.), air-drying and CO_2 drying	Netherlands	Incubation with methanol under dark conditions with stirring.	Polyphenols and flavonoids, chlorophylls, ascorbic acid, rosmarinic, chicoric, caftaric and caffeic	BTS radical cation scavenging activity, DPPH free radical scavenging activity assay	(Bušić et al., 2014)
O. basilicum seeds	India	Serial extraction using petroleum ether and methanol in Soxhlet extraction apparatus.	alkaloids, flavonoids, carbohydrates, tannins, terpenoids, Gamolenic acid, eicosatrienoic acids, nonanoic acid,	Antimicrobial, antioxidant And anticancer activities.	(Gajendiran et al., 2016)
Ocimum basilicum Leaves	Vietnam	Ethanol, methanol, acetone, and water used individually and stirring at 60°C for 40 minutes followed by centrifugation.	phenolic compounds	DPPH free radical activity	(Do et al., 2020)
Ocimum basilicum L. Leaves	Pakistan	Non-sequential extraction using n-hexane, dichloromethane, ethanol, and water. Extraction using rotary shaker followed by filtration and freeze drying.	Flavonoids, Alkaloids, Steroids, Tannins, Phenols, Saponins, Terpenoids and compounds identified as rosmarinic acid, ellagic acid, catechin, liquiritigenin, and umbelliferone.	Scavenging of hydrogen peroxide (H_2O_2), DPPH activity, ferric reducing antioxidant power (FRAP)	(Nadeem et al., 2022)

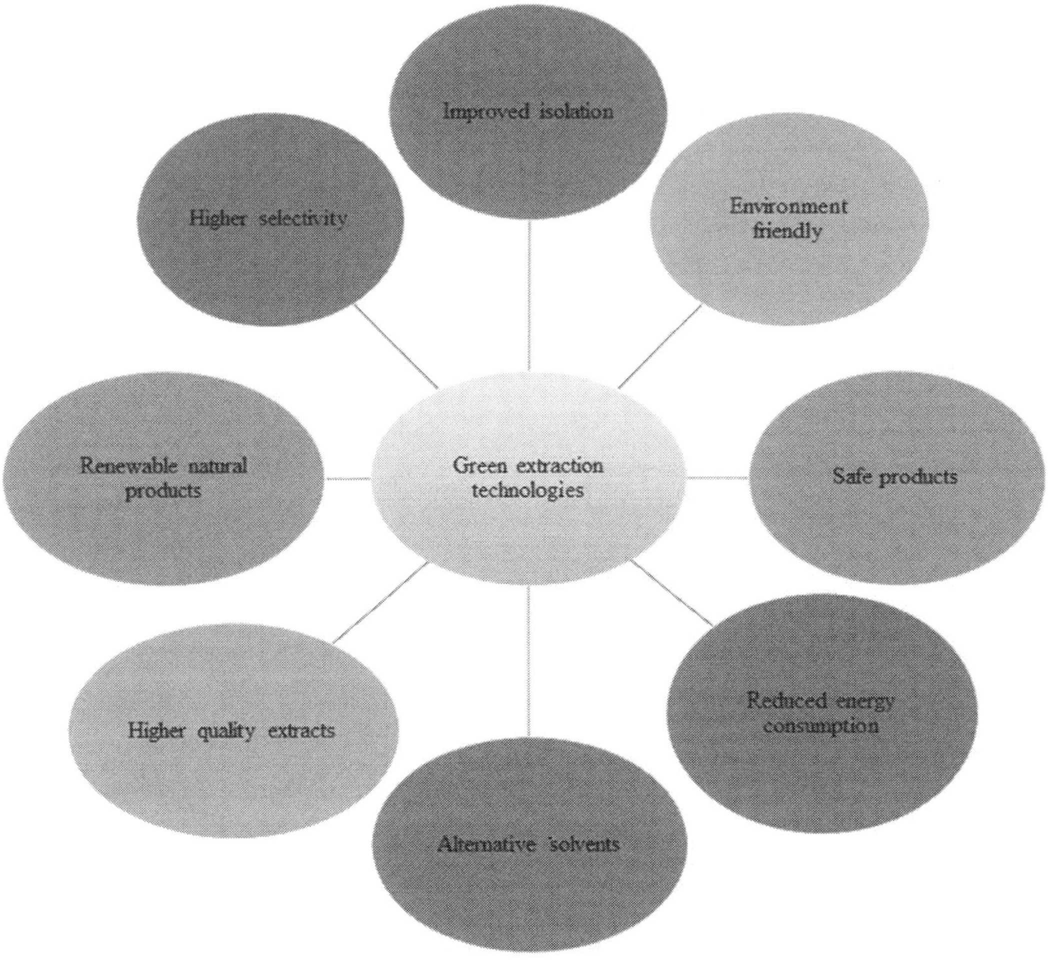

FIGURE 7.3 Benefits of green extraction technologies.

addition, the high-performance TLC (HPTLC) densitometry method has recently gained recognition due to its sensitive, precise, and cost-effectiveness, and that it shortens the time taken to process large samples. TLC-densitometry has been chosen by researchers over the popular HPLC method because of its greater sensitivity and capacity to process a large number of samples in a shorter period of time (Bhardwaj et al., 2020).

7.8.2 Green Technologies for the Extraction of Bioactive Compounds

7.8.2.1 Ultrasound-assisted Extraction (UAE)

Ultrasound-assisted extraction (UAE) is an innovative, low-cost, and efficient approach for increasing the rate of heat and mass transfer during extraction. UAE reduces the extraction period greatly while enhancing antioxidant abilities and extraction efficiency (Chemat & Khan, 2011). UAE is also suggested to be a promising sample preparation approach for a variety of applications (Pan et al., 2012). The improved performance of UAE can be attributed to sonic-aided cavitation, which induces cell wall breaches due to sonothermal bubble collapse and, as a consequence, an increased rate of mass transfer of cell components to a surrounding solvent (Xia et al., 2011). Upadhyay et al. (2015)

have compared conventional solvent extraction (CSE) approaches and UAE for the extraction of flavonoids and phenolics from the leaves of *Ocimum tenuiflorum* (basil leaves). Process factors (ethanol concentration, sonication time, and ultrasonic energy density) were found to have a significant association with total phenolic (TPC) and flavonoid content (FC) levels. UAE extracts has exhibited significantly higher extraction yields of flavonoids and phenolic compounds, and higher *in vitro* antioxidant activities (DPPH, reducing power, photochemiluminescent assay, Ferrous ion [Fe^{2+}] chelation activity), and thermo-oxidative stability than conventional solvent extracts. Similarly, kaempferol (KA) was extracted from basil leaves (*Ocimum basilicum*) using ultrasonication. Further, its yield was quantified using high-performance thin-layer chromatography in connection with ultraviolet detection (HPTLC-VIS).

Furthermore, the antioxidant activity of KA (IC_{50} = 0.68 μg/mL) was higher than that of the standard ascorbic acid (IC_{50} = 0.79 μg/mL) and butylated hydroxytoluene (BHT) (IC_{50} = 0.88 μg/mL) (Altemimi et al., 2020). In addition to extraction, ultrasound has the potential to successfully aromatize olive oils with basil, improving the sensorial and nutritional quality of the product. The UAE has sped up the diffusion of basil volatile chemicals into the oil, cutting processing time down to a few minutes (Veillet et al., 2010).

7.8.2.2 Pressurized Liquid Extraction (PLE)

Pressurized liquid extraction (PLE) is an environmentally-friendly extraction method which employs solvents at high pressure and temperatures (over their atmospheric boiling point). Under these conditions, the characteristics of the solvents changes, resulting in an enhanced mass-transfer rate (Herrero et al., 2015). PLE has already been demonstrated to be a potent extraction method for a wide range of bioactive compounds extracted from food by-products and medicinal herbs (Herrero et al., 2015; Hossain et al., 2011). Water, ethanol, or a mixture of the two, is commonly used to extract polar compounds in this technique. Pagano et al. (2018) have utilized ethyl lactate/ethanol along with water (water–ethanol mixtures, and water–ethyl lactate mixtures) as a solvent to extract bioactive compounds from dried distillation wastewaters of basil (*Ocimum basilicum* L.) through Hansen solubility parameters and PLE. This selective approach was utilized for the extraction of caffeic acid (CA) and rosmarinic acid (RA). Further, the extracts were quantified for caffeic acid (CA) and rosmarinic acid (RA) using the HPLC-UV system with gradient elution. At 50 °C, a water–ethanol mixture of 25:75 (% v/v) yields RA (23.90 ± 2.06 mg/g extract) with an extraction efficiency of 75.89 ± 16.03%, and absolute ethanol yields 2.42 0.04 mg/g extract with an extraction efficiency of 13.86 ± 4.96%. It has been recommended that dried basil distillation waste can be used as a new promising source of natural ingredients and/or functional constituents for nutraceutical, cosmetic, and food industries (Pagano et al., 2018).

7.8.2.3 Subcritical Water Extraction (SWE)

Subcritical water extraction (hydrolysis) is also known as pressurized hot water or hot-compressed water technique because it employs water at different temperatures and pressures, beyond its boiling temperature at 1 atm (>100 °C at 0.1 MPa) and under its critical point (374.5 °C at 22.06 MPa). In these subcritical environments, the aqueous medium generated hydronium (H_3O+) and hydroxide (OH) ions, increasing the reactivity of the water and allowing it to act as an acid, base, or double catalyst for hydrolysis (Rivas-Vela et al., 2021). The extraction of essential oil was carried out from Indonesia basil (kemangi) leaf using SWE. It was determined that product composition and yield was affected by extraction time and temperature. The presence of an antibacterial agent such as tricosane, eugenol, and 1,8 cineole were isolated from subcritical water at a temperature of 200 °C (Zullaikah et al., 2015). Water was used to extract basil and oregano leaves at temperatures of 100 and 150 °C in SWE. Terpenes (limonene, α-pinene, citronellol, and camphor) were extracted from basil and oregano using SWE in a range of traces to 65 μg/g herb (Yang et al., 2007).

7.8.2.4 Supercritical Fluid Extraction (SFE)

Carbon dioxide (CO_2) is employed as a non-polar solvent in the supercritical fluid extraction (SFE) technique at low temperatures and intermediate pressures. Small amounts of organic additives (cosolvents) of variable polarity are often employed to increase the solvating power of supercritical CO_2 (Romano et al., 2021). Romano et al. (2022) applied SFE for the extraction of bioactive compounds from basil leaves. The extraction conditions include pressure 30 MPa with 50 °C temperature of supercritical CO_2 for 2 h and the use of ethanol (10%) as a cosolvent. Two basil cultivars (Italiano Classico and Genovese) were submitted to the green extraction technique and demonstrated considerably higher antiradical activity (ABTS$^{•+}$ test), caffeic acid (1.69–1.92 mg/g), linalool (35–27%), and bergamotene (11–14%) contents than the control (100% ethanol). Using the novel technique, bioactive compounds were derived from basil extracts in a more environmentally-friendly way and lower consumption of alcohol. Coelho et al. (2018) conducted a similar experiment with the goal of comparing the extraction of bioactive compounds from basil (*Ocimum basilicum* L.) using various extraction methods, such as SFE (CO_2) (9.0 MPa, 40 °C; and 40.0 MPa, 40 °C), hydrodistillation (fresh plant), hydrodistillation (dry plant), and Soxhlet extraction (methanol or ethanol). Various approaches were applied for the study of the antioxidant activity of essential oils, such as 2,2-azino-bis (3-ethylbenzthiazoline-6-sulfonic acid [ABTS]), reducing power test, and 2,2-diphenyl-1-picrylhydrazyl (DPPH). In an ABTS assay, the SFE extract had a higher antioxidant capacity. The highest amount of phytol acetate 2 was observed in the SFE extracts compared with hydrodistillation oils. Furthermore, the essential oil (*O. basilicum*) was recognized as belonging to the chemotype Group 4, which is high in estragole and linalool.

7.8.2.5 Microwave-assisted Extraction (MAE)

Microwave-assisted extraction (MAE) is a crucial technique for separating bioactive chemicals from a variety of plant matrices. Due to various automation possibilities, solvent savings, diverse sample processing options, and quick contact time, MAE technologies have recently distinguished themselves from other extraction techniques. In MAE, the solvent enters the solid sample matrix, and the target chemicals exit via either nonpolar or polar solvents. Numerous variables affect extraction yield, including solvent type, liquid-to-solid ratio, time, and temperature (Şen et al., 2019). Şen et al. (2019) have applied MAE for the extraction of antioxidants from basil (*Ocimum basilicum* L.) using response surface methodology (RSM). The antioxidant activity of MAE extracts was revealed to be the most significantly influenced by the solvent concentration. The proposed technique showed a good accord between experimental and projected values. Upon comparing the extraction yield with classical heat extraction, MAE has exhibited almost a double recovery of the antioxidants from basil. Due to significant advantages such as the ability to extract multiple samples (12 samples at a time in the authors' system), to work beyond the boiling point using a closed vessel, to work quickly (in minutes), to automatically control temperature using power at varying voltages, to be simple to operate, and to have a high extraction efficiency, the MAE system offers advantages over traditional extraction approaches.

7.8.2.6 Pulsed Electric Field (PEF) Extraction

The application of a short duration pulse with high voltage into a food product placed between two electrodes promotes membrane permeability modification, and increasing extraction yield (Nowacka et al., 2019). Pulungan et al. (2020) investigated the impact of pulsed electric field (PEF) as a pre-treatment for the recovery of phenolic components from basil (*Ocimum americanum* L.) leaves (2020). Two factors, exposure time and voltage, were considered for the study; further, the extraction was done using maceration (using 80% ethanol at 50°C). PEF pre-treatment has a significant effect on the total phenolic components and antioxidant activity (IC_{50}) of basil extract. The optimum treatment yields a TPC of 260.50 mg GAE/g and an IC50 of 35.96 ppm at a voltage of 1500 Volts and an exposure time of 10 seconds. The TPC of the untreated sample (no PEF pre-treatment) is

175.75 mg GAE/g, and the IC_{50} is 116.25 ppm. According to the data, PEF treatment before basil leaf extraction raised TPC by 1.5 times and antioxidant activity by 3.2 times. SEM was used to confirm the presence of pores in the cell membrane caused by PEF, which may have helped in the release of phenolic substances (Pulungan et al., 2020).

Hendrawan et al. (2019) conducted a similar investigation on basil leaves and confirmed that pre-treatment period, electric field intensity variation, and the interaction of both variables all affect the total amount of phenols and flavonoids. The cumulative effect of two minutes of pre-treatment on an electric field of 3 kV cm-1 yielded the maximum flavonoid and phenolic compounds. PEF pre-treated samples has exhibited the highest yield of flavonoid and phenolic compounds. In general, the electroporation effect of PEF improved the level of extraction.

7.9 APPLICATIONS OF BASIL INGREDIENTS IN FOOD

Basil ingredients such as essential oils and basil seed gum are utilized in various food products. As in the previous sections we have discussed the biological role of essential oils; in addition, essential oils are also used in food formulations. Basil seed gum is utilized as a natural hydrocolloid. It is able to act as an emulsifier in cheese, enabling the production of firmer cheese at a lower cost (Hosseini-Parvar et al., 2015). Studies has shown novel stabilizer blends, in which the majority of the blend (>80%) made up of basil seed mucilage can produce a desirable ice cream (BahramParvar et al., 2012, 2013). Another study in 2016 showed that a 1:1 ratio of basil seed mucilage and guar gum had huge potential as a replacement for fat and a stabilizer for low-fat ice-creams (Javidi et al., 2016). This signifies the potential of basil seed mucilage as a promising food hydrocolloid and food texturing agent.

Essential oil in basil has many applications in the food manufacturing industries. It is a commonly used additive for meat products (i.e. sausages), dairy products (i.e. cheese), and alcoholic and non-alcoholic beverages and it can also be used alongside other herbs in many other food products, such as baked goods, confectionery, vinegars, pickled foods, sweets and condiments (Li & Chang, 2016). The incorporation of basil essential oil as a natural antioxidant into minced beef increased its stability against lipid oxidation and color during storage (Falowo et al., 2019). Similarly, when sprayed on the surface of sausages, it exhibits antifungal properties (Kocić-Tanackov et al., 2020; Saggiorato et al., 2012) while, *in vitro*, it had partial antibacterial properties (successful as antibacterial additive in all tested bacteria with the exception of *Pseudomonas aeruginosa*) (Gaio et al., 2015). In the production of cheese, aromatic compounds of basil essential oil improved the flavor and displayed antimicrobial properties without affecting the blooming of cheese (Licon et al., 2020). The nanoencapsulation of basil essential oil into food packaging has been studied with regard to its potential impact on food products. Films with basil essential oil incorporation retarded lipid oxidation and the microbial activity of food product better than those without such essential oil (Arfat et al., 2015) whereas nanoencapsulated basil essential oil could potentially replace conventional antimicrobials as it is able to provide safety, efficacy, and good physiochemical stability (da Silva Gündel et al., 2018).

7.10 CONCLUSION

Basil (*Ocimum basilicum* L.) is an annual plant of the mint family (*Lamiaceae*). Guven advances in the technological and extraction process, various bioactive compounds are derived from the basil plant. The plant's leaves are utilized to derive essential oils and bioactive compounds which have therapeutic values, including anti-cancer activity, antioxidant activity, radioprotective activity, immunomodulatory activity, anti-microbial activity, anti-inflammatory effects, anti-stress activity, anti-arthritic activity, anti-diabetic activity, and anti-pyretic activity, and being used as a prophylactic against cardiovascular disease. Not only essential oils, but also basil seed gums hold a bright future as a natural hydrocolloid in the food industry. These basil seed gums are utilized as an emulsifier and stabilizer, and as a fat substitute in many food products. Nevertheless, there is widespread scope for the utilization of basil ingredients in both medicinal and non-medicinal applications.

REFERENCES

Ahmed, H. M. (2016). Ethnopharmacobotanical study on the medicinal plants used by herbalists in Sulaymaniyah Province, Kurdistan, Iraq. *Journal of Ethnobiology and Ethnomedicine*, *12*(1), 1–17.

Al-Maskari, M. Y., Hanif, M. A., Al-Maskri, A. Y., & Al-Adawi, S. (2012). Basil: A natural source of antioxidants and neutraceuticals. In M. Mohamed Essa, A. Manickvasagan, and E. Sukumar (Eds.), *Natural Products and Their Active Compounds on Disease Prevention* (pp. 463–471). Nova Science Publishers, Inc.

Alomar, M. Y., & Al-Attar, A. M. (2019). Effect of basil leaves extract on liver fibrosis induced by thioacetamide in male rats. *International Journal of Pharmacology*, *15*(4), 478–485.

Altemimi, A. B., Mohammed, M. J., Yi-Chen, L., Watson, D. G., Lakhssassi, N., Cacciola, F., & Ibrahim, S. A. (2020). Optimization of Ultrasonicated Kaempferol Extraction from *Ocimum basilicum* Using a Box–Behnken Design and Its Densitometric Validation. *Foods*, *9*(10), 1379.

Amrani, S., Harnafi, H., Gadi, D., Mekhfi, H., Legssyer, A., Aziz, M., Martin-Nizard, F., & Bosca, L. (2009). Vasorelaxant and anti-platelet aggregation effects of aqueous *Ocimum basilicum* extract. *Journal of Ethnopharmacology*, *125*(1), 157–162.

Antonescu, I. A., Antonescu, A., Miere, F., Fritea, L., Teuşdea, A. C., Vicaş, L., Vicaş, S. I., Brihan, I., Domuţa, M., Zdrinca, M., Zdrinca, M., & Cavalu, S. (2021). Evaluation of Wound Healing Potential of Novel Hydrogel Based on *Ocimum basilicum* and Trifolium pratense Extracts. *Processes*, *9*(11), 2096.

Arfat, Y. A., Benjakul, S., Vongkamjan, K., Sumpavapol, P., & Yarnpakdee, S. (2015, 2015/10/01). Shelf-life extension of refrigerated sea bass slices wrapped with fish protein isolate/fish skin gelatin-zno nanocomposite film incorporated with basil leaf essential oil. *Journal of Food Science and Technology*, *52*(10), 6182–6193.

Arlett, C. F., Boettiger, D., Callander, R., Chalmers, D., Cooper, C., Cowell, J., Dean, S., Defais, M., DiGiovanni, J., & Edwards, A. (2012). *Chemical Carcinogenesis and Mutagenesis II*. Springer Science & Business Media.

BahramParvar, M., Mazaheri Tehrani, M., & Razavi, S. M. A. (2013). Effects of a Novel Stabilizer Blend and Bresence of K-carrageenan on Some Properties of Vanilla Ice Cream During Storage. *Food Bioscience*, *3*, 10–18.

BahramParvar, M., Razavi, S. M. A., & Mazaheri Tehrani, M. (2012). Optimising the ice cream formulation using basil seed gum (*Ocimum basilicum* L.) as a novel stabiliser to deliver improved processing quality. *International Journal of Food Science & Technology*, *47*(12), 2655–2661.

Bhardwaj, P., Banarjee, A., Jindal, D., Kaur, C., Singh, G., Kumar, P., Sharma, A., & Kumar, R. (2020). Validation of TLC-densitometry method for estimation of catechin in acacia catechu heartwood. *Pharmaceutical Chemistry Journal*, *54*(2), 184–189.

Bušić, A., Vojvodić, A., Komes, D., Akkermans, C., Belščak-Cvitanović, A., Stolk, M., & Hofland, G. (2014). Comparative evaluation of CO_2 drying as an alternative drying technique of basil (*Ocimum basilicum* L.)—The effect on bioactive and sensory properties. *Food Research International*, *64*, 34–42.

CABI. (2022). *Ocimum basilicum (basil)* (CABI Compendium 36858) [Data set]. CABI International. https://doi.org/10.1079/cabicompendium.36858

Castronuovo, D., Russo, D., Libonati, R., Faraone, I., Candido, V., Picuno, P., Andrade, P., Valentao, P., & Milella, L. (2019). Influence of shading treatment on yield, morphological traits and phenolic profile of sweet basil (*Ocimum basilicum* L.). *Scientia Horticulturae*, *254*, 91–98.

Chemat, F., & Khan, M. K. (2011). Applications of ultrasound in food technology: Processing, preservation and extraction. *Ultrasonics Sonochemistry*, *18*(4), 813–835.

Coelho, J., Veiga, J., Karmali, A., Nicolai, M., Pinto Reis, C., Nobre, B., & Palavra, A. (2018). Supercritical CO_2 extracts and volatile oil of basil (*Ocimum basilicum* L.) comparison with conventional methods. *Separations*, *5*(2), 21.

Darrah, H. H. (1980). *The cultivated basils*. Buckeye Print Company. Madison Wisconson.

Danesi, F., Elementi, S., Neri, R., Maranesi, M., D'Antuono, L. F., & Bordoni, A. (2008). Effect of cultivar on protection of cardiomyocytes from oxidative stress by essential oils and aqueous extracts of basil (*Ocimum basilicum L.*). *Journal of Agricultural and Food Chemistry*, *56*(21), 9911–9917.

da Silva Gündel, S., Velho, M. C., Diefenthaler, M. K., Favarin, F. R., Copetti, P. M., de Oliveira Fogaça, A., Klein, B., Wagner, R., Gündel, A., Sagrillo, M. R., & Ourique, A. F. (2018). Basil oil-nanoemulsions:

Development, cytotoxicity and evaluation of antioxidant and antimicrobial potential. *Journal of Drug Delivery Science and Technology*, *46*, 378–383.

Deligianni, E., Pizzi, E., Kavelaki, I., Siden-Kiamos, I., Sapienza, F. U., Fioravanti, R., Garzoli, S., Pace T., Ponzi, M., Ragno, R., & Curra, C. (2023). Screening of the activity of sixty essential oils against plasmodium early mosiquito stages in vitro and machine learning analysis reveal new putative inhibitors of malaria parasites. *International Journal for Paristology: Drugs and Drug resistance*, *23*, 87–93.

Devi, A. D., Devi, O. I., Singh, T. C., & Singh, E. (2014). A study of aromatic plant species especially in Thoubal district, Manipur, North East India. *International Journal of Scientific and Research Publications*, *4*, 2250–3153.

Dhama, K., Sharun, K., Gugjoo, M. B., Tiwari, R., Alagawany, M., Iqbal Yatoo, M., Thakur, P., Iqbal, H. M. N., Chaicumpa, W., Michalak, I., Elnesr, S. S., & Farag, M. R. (2021). A comprehensive review on chemical profile and pharmacological activities of *Ocimum basilicum*. *Food Reviews International*, 1–29.

Do, T. H., Truong, H. B., & Nguyen, H. C. (2020). Optimization of extraction of phenolic compounds from *Ocimum basilicum* leaves and evaluation of their antioxidant activity. *Pharmaceutical Chemistry Journal*, *54*(2), 162–169.

Doctor, N., Parker, G., Vang, K., Smith, M., Kayan, B., & Yang, Y. (2020). Stability and extraction of vanillin and coumarin under subcritical water conditions. *Molecules*, *25*(5), 1061.

Falowo, A. B., Mukumbo, F. E., Idamokoro, E. M., Afolayan, A. J., & Muchenje, V. (2019). Phytochemical constituents and antioxidant activity of sweet basil (*Ocimum basilicum* L.) essential oil on ground beef from boran and nguni cattle. *International Journal of Food Science*, *2019*, 2628747.

Fang, J.-Y., Leu, Y.-L., Hwang, T.-L., & Cheng, H.-C. (2004). Essential oils from sweet basil (*Ocimum basilicum*) as novel enhancers to accelerate transdermal drug delivery. *Biological and Pharmaceutical Bulletin*, *27*(11), 1819–1825.

Fathiazad, F., Matlobi, A., Khorrami, A., Hamedeyazdan, S., Soraya, H., Hammami, M., Maleki-Dizaji, N., & Garjani, A. (2012). Phytochemical screening and evaluation of cardioprotective activity of ethanolic extract of *Ocimum basilicum* L. (basil) against isoproterenol induced myocardial infarction in rats. *DARU Journal of Pharmaceutical Sciences*, *20*(1), 87.

Fernandes, F., Pereira, E., Círić, A., Soković, M., Calhelha, R. C., Barros, L., & Ferreira, I. C. (2019). *Ocimum basilicum* var. purpurascens leaves (red rubin basil): A source of bioactive compounds and natural pigments for the food industry. *Food & function*, *10*(6), 3161–3171.

Filip, S. (2017). Basil (*Ocimum basilicum* L.) a source of valuable phytonutrients. *International Journal of Clinical Nutrition & Dietetics*, *3*, 118.

Gaio, I., Saggiorato, A. G., Treichel, H., Cichoski, A. J., Astolfi, V., Cardoso, R. I., Toniazzo, G., Valduga, E., Paroul, N., & Cansian, R. L. (2015). Antibacterial activity of basil essential oil (*Ocimum basilicum* L.) in Italian-type sausage. *Journal für Verbraucherschutz und Lebensmittelsicherheit*, *10*(4), 323–329.

Gajendiran, A., Thangaraman, V., Thangamani, S., Ravi, D., & Abraham, J. (2016). Antimicrobial, antioxidant and anticancer screening of *Ocimum basilicum* seeds. *Bulletin of Pharmaceutical Research*, *6*(3), 114–119.

Gao, X., Cassidy, A., Schwarzschild, M., Rimm, E. B., & Ascherio, A. (2012). Habitual intake of dietary flavonoids and risk of Parkinson disease. *Neurology*, *78*(15), 1138–1145.

Golak, B. C., Prabhat, K., Bhabani, S. N., Sangram, K. P., & Tripathy, S. (2010). Phytochemical investigation and evaluation of analgesic activity of leafy extracts of various *Ocimum* (tulsi) species. *The Indian Pharmacist*, *8*, 67–70.

Grayer, R. J., Kite, G. C., Veitch, N. C., Eckert, M. R., Marin, P. D., Senanayake, P., & Paton, A. J. (2002). Leaf flavonoid glycosides as chemosystematic characters in *Ocimum*. *Biochemical Systematics and Ecology*, *30*(4), 327–342.

Harnafi, H., Caid, H. S., el Houda Bouanani, N., Aziz, M., & Amrani, S. (2008). Hypolipemic activity of polyphenol-rich extracts from *Ocimum basilicum* in Triton WR-1339-induced hyperlipidemic mice. *Food Chemistry*, *108*(1), 205–212.

Hashemi, N., Ommi, D., Kheyri, P., Khamesipour, F., Setzer, W. N., & Benchimol, M. (2021). A review study on the anti-*trichomonos* activities of medicinal plants. *International Journal for Paristology: Drugs and Drug resistance*, *15*, 92104.

Hendrawan, Y., Sabrinauly, S., Hawa, L. C., Argo, B. D., & Rachmawati, M. (2019). Analysis of the phenol and flavonoid content from basil leaves (*Ocimum americanum* L) extract using pulsed electric field (PEF) pre-treatment. *Agricultural Engineering International: CIGR Journal*, *21*(2), 149–158.

Herrero, M., del Pilar Sánchez-Camargo, A., Cifuentes, A., & Ibáñez, E. (2015). Plants, seaweeds, microalgae and food by-products as natural sources of functional ingredients obtained using pressurized liquid extraction and supercritical fluid extraction. *TrAC Trends in Analytical Chemistry*, *71*, 26–38.

Hossain, M., Barry-Ryan, C., Martin-Diana, A. B., & Brunton, N. (2011). Optimisation of accelerated solvent extraction of antioxidant compounds from rosemary (*Rosmarinus officinalis* L.), marjoram (*Origanum majorana* L.) and oregano (*Origanum vulgare* L.) using response surface methodology. *Food Chemistry, 126*(1), 339–346.

Hosseini-Parvar, S. H., Matia-Merino, L., & Golding, M. (2015). Effect of basil seed gum (BSG) on textural, rheological and microstructural properties of model processed cheese. *Food Hydrocolloids, 43*, 557–567.

Idu, M., Umweni, A., Odaro, T., & Ojelede, L. (2009). Ethnobotanical Plants Used for Oral Healthcare Among the Esan Tribe of Edo State, Nigeria. *Ethnobotanical leaflets, 2009*(4), 15.

Jadoon, S., Karim, S., Bin Asad, M. H. H., Akram, M. R., Khan, A. K., Malik, A., Chen, C., & Murtaza, G. (2015). Anti-aging potential of phytoextract loaded-pharmaceutical creams for human skin cell longevity. *Oxidative Medicine and Cellular Longevity, 2015*, 709628.

Jahejo, A. R., Rajput, N., Wen-Xia, T., Naeem, M., Kalhoro, D. H., Kaka, A., Niu, S., & Jia, F. J. (2019). Immunomodulatory and growth promoting effects of basil (*Ocimum basilicum*) and ascorbic acid in heat stressed broiler chickens. *Pakistan Journal of Zoology, 51*(3), 801.

Javidi, F., Razavi, S. M. A., Behrouzian, F., & Alghooneh, A. (2016). The Influence of Basil Seed Gum, Guar Gum and Their Blend on the Rheological, Physical and Sensory Properties of Low Fat Ice Cream. *Food Hydrocolloids, 52*, 625–633.

Kadan, S., Saad, B., Sasson, Y., & Zaid, H. (2016). In vitro evaluation of anti-diabetic activity and cytotoxicity of chemically analysed *Ocimum basilicum* extracts. *Food Chemistry, 196*, 1066–1074.

Kalamartzis, L., Menexes, G., Georgiou, P., & Dordas, C. (2020). Effect of water stress on the physiological characteristics of five basil (*Ocimum basilicum* L.) cultivars. *Agronomy, 10*(7), 1029.

Kéita, S. M., Vincent, C., Schmit, J. P., & Bélanger, A. (2000). Essential oil composition of *Ocimum basilicum* L., *O. gratissimum* L. and *O. suave* L. in the Republic of Guinea. *Flavour and Fragrance Journal, 15*(5), 339–341.

Kocić-Tanackov, S., Dimić, G., Đerić, N., Mojović, L., Tomović, V., Šojić, B., Đukić-Vuković, A., & Pejin, J. (2020). Growth control of molds isolated from smoked fermented sausages using basil and caraway essential oils, in vitro and in vivo. *Food Science and Technology, 123*, 109095.

Legerská, B., Chmelová, D., & Ondrejovič, M. (2020). TLC-Bioautography as a fast and cheap screening method for the detection of α-chymotrypsin inhibitors in crude plant extracts. *Journal of Biotechnology, 313*, 11–17.

Li, Q. X., & Chang, C. L. (2016). Chapter 25 - Basil (*Ocimum basilicum* L.) Oils. In V. R. Preedy (Ed.), *Essential Oils in Food Preservation, Flavor and Safety* (pp. 231–238). Academic Press.

Licon, C. C., Moro, A., Librán, C. M., Molina, A. M., Zalacain, A., Berruga, M. I., & Carmona, M. (2020). Volatile Transference and Antimicrobial Activity of Cheeses Made with Ewes' Milk Fortified with Essential Oils. *Foods, 9*(1).

Ma, N., & Le, N. (2021). Extraction optimization of total phenolics from Thai basil (*Ocimum basilicum* var. thyrsiflora) leaves and bioactivities of the extract. *AIP Conference Proceedings, 2406*(1), 050002.

Manosroi, A., & Manosroi, J. (2003). Free radical scavenging and tyrosinase inhibition activity of aromatic volatile oil from Thai medicinal plants for cosmetic uses. *Acta Horticulturae, 680*, 97–100.

Marotti, M., Piccaglia, R., & Giovanelli, E. (1996). Differences in essential oil composition of basil (*Ocimum basilicum* L.) Italian cultivars related to morphological characteristics. *Journal of Agricultural and Food Chemistry, 44*(12), 3926–3929.

Marwat, S. K., Khan, M. S., Ghulam, S., Anwar, N., Mustafa, G., & Usman, K. (2011). Phytochemical constituents and pharmacological activities of sweet Basil-*Ocimum basilicum* L. (*Lamiaceae*). *Asian Journal of Chemistry, 23*(9), 3773.

Muhammad, A. H., Masoud, Y. A.-M., Awatif, A.-M., Ahmed, A.-S., Ahmed, Y. A.-M., & Jamal, N. A.-S. (2011). Essential oil composition, antimicrobial and antioxidant activities of unexplored Omani basil. *Journal of Medicinal Plants Research, 5*(5), 751–757.

Muneefa, K., Doss, V. A., & Sowndarya, R. (2017). Beneficial effect of hydroethanolic extract of *Ocimum basilicum* L on enzymic and non enzymic antioxidant in depression induced rats. *Journal of Medicinal Plants Studies, 5*(3), 185–188.

Nadeem, H. R., Akhtar, S., Sestili, P., Ismail, T., Neugart, S., Qamar, M., & Esatbeyoglu, T. (2022). Toxicity, antioxidant activity, and phytochemicals of basil (*Ocimum basilicum* L.) Leaves cultivated in southern Punjab, Pakistan. *Foods, 11*(9), 1239.

Newman, D. J., & Cragg, G. M. (2016). Natural products as sources of new drugs from 1981 to 2014. *Journal of Natural Products, 79*(3), 629–661.

Nowacka, M., Tappi, S., Wiktor, A., Rybak, K., Miszczykowska, A., Czyzewski, J., Drozdzal, K., Witrowa-Rajchert, D., & Tylewicz, U. (2019, Jul 5). The impact of pulsed electric field on the extraction of bioactive compounds from beetroot. *Foods, 8*(7), 244.

Opalchenova, G., & Obreshkova, D. (2003). Comparative studies on the activity of basil—an essential oil from *Ocimum basilicum* L.—against multidrug resistant clinical isolates of the genera *Staphylococcus, Enterococcus* and *Pseudomonas* by using different test methods. *Journal of Microbiological Methods*, *54*(1), 105–110.

Pagano, I., Sánchez-Camargo, A. d. P., Mendiola, J. A., Campone, L., Cifuentes, A., Rastrelli, L., & Ibañez, E. (2018). Selective extraction of high-value phenolic compounds from distillation wastewater of basil (*Ocimum basilicum* L.) by pressurized liquid extraction. *Electrophoresis*, *39*(15), 1884–1891.

Pan, G., Yu, G., Zhu, C., & Qiao, J. (2012). Optimization of ultrasound-assisted extraction (UAE) of flavonoids compounds (FC) from hawthorn seed (HS). *Ultrasonics Sonochemistry*, *19*(3), 486–490.

Pedro, A. C., Moreira, F., Granato, D., & Rosso, N. D. (2016). Extraction of bioactive compounds and free radical scavenging activity of purple basil (*Ocimum basilicum* L.) leaf extracts as affected by temperature and time. *Anais da Academia Brasileira de Ciências*, *88*, 1055–1068.

Perez, R., Navarro, T., & Lorenzo, C. D. (2007). HS–SPME analysis of the volatile compounds from spices as a source of flavour in "Campo Real" table olive preparations. *Flavour and Fragrance Journal*, *22*(4), 265–273.

Peschel, W., Sánchez-Rabaneda, F., Diekmann, W., Plescher, A., Gartzía, I., Jiménez, D., Lamuela-Raventos, R., Buxaderas, S., & Codina, C. (2006). An industrial approach in the search of natural antioxidants from vegetable and fruit wastes. *Food Chemistry*, *97*(1), 137–150.

Prabu, S. L., & Suriyaprakash, T. (2012). *Extraction of Drug from the Biological Matrix: A Review*. IntechOpen.

Pratama, H. N., Dwiyanti, R. D., & Muhlisin, A. (2020). Growth of Malassezia furfur in Media with The Addition of Basil (*Ocimum basilicum* Linn) Powder. *Tropical Health and Medical Research*, *2*(1), 26–33.

Pulungan, M., Purwaningsih, I., & Sita, P. (2020). Extraction of phenolic compounds from basil (*Ocimum americanum* L.) leaves with pretreatment using pulsed electric field (PEF). *IOP Conference Series: Earth and Environmental Science*, *475*(1), 012056.

Rao, K. S., Chaudhury, P. K., & Pradhan, A. (2010). Evaluation of anti-oxidant activities and total phenolic content of Chromolaena odorata. *Food and Chemical Toxicology*, *48*(2), 729–732.

Rasul, A., & Akhtar, N. (2011). Formulation and in vivo evaluation for anti-aging effects of an emulsion containing basil extract using non-invasive biophysical techniques. *DARU: Journal of Faculty of Pharmacy, Tehran University of Medical Sciences*, *19*(5), 344.

Rice-Evans, C., Miller, N., & Paganga, G. (1997). Antioxidant properties of phenolic compounds. *Trends in Plant Science*, *2*(4), 152–159.

Rivas-Vela, C. I., Amaya-Llano, S. L., Castaño-Tostado, E., & Castillo-Herrera, G. A. (2021). Protein hydrolysis by subcritical water: A new perspective on obtaining bioactive peptides. *Molecules*, *26*(21), 6655.

Romano, R., Aiello, A., Meca, G., De Luca, L., Pizzolongo, F., & Masi, P. (2021). Recovery of bioactive compounds from walnut (*Juglans regia* L.) green husk by supercritical carbon dioxide extraction. *International Journal of Food Science & Technology*, *56*(9), 4658–4668.

Romano, R., De Luca, L., Aiello, A., Pagano, R., Di Pierro, P., Pizzolongo, F., & Masi, P. (2022). Basil (*Ocimum basilicum* L.) leaves as a source of bioactive compounds. *Foods*, *11*(20), 3212.

Sagbo, I. J., Kola, E., & Otang-Mbeng, W. (2020). Phytochemical screening and gas chromatography-mass spectrometry analysis of ethanol extract of *Scambiosa columbabria* L. *Pharmacognosy Research*, *12*(1), 35–39.

Saggiorato, A. G., Gaio, I., Treichel, H., de Oliveira, D., Cichoski, A. J., & Cansian, R. L. (2012). Antifungal activity of basil essential oil (*Ocimum basilicum* L.): Evaluation in vitro and on an italian-type sausage surface. *Food and Bioprocess Technology*, *5*(1), 378–384.

Şen, F. B., Aşçı, Y. S., Bekdeşer, B., Bener, M., & Apak, R. (2019). Optimization of Microwave-Assisted Extraction (MAE) for the isolation of antioxidants from basil (*Ocimum basilicum* L.) by response surface methodology (RSM). *Analytical Letters*, *52*(17), 2751–2763.

Sestili, P., Ismail, T., Calcabrini, C., Guescini, M., Catanzaro, E., Turrini, E., Layla, A., Akhtar, S., & Fimognari, C. (2018). The potential effects of *Ocimum basilicum* on health: A review of pharmacological and toxicological studies. *Expert Opinion on Drug Metabolism & Toxicology*, *14*(7), 679–692.

Shahrajabian, M. H., Sun, W., & Cheng, Q. (2020). Chemical components and pharmacological benefits of Basil (*Ocimum basilicum*): A review. *International Journal of Food Properties*, *23*(1), 1961–1970.

Siddiqui, B. S., Bhatti, H. A., Begum, S., & Perwaiz, S. (2012). Evaluation of the antimycobacterium activity of the constituents from *Ocimum basilicum* against *Mycobacterium tuberculosis*. *Journal of Ethnopharmacology*, *144*(1), 220–222.

Simon, J. E., Morales, M. R., Phippen, W. B., Vieira, R. F., & Hao, Z. (1999). Basil: A source of aroma compounds and a popular culinary and ornamental herb. *Perspectives on New Crops and New Uses*, *16*, 499–505.

Simon, J. E., Quinn, J., & Murray, R. G. (1990). Basil: A source of essential oils. In J. Janick and J. E. Simon (Eds.), *Advances in New Crops: Proceedings of the First National Symposium'new Crops: Research, Development, Economics*'. Indianapolis Indiana.

Solanki, R., Mathur, V., Purohit, S., & Mathur, M. (2017). Evaluation of wound healing activity of ethanolic extract of *Ocimum basilicum* and *Aegle marmelos* leaves in male albino rats. *International Journal of Drug Research and Technology*, *2*(4).

Tongnuanchan, P., & Benjakul, S. (2014). Essential oils: Extraction, bioactivities, and their uses for food preservation. *Journal of Food Science*, *79*(7), R1231–R1249.

Umamageswar, A., & Kudagi, B. (2015). Anti-inflammatory and analgesic properties of *Ocimum* sanctum: A comparative study using animal models. *International Journal of Basic & Clinical Pharmacology*, *4*(5), 981–986.

Upadhyay, R., Nachiappan, G., & Mishra, H. N. (2015). Ultrasound-assisted extraction of flavonoids and phenolic compounds from *Ocimum tenuiflorum* leaves. *Food Science and Biotechnology*, *24*(6), 1951–1958.

USDA, 2020. *The PLANTS database*. United States Department of Agriculture. https://plants.usda.gov/plantprofile/OCIMU

Veillet, S., Tomao, V., & Chemat, F. (2010). Ultrasound assisted maceration: An original procedure for direct aromatisation of olive oil with basil. *Food Chemistry*, *123*(3), 905–911.

Vlase, L., Benedec, D., Hanganu, D., Damian, G., Csillag, I., Sevastre, B., Mot, A. C., Silaghi-Dumitrescu, R., & Tilea, I. (2014). Evaluation of antioxidant and antimicrobial activities and phenolic profile for *Hyssopus officinalis*, *Ocimum basilicum* and *Teucrium chamaedrys*. *Molecules*, *19*(5), 5490–5507.

Widjaja, S. S., Rusdiana, & Savira, M. (2019). Glucose lowering effect of basil leaves in diabetic rats. *Open Access Macedonian Journal of Medical Sciences*, *7*(9), 1415–1417.

Wojdyło, A., Oszmiański, J., & Czemerys, R. (2007). Antioxidant activity and phenolic compounds in 32 selected herbs. *Food Chemistry*, *105*(3), 940–949.

Xia, E. Q., Ai, X. X., Zang, S. Y., Guan, T. T., Xu, X. R., & Li, H. B. (2011). Ultrasound-assisted extraction of phillyrin from *Forsythia suspensa*. *Ultrasonics Sonochemistry*, *18*(2), 549–552.

Yang, Y., Kayan, B., Bozer, N., Pate, B., Baker, C., & Gizir, A. M. (2007). Terpene degradation and extraction from basil and oregano leaves using subcritical water. *Journal of Chromatography A*, *1152*(1), 262–267.

Zahra, K., Khan, M. A., & Iqbal, F. (2015). Oral supplementation of *Ocimum basilicum* has the potential to improves the locomotory, exploratory, anxiolytic behavior and learning in adult male albino mice. *Neurological Sciences*, *36*(1), 73–78.

Złotek, U., Mikulska, S., Nagajek, M., & Swieca, M. (2016). The effect of different solvents and number of extraction steps on the polyphenol content and antioxidant capacity of basil leaves (*Ocimum basilicum* L.) extracts. *Saudi Journal of Biological Sciences*, *23*(5), 628–633.

Zullaikah, S., Clarizka, D. C., Fulanah, D., Fitri, L., & Refit, W. Y. (2015). *Subcritical Water Extraction of Essential Oils from Indonesia Basil (Kemangi) Leaf: Effects of Temperature and Extraction Time on Yield and Product Composition*. Seminar Nasional Teknik Kimia Kejuangan. http://jurnal.upnyk.ac.id/index.php/kejuangan/article/view/475/435

8 Wound Wort/Common Self-Heal (*Prunella vulgaris*)

Santwana Palai and Raees ul Haq

8.1 INTRODUCTION

Prunella vulgaris L. (*labiatae*) is more commonly referred to as common self-heal. It was extensively used in both European and Chinese medication to treat fever, sore throats, and heal wounds. The perennial plant is a common herb that grows wild in Kashmir Valley. The leaves of *Prunella vulgaris* L. have a notched edge and a stem which is 1 to 2 feet in length. The flowers appearing at the stem's apex are purple (Figure 8.1). Mostly from June through August, the flowers are in full bloom. The herb is of great medical significance, and it is a necessary component of traditional Unani compound medicine used to treat headaches, colds, and sore throats (Gazala et al., 2016).

During the long winters in Kashmir, the herb is boiled and used as a brain tonic in Unani medicine. Its steam is inhaled, clearing mucus and easing headaches. The herb is one of the components of a composite traditional medicine used in Kashmir to bathe pregnant mothers after giving birth. This herb's aqueous extract is now being tested in clinical trials to treat herpetic keratitis. It proves highly beneficial for health, being used as a diuretic, hypotensive, antibiotic, antiseptic, antirheumatic, antipyretic, and antibacterial agent. It is applied to cure sores, ulcers, and cuts (Rasool et al., 2009).

Phenylpropanoids, organic acids, flavonoids, volatile oils, triterpenoids, sterols, and polysaccharides are just a few of the diverse chemical components to be found in *P. vulgaris*. The two main active components are terpenoids and flavonoids. Modern pharmacological research has revealed its numerous pharmacological actions, including liver protection, antibacterial, anti-hypertension, hypolipidemic, hypoglycemic, antiviral, anti-inflammatory, anti-tumor, and immunological modulation (Ksouri et al., 2012).

Due to restrictions in some particular studies, the current research on *P. vulgaris* is still relatively limited. As a result, this chapter's descriptions of botany, ethnopharmacology, traditional uses, phytochemistry, pharmacological effects, and toxicity can benefit readers, particularly students and researchers working in this sector. It can help in future studies on *P. vulgaris*' biological activities and medical uses.

8.2 OCCURRENCE

The herbaceous perennial *P. vulgaris* is thought to have originated in Europe, Eurasia, and America, but the plant has since spread widely throughout all continents. It was spread over the world in the 1800s and has since been identified as invasive in New Zealand and on numerous Pacific islands. It is believed to be endangering native plants and habitats on the Hawaiian Islands, along with other invasive species. Although it has been purposefully planted as an ornamental in some locations, it may be invasive in lawns in its natural habitat and is frequently regarded as a weed (Patel & Sharma, 2019).

DOI: 10.1201/9781003292340-8

FIGURE 8.1 *Prunella vulgaris* in the meadows of Kashmir.

8.3 SYSTEMATIC CLASSIFICATION

Domain	Eukaryota
Kingdom	Plantae
Phylum	Spermatophyta
Subphylum	Angiospermae
Class	Dicotyledonae
Order	Lamiales
Family	Lamiaceae
Genus	Prunella
Species	*Prunella vulgaris*

8.4 PHYTOCHEMICAL OVERVIEW

Since *Prunella vulgaris* contains a substantial amount of carbohydrates, protein, and fat, over the years it has been widely used for its good taste and nutritional value. In addition to high levels of proteins, lipids, and carbohydrates, the plant has a high concentration of rosmarinic acid and also prunellin, an anti-HIV polysaccharide which increases the plant's therapeutic utility. With its various species and populations, this herb has a variety of bioactive compounds that vary according to climatic zones and population sizes. The higher kind of this priceless plant has more of the phytochemicals crucial for both medicinal and therapeutic purposes. Alkaloids were also found in the herb as a secondary metabolite in addition to saponins. These serve a variety of purposes, but their analgesic, antispasmodic, and antibacterial properties are among their most important ones. The plant is also abundant in tannins and phenolics (Zhang et al., 2020). The phytochemicals present include the following: Prunellin, rosmarinic acid, oleanolic acid, ursolic acid, β-amyrin, pruvuloside, luteolin, kaemferol, betulinic acid (detailed in Figure 8.2); rutin, quercetin, caffeic acid, euscaphic acid, homoorientin, cynaroside, hesperidin, quercetin (Figure 8.3); and wogonin, *p*-hydroxycinnamic acid, umbelliferone, esculetin, methyl rosmarinate, ethyl rosmarinate and scopoletin (Figure 8.4). These allow it to contribute to a wide range of different pharmacological properties.

Mainly, three types of triterpenoids are present in *Prunella vulgaris*: ursane, lupane, and oleanane. Triterpenoids are the major phytoactive ingredients showing anti-diabetes, anti-inflammatory, anti-cancer, antibacterial, anti-viral and hepatoprotective effects (Li et al., 2019).

The presence of the phenolic component rosmarinic acid is confirmed. Of the 87 distinct triterpenoid components, oleanolic acid, ursolic acid, and β-amyrin all possess antihypertensive properties. The saponins present extend anti-cardiac arrhythmic properties. The 3-hydroxyolean-5,12-diene present in spikes significantly displayed cytotoxic action against human colon adenoblastoma (HT29). The cytotoxic effects of ursolic acid on SK-MEL-2, SK-OV-3, HCT15 and A549 cells were mild. *P. vulgaris* phytochemicals also include prunellin, betulinic acid, and 2a,3a-dihydroxyurs-12-en-28-oic acid, ursolic acid, etc. (Pan et al., 2022).

8.5 MEDICINAL USES

Self-heal (*Prunella vulgaris* L.) was principally employed to treat fevers, throat problems, and wounds. It has vast therapeutic applications due to its high level of rosmarinic acid. The presence of phenolic compounds demonstrates the antimicrobial properties of the plants. Typhoid fever and other bacterial illnesses, particularly those caused by *Escherichia coli, Bacillus typhi, Pseudomonas*, and *Mycobacterium tuberculosis*, can also be effectively treated through the use of the herb. The polysaccharide prunellin exerts antiviral action. The north Indian traditional healers advise inhaling this herb's steam to eliminate mucus from the chest and brain cavity, thereby lessening chest infections and easing headaches. One month after giving birth, expectant women are bathed in its decoction (Guru et al., 2021). Additionally, it serves as a disinfectant, removes overall body weakness in mothers, and acts as a brain tonic. The herb also includes saponins, which help red blood cells coagulate and are used to control bleeding, cure wounds, and treat ulcers. Alkaloids are present which have bactericidal, antispasmodic, and analgesic properties. The astringency of herbs owing to its high tannin content helps to accelerate the healing of wounds and to ease inflamed mucous membranes. *P. vulgaris* is also found to have outstanding anti-herpes simplex virus and anti-HIV effects. The plant prevents allergic responses and anaphylactic shock. It guards against lipid peroxidation in kidney, brain homogenates, and rat red blood cell (RBC). Its immunomodulatory effects are helpful in the treatment of cancer, diabetes, herpes and other inflammatory conditions (Patel et al., 2021; Kapoor et al., 2017).

8.6 APPLICATION OF ACTIVE INGREDIENT

The modern phytochemical understanding of this plant's ability to self-heal and pharmacological studies of its extracts and pure components can be employed to treat ailments (Figure 8.5).

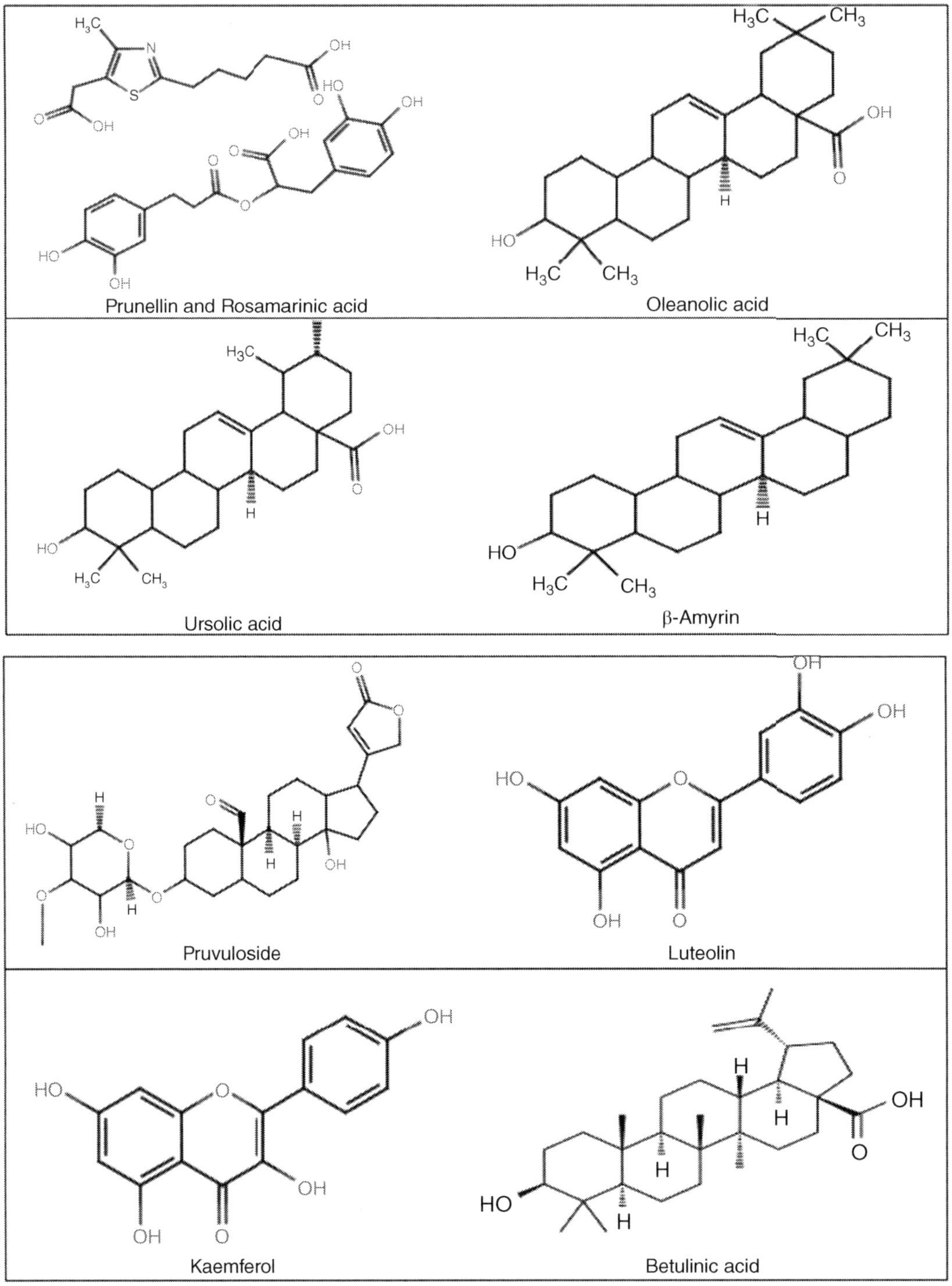

FIGURE 8.2 Structure of some of the phytochemicals of *Prunella vulgaris* (Prunellin, rosmarinic acid, oleanolic acid, ursolic acid, β-amyrin, pruvuloside, luteolin, kaemferol, betulinic acid).

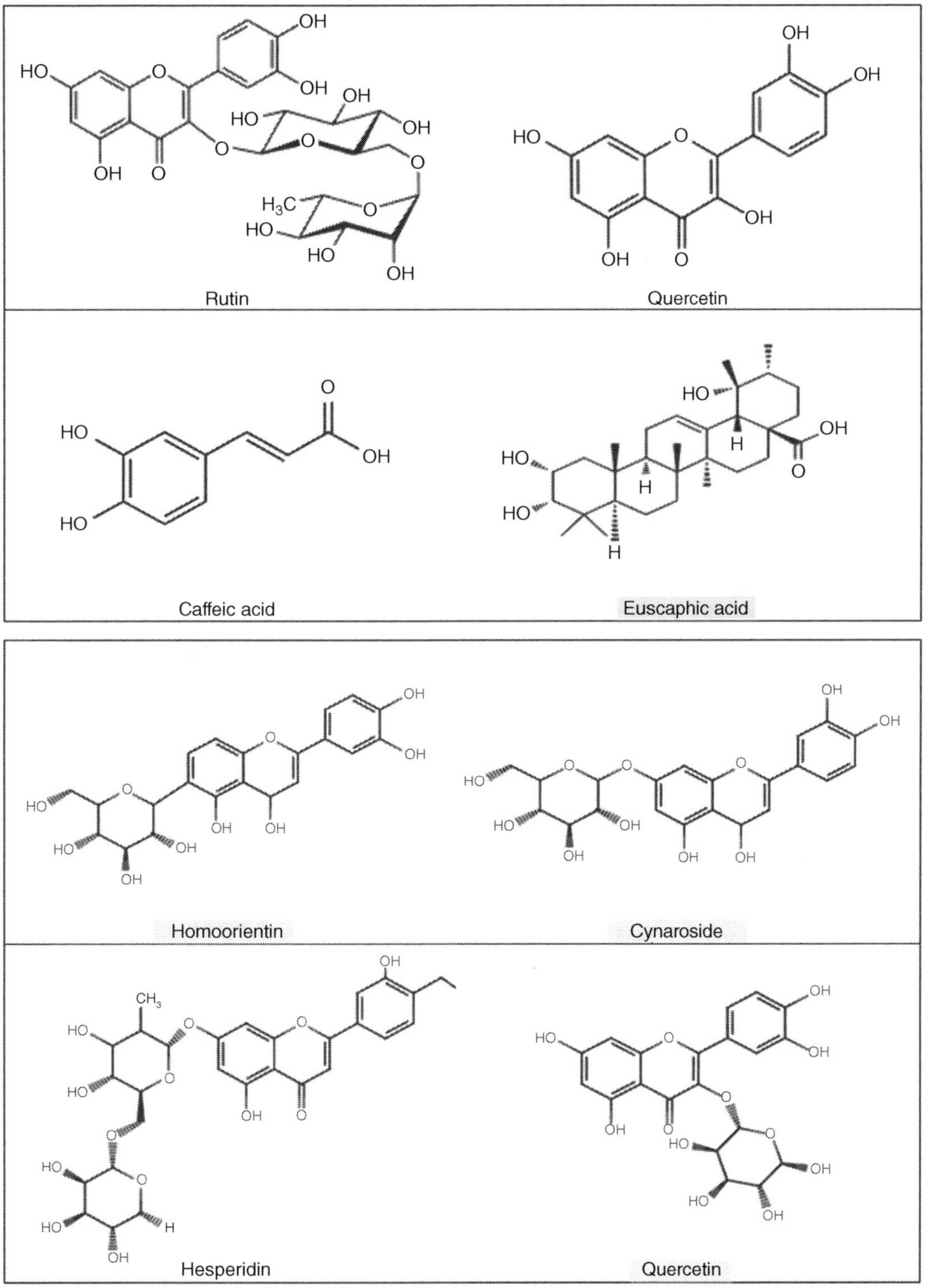

FIGURE 8.3 Chemical structure of rutin, quercetin, caffeic acid, euscaphic acid, homoorientin, cynaroside, hesperidin and quercetin found in *Prunella vulgaris*.

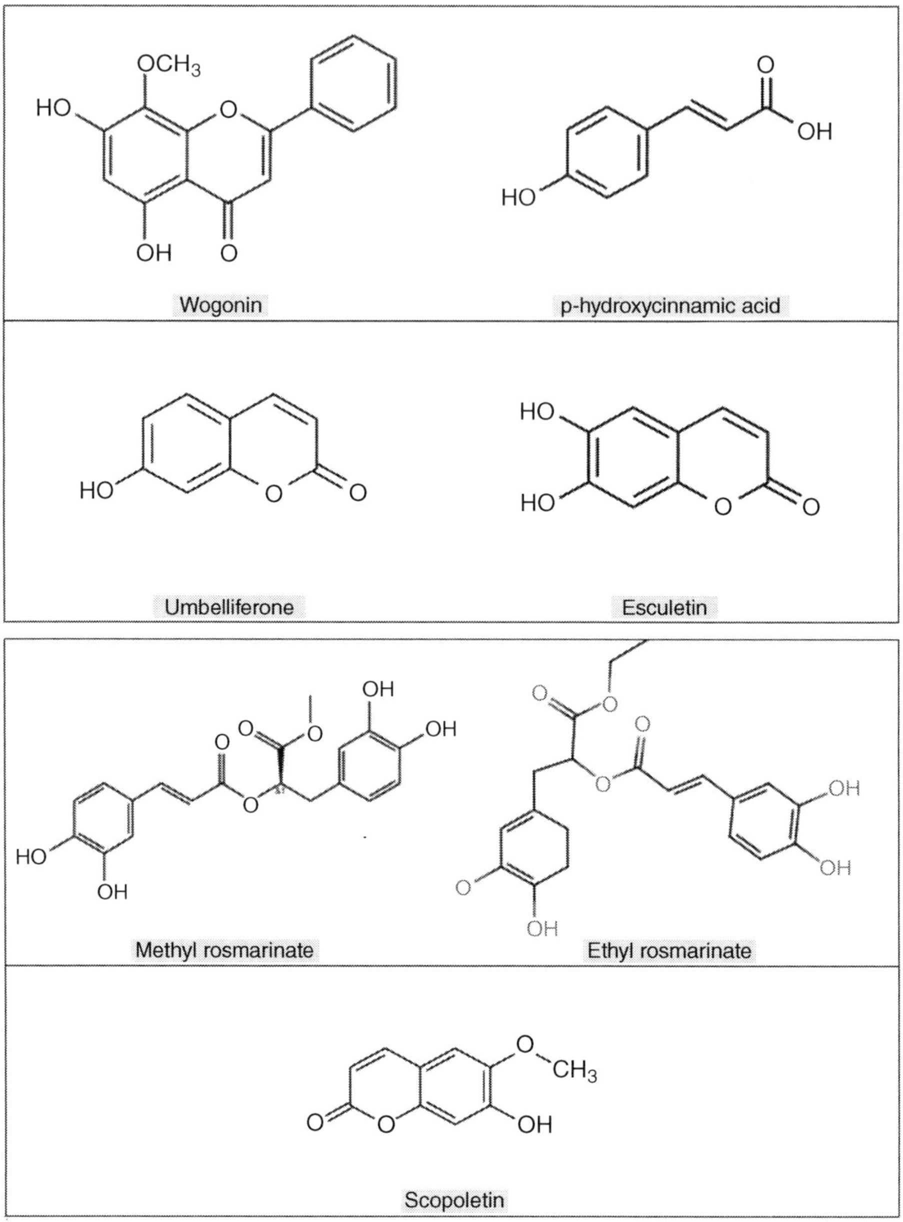

FIGURE 8.4 Wogonin, *p*-hydroxycinnamic acid, umbelliferone, esculetin, methyl rosmarinate, ethyl rosmarinate and scopoletin structure.

8.6.1 Anti-oxidant Effects

Prunella vulgaris has a clear antioxidant potential in vitro, enabling the food and pharmaceutical industries to use it more frequently as a natural antioxidant. In vitro testing revealed that the flavonoids in *Prunella vulgaris* have strong antioxidant activity and could effectively scavenge 2,2diphenyl-1-picrylhydrazyl (DPPH) and hydroxyl (OH) radicals (Hwang et al., 2013). The antioxidant activity varied amongst *Prunella vulgaris* extracts. The lipid peroxidation inhibition, DPPH,

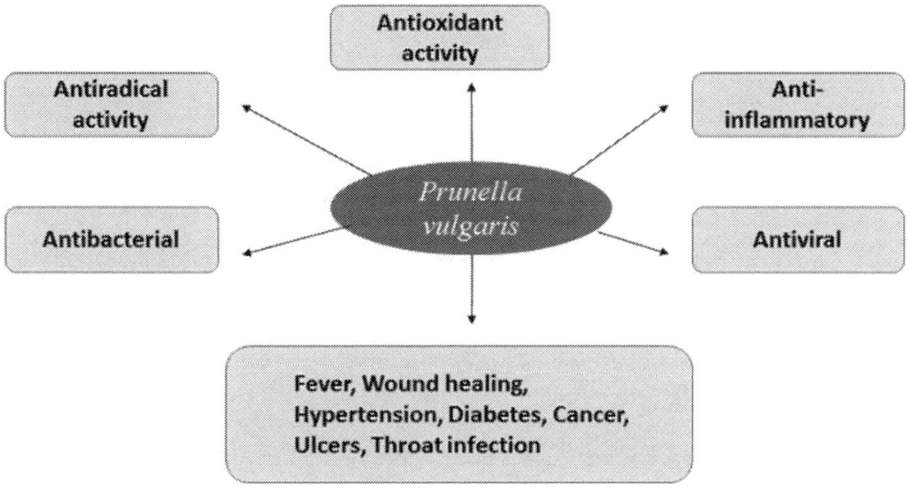

FIGURE 8.5 Pharmacological activity of *Prunella vulgaris*.

and OH scavenging activity of the 70% ethanol extract were all relatively strong (Zhu et al., 2018). The most effective O_2 scavenger was the aqueous extract. The *Prunella vulgaris* extract in 50% ethanol demonstrated strong antiradical, antioxidant, and minimal hydrolytic action. *Prunella vulgaris* aqueous extract's ethyl acetate fraction demonstrated a strong scavenging capacity and a moderately high amount of total flavonoids, phenolic acids, and triterpenoids (Feng et al., 2010).

In mice suffering from acute restraint stress damage, *Prunella vulgaris* demonstrated a protective effect. In mouse brain tissue, it may drastically lower the levels of H_2O_2, MDA, and protein carbonyl while boosting superoxide dismutase (SOD) activity. It also demonstrated that *Prunella vulgaris*' total flavonoids, which outperformed its total triterpenoids and polysaccharides, had the strongest antioxidant activity (Ahn et al., 2018).

The *Prunella vulgaris* extract in 60% ethanol demonstrated considerable in vitro antioxidant activity and can dramatically slow tumor growth by boosting SOD action and lowering MDA levels in the blood of tumor-bearing mice. *Prunella vulgaris* contains significant bioactive substances such as quercetin, rutin, caffeic acid, rosmarinic acid, and so on, which are essential for antioxidant activity (Pan et al., 2022).

8.6.2 Anti-inflammatory Effects

In response to microbial invasion, the polysaccharide portion of *Prunella vulgaris* demonstrates both anti-inflammatory and immunostimulatory actions. Due to its anti-inflammatory activity, rosmarinic acid in *Prunella vulgaris* decreased TNF-α, IL-1, IL-6, and iNOS production on LPS-persuaded inflammation of gingival fibroblasts in humans. The total triterpenoids from *Prunella vulgaris* were able to drastically reduce the gene expression of Jak2 and Stat3, showing some anti-inflammatory effects linked to the Jak/Stat system, as well as the production of TNF-α, PGE2, and IL-6 in LPS-enthused RAW264.7 cells (Han et al., 2009).

Due to an anti-inflammatory mechanism associated with the NLRP3/Caspase1/IL-1 pathway, the *Prunella vulgaris* aqueous extract could suggestively decrease inflammatory proteins NLRP3, Caspase-l, and IL-1 expression in corneal tissue and recover the inflammatory abrasions of cornea of rats in allergic conjunctivitis (Li et al., 2019; Ahmad et al., 2020).

Prunella vulgaris and *Fallopia japonica* extract might both be able to treat myopia. The combination condensed the expression of TNF-α, IL-6, and IL-8, and abridged inflammation via reducing the phosphorylation of AKT and the NFκ-B path. Additionally, they increased type I collagen expression in the monocular form deprivation-induced hamster model while decreasing myopia-related TGF-α, MMP-2, and NFκ-B expression (Lin et al., 2021).

In experimental autoimmune thyroiditis rats, *Prunella vulgaris* demonstrated anti-inflammatory and immunoregulatory effects. *Prunella vulgaris* therapy reduced the thyroiditis inflammation score, thyroid volume, and blood thyroglobulin antibody points in experimental autoimmune thyroiditis rats. Studies on the mechanisms involved showed that *Prunella vulgaris* could reduce experimental autoimmune thyroiditis by preventing a HMGB1/TLR9 signalling pathway (Guo et al., 2021).

8.6.3 Antibacterial Effect

Prunella vulgaris can be utilized in the study and creation of antibacterial drugs. By using the broth microdilution method, the minimal inhibitory concentration of *Prunella vulgaris* extract was identified. To obtain the active fractions, column chromatography was also applied to the extract. The active fraction was exposed to spectral analysis utilising the 1H 13C nuclear magnetic resonance, Fourier-transform infrared spectroscopy, and mass spectroscopy for the discovery and isolation of the antibacterial molecule. Based on the molecular formula, functional group analysis and molecular mass, the isolated molecule from the active fraction of *Prunella vulgaris* flower extract was recognized as [(2-(E)-3-(3-4-dihydroxyphenyl) acryloyloxy)-3-(3, dihydroxyphenyl) propanoic acid (Rosmarinic acid)] (Dar et al., 2022).

Both in vitro and in vivo investigations using the herbal extract of *Prunella vulgaris* revealed considerable antibacterial activity. *Prunella vulgaris* plant extracts in methanolic and petroleum ether were tested in vitro for their ability to combat *S. aureus, B. subtilis, S. typhi*, and *E. coli*. The rats were administered the antibiotic ofloxacin and plant extracts for controlled in vivo trials employing the *E. coli*-induced peritonitis in rat models. When the effectiveness of plant extracts and ofloxacin were examined, the methanol extracts demonstrated similar results in removing the pathogen from the systems of animals. Compared to methanol, petroleum extracts and ether extracts have lower levels of antibacterial activity (Komal et al., 2018).

8.6.4 Antiviral Effects

Prunella vulgaris contains powerful antiviral compounds which are effective against the influenza, HSV, and polysaccharide-based viruses. These antiviral compounds of *Prunella vulgaris* specifically targeted the obstruction of the life cycle stages of the virus. *Prunella vulgaris* compounds operated primarily through hindering the attachment and access of virions into the host cells and reducing the activities of integrases, proteases, reverse transcriptases, and glycohydrolases (Mak & Walsh, 2022).

8.6.4.1 HIV

Extracts of *Prunella vulgaris* inhibited HIV infection in vitro (Yao et al., 1992). In vitro, a spike extract of *Prunella vulgaris* suppresses HIV replication at reverse transcription and can be absorbed from the colon in vivo (Kageyama et al., 2000). Using multiple cell lines, a screening method evaluated the anti-HIV efficacy of extracts from plants like *Prunella vulgaris*. Preincubation of HIV-1 viruses with *Prunella vulgaris* extract revealed that the attachment was severely inhibited. Polymerase chain reaction (PCR) was used to verify that infected cells treated with the extract did not contain any HIV-1 proviral DNA. The extract demonstrated that *Prunella vulgaris* prevented viral addon to CD4 cells by inhibiting the bondage of isolated HIV envelope glycoprotein gp120 to CD4 cell receptors (Oh et al., 2011).

The aqueous extracts of herbs such as *Prunella vulgaris* were examined against HIV-1. The stoppage of HIV-1 reverse transcriptase, protein glycosylation, and HIV-1 envelope glycoprotein gp120's attachment near immobilized CD4 receptors is recognized as the anti-viral actions of *Prunella vulgaris* during the addon and duplication stages (Collins et al., 2009).

8.6.4.2 Herpes Simplex Virus

Plaque reduction assays were used to measure anti-HSV activity in RC-37 cells (Nolkemper et al., 2006). Through the application of the time-of-addition method, all isolated chemicals of *Prunella vulgaris* showed an impact on HSV-1 and herpes simplex virus type 2 (HSV-2) before attachment without impact on intracellular viral proliferation. The anti-HSV-1 properties of many herbs were also assessed using both in vitro and clinical trials, and both found *Prunella vulgaris* to be extremely effective when used both therapeutically and as a preventative strategy in clinical settings (Mak & Walsh, 2022).

The aqueous polysaccharide fraction of *Prunella vulgaris* worked well against HSV-1 and HSV-2 through plaque reduction assay, establishing their antiviral properties (Tapondjou et al., 2003; Zhang, 2003). Pre-treatment of HSV-1 viruses with the polysaccharide prevented infection. The polysaccharide added postinfection decreased the amount of intracellular viruses. Antiviral effects were observed both during the early stages of attachment and after entrance. After *Prunella vulgaris* polysaccharide fraction treatment, the expression of HSV-1 and HSV-2 antigens within their host Vera cells was observed using flow cytometry. The polysaccharide from *Prunella vulgaris* decreased expression and worked well against HSV-1 and HSV-2 viruses (Chiu et al., 2004).

Two triterpenes of *Prunella vulgaris* were responsible for the anti-HSV-1 actions. Plaque reduction assay was performed with Vero cells as host cells. The anti-HSV properties of *Prunella vulgaris* isolated polysaccharides performed with methods like ethanol precipitation showed the blocking of the binding of HSV-1 viruses to Vero cells and prevent their penetration in vitro and in vivo studies using guinea pigs and mice (Khan et al., 2005).

In addition, *Prunella vulgaris* polysaccharide exhibited repressive effects on both HSV-1 and HSV-2 using plaque reduction assays. Using both in vivo experiments with rainbow trout and in vitro studies with *epithelioma papulosum cyprini* cells, *Prunella vulgaris* owing to ursolic acid demonstrated highly effective anti-IHNV (a fish viral pathogen) activity (Cai et al., 2017).

A conformation-specific procedure was developed for the monoclonal antibody NC-1 which recognized small chemical compounds that prevented the formation of HIV-1 virions' glycoprotein gp41 six-helix bundles. This process shaped the capability of attachment inhibition of *Prunella vulgaris* (Liu et al., 2002).

8.7 PHARMACOLOGICAL APPLICATIONS

8.7.1 Hypertension

Prunella vulgaris is a traditional remedy for hypertension, hyperglycemia, and hyperlipidemia, and clinical studies have supported its effectiveness in treating these conditions. Its pharmacological effects can be further explored and studied by researchers (Yu et al., 2021). The rat thoracic aorta may be relaxed by the total alkaloids compound of *Prunella vulgaris* in dose-dependent manner and the vasodilation effect was endothelium-independent. The suppression of intracellular calcium release and external calcium influx may be connected to the process (Que et al., 2015). *Prunella vulgaris* and *Uncaria rhynchophylla* work together to lower the systolic blood pressure of spontaneously hypertensive rats and regulate the hypertension-related cardiac hypertrophy. The mechanism concerned involves the lowering of ANG II and ET levels in serum and the elevation of CGRP levels (Mohammed et al., 2022). Through its impact on body weight and serum lipids of obese C57BL/6J male mice caused by excess sugar excess fat diet in vivo, the viability of *Prunella vulgaris* in treating obesity was studied. *Prunella vulgaris* could significantly lower TG, TC, and LDL while raising HDL (Deng et al., 2021).

8.7.2 Diabetes

Prunella vulgaris aqueous extract blocked the mRNA expression of SGLT-1, α-glucosidase, Na^+ -K^+ -ATPase and GLUT-2 in Caco-2 cells to postpone carbohydrate breakdown and impact glucose absorption, thereby delaying blood sugar spikes after meals (Wu et al., 2010). *Prunella vulgaris* has improved hypoglycemic potential. In diabetic ICR mice, the continuing administration of *Prunella vulgaris* ethanol extract may reduce weight loss and polydipsia. The cholesterol, serum triglyceride, and LDL content of diabetic mice could be considerably decreased and increased by doses of *Prunella vulgaris* extract (8, 4 g/kg) according to the findings (Li et al., 2006).

Prunella vulgaris extract in 80% ethanol contain four pancreatic lipase inhibitors: rutin, caffeic acid, hesperidin, and ursolic acid (Dabur et al., 2018; Chen et al., 2018). Many extracts and monomer compounds exhibited likely antagonistic actions on pancreatic lipase during in vitro screening. *Prunella vulgaris* methanol extract showed 74.7% inhibitory impact on pancreatic lipase when compared to quercetin and luteolin extracts (Zheng et al., 2010). *Prunella vulgaris* stops the development of diabetic atherosclerosis. In HFHC diet-db/db mice, *Prunella vulgaris* aqueous extract decreased hyperglycemia and diabetic vascular dysfunction (Hwang et al., 2012).

8.7.3 Cancer

Thyroid cancer has a somewhat better prognosis than other cancers because of its low potential for malignancy, yet its unchecked proliferation still occurs. *Prunella vulgaris* extracts prevented TPC-1 and SW579 cells from proliferating and migrating shown by the wound healing, cell growth curve, cell cycle analysis, colony formation assay, and migration assays. In a subcutaneous xenotransplanted tumor model, the antitumor effect of *Prunella vulgaris* was confirmed. MKI67, PCNA, CTNNB1, and CDH1 expression in xenograft tissue both with and without *Prunella vulgaris* treatment was screened by qRT-PCR and the expressions were confirmed by Western blot. *Prunella vulgaris* could prevent thyroid cancer cells from proliferating and migrating both in vivo and in vitro by regulating the expression of MKI67, PCNA, and CDH1. *Prunella vulgaris* is likely to be established as novel anti-thyroid cancer medication (Chang et al., 2021).

In the case of lung cancer chemoprevention studies, the chemical compositions of ethanolic extract *Prunella vulgaris* and oleanolic acid (its primary active ingredient) were determined using high-performance liquid chromatography, high-performance thin layer chromatography, and liquid chromatography-mass spectrometry. No appreciable differences were seen in terms of cell viability between the ethanolic extracts of *Prunella vulgaris* and oleanolic acid in lung adenocarcinoma SPC-A-1 cells in MTT assays. Acridine orange-ethidium bromide fluorescence detection was used to measure the apoptosis of SPC-A-1 cells. Additionally, immunocytochemistry experiments revealed higher Bax and Bad protein expression and lower Bcl-2 expression. Oleanolic acid therapy increased the Bax/Bcl-2 ratio. It caused lung cancer cells to undergo apoptosis down-regulating Bcl-2 expression, and up-regulating Bax and Bad expression (Feng et al., 2011).

Prunella vulgaris might have anti-cancer effects on breast CAFs through the prevention of bFGF expression, which would obstruct the development of breast cancer SKBr-3 cells. By causing apoptosis and stopping the cell cycle, PVP Polysaccharide prevented CAFs from surviving and migrating. In this, bFGF decreased the G0 phase, protected CAFs from apoptosis, and encouraged CAF migration and proliferation. *Prunella vulgaris* might negate the effects of bFGF on CAFs. It is possible to indirectly reduce the viability of breast cancer SKBr-3 cells by cultivating them with CAFs-CM and PVP. Additionally, *Prunella vulgaris* decreased the secretion of the bFGF's protein and mRNA in CAFs (Hao et al., 2020).

8.7.4 Wounds and Ulcers

Prunella vulgaris show notable anti-inflammatory and wound-healing effects. The actions of the methanolic extract and sub-extracts of *Prunella vulgaris* on wound healing were examined through wound models comprising linear and circular excision in rats and mice. The acetic acid-persuaded capillary permeability trial was undertaken to assess its anti-inflammatory effects. The ethyl acetate sub-extract with greatest activity was used for isolation investigations. By using spectroscopic techniques (MS, IR, UV, 1H-NMR, 13C-NMR, 2D-NMR), the chemicals' structures were determined to be rosmarinic acid, ursolic acid, methyl arginolate, methyl 3-epimaclinate, chlorogenic acid, and ethyl rosmarinate. The wound-healing processes of these pure compounds was examined through an inhibition of enzymes such as collagenase, hyaluronidase, and elastase. The anti-inflammatory and wound-healing properties is attributed to the presence of rosmarinic acid, chlorogenic acid, and ursolic acid (Akkol et al., 2022).

8.8 FUTURE LINE OF WORK

There is a need for additional study and development on the biological impacts and the structure–activity connections and purification of phytochemicals. The quality control of *Prunella vulgaris* needs to be thoroughly assessed using chromatographic fingerprinting to gather data on chemical composition and conduct correlation research with regard to its effectiveness and safety assessment.

8.9 CONCLUSION

The medicinal properties of *Prunella vulgaris* L. have been used traditionally to treat a variety of infectious disorders. The phytochemical composition and broad pharmacological effects of *Prunella vulgaris* are presented to justify its medicinal purposes. The investigation of *Prunella vulgaris* phytocomponents with the advancements in science and technology will give rise to some useful, effective, novel, environmentally friendly, and economically viable options. With its distinctive advantages, *Prunella vulgaris* has a great deal of potential for drug resource development for clinical use. It also has a chance for development and for its use in the beverage and health products industries.

REFERENCES

Ahmad, G., Masoodi, M. H., Tabassum, N., & Mir, R. A. (2020). Phytochemical Analysis and Anti-inflammatory Activity of Various Extracts Obtained from Floral Spikes of *Prunella vulgaris* L. *Jordan Journal of Pharmaceutical Sciences*, *13*(1), 41–52.

Ahn, E. Y., Lee, Y. J., Park, J., Chun, P., & Park, Y. (2018). Antioxidant potential of *Artemisia capillaris, Portulaca oleracea*, and *Prunella vulgaris* extracts for biofabrication of gold nanoparticles and cytotoxicity assessment. *Nanoscale research letters*, *13*(1), 1–14.

Akkol, E. K., Renda, G., İlhan, M., & Bektaş, N. Y. (2022). Wound healing acceleration and anti-inflammatory potential of *Prunella vulgaris* L.: From conventional use to preclinical scientific verification. *Journal of Ethnopharmacology*, *15*, 115411.

Cai, S., Yang, Y., Wu, R., Yang, Y., & Tan, H. (2017). A pharmacodynamic study on the anti-herpes simplex virus (HSV) activity of the polysaccharides and gels from *Prunella vulgaris* L. *World Science and Technology-Modernization of Traditional Chinese Medicine*, *12*, 247–253.

Chang, C. H., Lin, C. P., MacDonald, I., Chiu, T. W., & Huang, S. T. (2021). Resolution of a vocal fold polyp treated with Chinese herbal medicine: One case report with literature review. *Journal of Herbal Medicine*, *29*, 100486.

Chen, T. G., Li, L. Y., Wei, Y. R., & Zhang, L. W. (2018). Screening, identification and activity evaluation of pancreatic lipase inhibition in *Prunella vulgaris*. *China Journal of Chinese Materia Medica*, *43*(23), 4665–4671.

Chiu, L. C. M., Zhu, W., & Ooi, V. E. C. (2004). A polysaccharide fraction from medicinal herb *Prunella vulgaris* downregulates the expression of herpes simplex virus antigen in Vero cells. *Journal of Ethnopharmacology*, *93*(1), 63–68.

Collins, N. H., Lessey, E. C., DuSell, C. D., McDonnell, D. P., Fowler, L., Palomino, W. A., & Lessey, B. A. (2009). Characterization of antiestrogenic activity of the Chinese herb, *Prunella vulgaris*, using in vitro and in vivo (Mouse Xenograft) models. *Biology of reproduction*, *80*(2), 375–383.

Dabur, R., Sharma, B., & Mittal, A. (2018). Mechanistic approach of anti-diabetic compounds identified from natural sources. *Chemical Biology Letters*, *5*(2), 63–99.

Dar, K. A., Senthilmurugan, S., Ali, S., AlSadoon, M. K., & Paray, B. A. (2022). Studies on the isolation and identification of the antibacterial compound from *Prunella vulgaris* L. flower extract. *Journal of King Saud University-Science*, *16*, 102324.

Deng, J., Li, L., Lin, L. M., Li, Y. M., Xia, B. H., & Liao, D. F. (2021). Metabolic mechanism of *Prunella vulgaris* in treatment of ethanol-induced oxidative stress in rats based on metabonomics. *China Journal of Chinese Materia Medica*, *46*(7), 1813–1821.

Feng, L., Au-Yeung, W., Xu, Y. H., Wang, S. S., Zhu, Q., & Xiang, P. (2011). Oleanolic acid from *Prunella vulgaris* L. induces SPC-A-1 cell line apoptosis via regulation of Bax, Bad and Bcl-2 expression. *Asian Pacific Journal of Cancer Prevention*, *12*(2), 403.

Feng, L., Jia, X., Zhu, M. M., Chen, Y., & Shi, F. (2010). Antioxidant activities of total phenols of *Prunella vulgaris* L. in vitro and in tumor-bearing mice. *Molecules*, *15*(12), 9145–9156.

Gazala, Q., Murtaza, I., Ara, S., Qazi, H., Geelani, S. M., & Amir, S. (2016). Characterization and antimicrobial activity of some natural dye yielding plant species of Kashmir Valley. *Journal of Industrial Pollution Control*, *32*(2), 518–528.

Guo, Q., Qu, H., Zhang, H., & Zhong, X. (2021). *Prunella vulgaris* L. attenuates experimental autoimmune thyroiditis by inhibiting HMGB1/TLR9 signaling. *Drug Design, Development and Therapy*, *15*, 4559.

Guru, A., Dwivedi, P., Kaur, P., & Pandey, D. K. (2021). Exploring the role of elicitors in enhancing medicinal values of plants under in vitro condition. *South African Journal of Botany*, *149*, 1029–1043.

Han, E. H., Choi, J. H., Hwang, Y. P., Park, H. J., Choi, C. Y., Chung, Y. C., & Jeong, H. G. (2009). Immunostimulatory activity of aqueous extract isolated from *Prunella vulgaris*. *Food and Chemical Toxicology*, *47*(1), 62–69.

Hao, J., Ding, X. L., Yang, X., & Wu, X. Z. (2020). *Prunella vulgaris* polysaccharide inhibits growth and migration of breast carcinoma-associated fibroblasts by suppressing expression of basic fibroblast growth factor. *Chinese Journal of Integrative Medicine*, *26*(4), 270–276.

Hwang, S. M., Kim, J. S., Lee, Y. J., Yoon, J. J., Lee, S. M., Kang, D. G., & Lee, H. S. (2012). Anti-diabetic atherosclerosis effect of *Prunella vulgaris* in db/db mice with type 2 diabetes. *The American journal of Chinese Medicine*, *40*(05), 937–951.

Hwang, Y. J., Lee, E. J., Kim, H. R., & Hwang, K. A. (2013). In vitro antioxidant and anticancer effects of solvent fractions from *Prunella vulgaris* var. lilacina. *BMC Complementary and Alternative Medicine*, *13*(1), 1–9.

Kageyama, S., Kurokawa, M., & Shiraki, K. (2000). Extract of *Prunella vulgaris* spikes inhibits HIV replication at reverse transcription in vitro and can be absorbed from intestine in vivo. *Antiviral Chemistry and Chemotherapy*, *11*(2), 157–164.

Kapoor, R., Sharma, B., & Kanwar, S. S. (2017). Antiviral phytochemicals: an overview. *Biochemistry and Physiology*, *6*(2), 220.

Khan, M. T. H., Ather, A., Thompson, K. D., & Gambari, R. (2005). Extracts and molecules from medicinal plants against herpes simplex viruses. *Antiviral Research*, *67*(2), 107–119.

Komal, S., Kazmi, S. A. J., Khan, J. A., & Gilani, M. M. (2018). Antimicrobial activity of *Prunella vulgaris* extracts against multi-drug resistant *Escherichia coli* from patients of urinary tract infection. *Pakistan Journal of Medical Sciences*, *34*(3), 616–620.

Ksouri, R., Ksouri, W. M., Jallali, I., Debez, A., Magné, C., Hiroko, I., & Abdelly, C. (2012). Medicinal halophytes: potent source of health promoting biomolecules with medical, nutraceutical and food applications. *Critical Reviews in Biotechnology*, *32*(4), 289–326.

Li, B. Y., Hu, Y., Li, J., Shi, K., Shen, Y. F., Zhu, B., & Wang, G. X. (2019). Ursolic acid from *Prunella vulgaris* L. efficiently inhibits IHNV infection in vitro and in vivo. *Virus Research*, *273*, 197741.

Li, Y., Ji, B. P., Zheng, J., Li, B., & Zhang, X. F. (2006). Effects of *Prunella vulgaris* L. extracts on blood glucose and blood lipid in streptozotocin-induced diabetic ICR mice. *Food Science*, *27*(6), 212–215.

Lin, C. H., Chen, C. S., Wang, Y. C., Lin, E. S., Chang, C. Y., Chen, J. J. Y., & Wan, L. (2021). *Fallopia japonica* and *Prunella vulgaris* inhibit myopia progression by suppressing tAKT and NFκB mediated inflammatory reactions. *BMC Complementary Medicine and Therapies*, *22*, 271.

Liu, S., Jiang, S., Wu, Z., Lv, L., Zhang, J., Zhu, Z., & Wu, S. (2002). Identification of inhibitors of the HIV-1 gp41 six-helix bundle formation from extracts of Chinese medicinal herbs *Prunella vulgaris* and *Rhizoma cibotte*. *Life Sciences*, *71*(15), 1779–1791.

Mak, W. C. K., & Walsh, S. (2022). The anti-viral activity of *Prunella vulgaris*: A narrative review. *Integrative Medicine Reports*, *1*(1), 107–122.

Mohammed, S. A., Liu, H., Baldi, S., Chen, P., Lu, F., & Liu, S. (2022). GJD modulates cardiac/vascular inflammation and decreases blood pressure in hypertensive rats. *Mediators of Inflammation*, *2022*, 7345116.

Nolkemper, S., Reichling, J., Stintzing, F. C., Carle, R., & Schnitzler, P. (2006). Antiviral effect of aqueous extracts from species of the *Lamiaceae* family against Herpes simplex virus type 1 and type 2 in vitro. *Planta Medica*, *72*(15), 1378–1382.

Oh, C., Price, J., Brindley, M. A., Widrlechner, M. P., Qu, L., McCoy, J. A., & Maury, W. (2011). Inhibition of HIV-1 infection by aqueous extracts of *Prunella vulgaris* L. *Virology Journal*, *8*(1), 1–10.

Pan, J., Wang, H., & Chen, Y. (2022). *Prunella vulgaris* L.–A review of its ethnopharmacology, phytochemistry, quality control and pharmacological effects. *Frontiers in Pharmacology*, *13*, 903171.

Patel, J. K., & Sharma, M. K. (2019). A review on ethno botanical, photochemistry, bioactivities and medicinal potential of *Prunella vulgaris* (mint family/common self-heal). *East African Scholars Journal of Biotechnology and Genetics*, *1*(5), 99–105.

Patel, J. K., Joshi, C. K., & Sharma, M. K. (2021). *Prunella vulgaris* L. potentiates answer to the emergence of dreaded antibiotic resistance. *Journal of Pure and Applied Microbiology*, *5*(3), 1429–1441.

Que, S., Lou, Y., & Hou, J. (2015). Vasodilation effect and underlying mechanism of active fraction from compound *Prunella vulgaris* L. in isolated thoracic aorta of rats. *China Pharmacist*, *12*, 884–887.

Rasool, R., Kamili, A. N., Ganai, B. A., & Akbar, S. (2009). Effect of BAP and NAA on shoot regeneration in *Prunella vulgaris*. *Journal of Natural Sciences and Mathematics*, *3*(1), 21–26.

Tapondjou, L. A., Lontsi, D., Beibam Luc, S., Jongwon, C., Kyung-Tae, L., Hyun-Ju, J., & Hee-Juhn, P. (2003). In Vivo anti-nociceptive and anti-inflammatory effect of the two triterpenes, ursolic acid and 23-hydroxy-ursolic acid, from *Cussonia bancoensis*. *Archives of Pharmacal Research*, *26*(2), 143–146.

Wu, H., Ha, T., & Gao, M. (2010). *Prunella vulgaris* L. extract decreasing mRNA expressions of α-glucosidase SGLT-1, GLUT-2 and Na+-K+-ATPase in Caco-2 cells. *Chinese Journal of Biochemical Pharmaceutics*, *31*(6), 373–376.

Yao, X. J., Wainberg, M. A., & Parniak, M. A. (1992). Mechanism of inhibition of HIV-1 infection in vitro by purified extract of *Prunella vulgaris*. *Virology*, *187*(1), 56–62.

Yu F, Zhang L, Ma R, Liu C, Wang Q, and Yin D. (2021). The antitumour effect of *Prunella vulgaris* extract on thyroid cancer cells in vitro and in vivo. *Evidence-Based Complementary and Alternative Medicine*, *2021*, 8869323.

Zhang, Y. (2003). *Studies on the anti-herpes simplex virus (HSV) constituents from a Chinese herbal medicine, Prunella vulgaris* (3095430) [Doctoral thesis, The Chinese University of Hong Kong]. CUHK Electronic Theses and Dissertations Collection, The Chinese University of Hong Kong Library. https://repository.lib.cuhk.edu.hk/sc/item/cuhk-343273

Zhang, Z., Zhou, Y., Lin, Y., Li, Y., Xia, B., Lin, L., & Liao, D. (2020). GC–MS-based metabolomics research on the anti-hyperlipidaemic activity of *Prunella vulgaris* L. polysaccharides. *International Journal of Biological Macromolecules*, *159*, 461–473.

Zheng, C. D., Duan, Y. Q., Gao, J. M., & Ruan, Z. G. (2010). Screening for anti-lipase properties of 37 traditional Chinese medicinal herbs. *Journal of the Chinese Medical Association*, *73*(6), 319–324.

Zhu, M., Zhan, J. L., Gao, Z. Y., Fan, J. F., & Zhao, E. L. (2018). Study on ultrasonic-assisted extraction and antioxidant activities of polyphone from *Prunella vulgaris* L. *China Food Additives 12*, 136–142.

9 Himalayan Rhubarb (*Rheum emodi*)

Pratiksha, Elizabeth Thomas, and C. G. Harshitha

9.1 INTRODUCTION

Indian Ayurveda medicine has used plants for the preparation of traditional medicines since approximately 1000 BC (Pandith et al., 2018). Over the years, herbal medicines have gained importance because of their therapeutic properties; this has promoted the development of a number of herbal drug manufacturing companies (Wasim et al., 2013). These medicinal plants have been proven effective in curing deleterious infections and chronic illnesses, improving overall holistic health (Khattak et al., 2020). At present, about 45% of the best-selling drugs are produced using natural products and their derivatives (Pandith et al., 2018).

Rheum emodi, or Himalayan rhubarb, is a common herb used around the world because of its medicinal properties and which, therefore, often goes by the name "wondrous drug". This perennial plant belongs to the Polygonaceae family. Due to its tonic, laxative, and diuretic effects, since ancient times, rhubarb has been found to be helpful in treating fever, diarrhea, menstrual disorder, cough, indigestion, bacterial infection, liver disorders, and jaundice, among other ailments. The dried rhizome, roots and petiole of the plant are generally utilized for medicinal purposes and are collected when it is in season, from October to November. It is known by different names in different languages such as Indian rhubarb or Himalayan rhubarb in English, *Pita* in Sanskrit, *Rewand* in Arabic, *Pambhaak* in Kashmiri, *Rewandchini* in Persian, *Amlaparni* in Ayurvedic, *Nattu-revalchini* in Telugu, etc., as shown in Table 9.1 (Rehman et al., 2014). It is also considered to be an important foodstuff for the people of Kashmir. Despite being classified as a vegetable, rhubarb is often considered a fruit because of its use in jams, jellies, sauces, crumbles, pies etc. The edible parts of the plant are the leaves, stalks, rhizome, and flowers, which are consumed either raw or cooked by boiling in salt and pepper (Kounsar et al., 2011). These are mainly consumed during the winters. Before consuming, the rhizomes of the plant are first cut into small pieces and are dried, either naturally in the sun or artificially (Nazir et al., 2013).

9.2 OCCURRENCE

As the name suggests, Himalayan rhubarb, or *Rheum emodi*, is a perennial herb native to Himalayan region and is widespread from Kashmir to Sikkim. It is distributed at an elevation of 2800–3000m and well adapted to colder climates; it therefore grows well in the sub-tropical and temperate Himalayan region (Radhika et al., 2010; Rumpunen & Henriksen, 1999). The origin of rhubarb dates back to ancient times, when it was first imported by the Romans from the unknown barbarian lands across the Rha River, from where it derives the name *Rha barbarum*, eventually being called *rhabarbarum* (Zargar et al., 2011). In 1753, Carl Linnaeus for the first time reported on the "*Rheum*" genus, which belongs to the family Polygonaceae (Wani et al., 2022).

For its medicinal use, the cultivation of rhubarb dates back to around 5000 years. For the proper cultivation of rhubarb, the permeable, humus rich and well-drained soil of the alpine and sub-alpine areas of Himalaya is preferred (Singh et al., 2017; Zargar et al., 2011). The dark red-purple-colored flowers 3 mm in diameter protrudes from the dense, clustered stalks of the plant having a length of

 DOI: 10.1201/9781003292340-9

TABLE 9.1
Different Names of *Rheum emodi* or Rhubarb as per Different Languages

Language	Name
Latin	*Rheum emodi*
English	*Himalayan Rhubarb/ Indian Rhubarb*
French	*Rhubarb de parse*
Persian	*Rewandchini*
Arabic	*Rewand*
Ayurvedic	*Amlaparni*
Sanskrit	*Pita*
Kashmiri	*Pambhaak/Pamchallan*
Tamil	*Nattu-ireval-chini*
Telugu	*Nattu-revalchini*
Unani	*Revandchini*

Source: Rehman et al., 2014; Daniyal et al., 2019.

about 30 cm (Pandith et al., 2014). The purple color of the flowers is the result of the presence of the pigment anthocyanin (Rana and Duhan, 2018). June and July is the flowering season, whereas fruits develop in the month of September and October. Its fruits are purple, angled and about 11–12 mm in diameter. This plant grows well in shady areas, on rocks and near rivulets (Daniyal et al., 2019; Zargar et al., 2011). Broad oval leaves are around 40 cm in length, having a long and scabby petiole which ranges from 30 to 45 cm in length. Its hard stems carry brownish streaks. The plant roots are thick and strong with a rough texture and are dark brown in color, whereas the stems are hollow (Daniyal et al., 2019; Singh et al., 2017). Because of its use in folk medicine, rhubarb has been grown in Germany, China, Tibet and some of the European countries. It is also cultivated in Assam because of its edible characteristics (Daniyal et al., 2019).

9.3 SYSTEMATIC CLASSIFICATION

The botanical name of rhubarb is *Rheum rhabarbarum*. Approximately 60 perennial species of *Rheum* genus are distributed all over the globe. Of these, some species possess medicinal properties, while others are used for culinary and decoration purposes. Around ten species of *Rheum emodi* can be found in India (Rokaya et al., 2012; Rehman et al., 2014). Himalayan rhubarb, Chinese rhubarb, and rhapontic rhubarb are the three main types of the plant (Singh et al., 2017) (Figure 9.1).

As stated earlier in the chapter, the rhubarb plant belongs to the family polygonaceae. *Rheum australe, R. undulatum, R. rhaponticum, R. palmatum, R. ribes, R. nobile, R. moorcroftianum, R. webbianum, R. speciformae, R. tibetana* are among the major rhubarb varieties which has been studied. *Rheum nobile*, commonly known as Sikkim rhubarb, is limited to the eastern region of Himalayas (Tayade et al., 2012). The *Rheum webbanium* variety of rhubarb is mainly restricted to the alpine regions of the Himalayas. It grows well in the cold and arid high-altitude regions of Ladakh Himalaya (Wani et al., 2022; Tayade et al., 2012) and can tolerate extreme cold conditions because its rhizomes are able to penetrate underground, sprouting only when favorable conditions prevail, giving rise to juvenile shoots (Wani et al., 2022). *Rheum australe* D. Don is another, less explored variety of rhubarb, one which has not been domesticated yet due to an absence of strategies and which is generally harvested in the wild. Bezzi (1987) and Voltolina (1987) have provided data on the cultivation and manuring of *Rheum palmatum* and germination, content of active principles and yield of plant materials (fresh as well as dry forms) in *Rheum undulatum*, respectively (Jeelani, 2015). The Gosaikunda

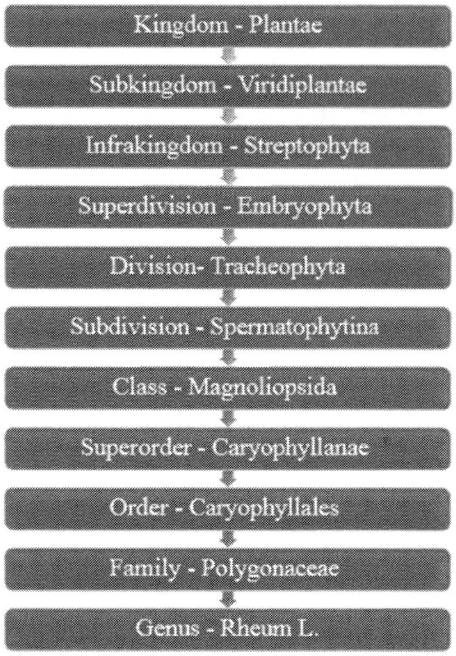

FIGURE 9.1 Taxonomic hierarchy of *Rheum* L. (Integrated Taxonomic Information System-Report, 2011.)

area of Nepal is one of the natural habitats for the growing of *Rheum acuminatum* and *Rheum australe*. It can grow up to an elevation of 3200–5200m. Another variant, *Rheum raponticum*, is also known as Siberian rhubarb (Kolodziejczyk-Czepas & Liudvytska, 2021).

9.4 NUTRITIONAL ATTRIBUTES OF THE PLANT

Because of to its medicinal properties, Himalayan rhubarb is being used by a large number of pharmaceutical industries. Ayurvedic medicines such as kankayan vati, chinimco and arjin tablets all contain Indian rhubarb as an important ingredient in their formulations (Singh et al., 2017). Studies have shown the anticancer and antioxidant potential of the extracts obtained from Rheum rhizomes (Radhika et al., 2010). In Japanese and Chinese traditional medicines, the rhizomes of the rhubarb plant has been found to be effective in treating syndromes involved in blood stagnation, such as diabetes, ischemia, inflammation, and atherosclerosis (Park and Lee, 2021). The diabetic property of the rhizome is because of the stilbene compounds present in it and has been found effective in preventing kidney and liver damage by scavenging and inhibiting free radical formation (Radhika et al., 2010). The pharmacological properties of the plant have been exploited in the treatment of blood fat regulation, hepatitis and cancer (Nazir et al., 2013). A rich source of dietary fiber, rhubarb also has lipid-lowering properties. It controls the development of cholesterol through the inhibition of squalene epoxidase (Tayade et al., 2012). Dried and ground rhubarb stalk fibre has been proven effective in reducing cholesterol level, total serum cholesterol, and low density lipoproteins (LDL) without affecting high-density lipoproteins (HDL) (Goel et al., 1997). The nutritional composition of rhubarb is shown in Table 9.2.

9.5 PHYTOCHEMICAL OVERVIEW OF THE PLANT

Singh and Chaturvedi (2018) performed a phytochemical profiling of *Rheum emodi* and found that carbohydrates, proteins, phenols, alkaloids, flavonoids, anthraquinones, quinones, and glycosides

TABLE 9.2
Nutritional Composition of Rhubarb, Raw

Nutrients	Amount	% RDA
Energy	21 kJ	1
Protein	0.9 g	2
Total lipid	0.2 g	1
Total dietary fiber	1.8 g	5
Carbohydrates	4.54 g	3
Sugar	1.1 g	
Water	93.6 g	
Ash	0.76 g	

Source: USDA, 2018; Mehra & Jain, 2020.

are the major phytochemicals present in the plant. Methanol was used as a solvent for extracting 95 bioactive compounds in total, of which the rhizome contained 31 compounds, the fruit contained 38 compounds, the leaf contained 27 compounds, and the callus 47 compounds (Singh & Chaturvedi, 2019). Free anthraquinones and their glycosides are the major phytoconstituents present in Indian rhubarb (Zargar et al., 2011). Emodin, chrysophanol, rhein, aloe-emodin, and physcion are the major anthraquinones present in the plant, and these are to be found present in every species of rhubarb (Singh & Chaturvedi, 2019; Tayade et al., 2012). Chrysophanol (43.97%) and physcion (3.23%) are mainly present in the rhizome, whereas the fruit contains only physcion (4.66%) (Singh & Chaturvedi, 2019). Roots of the plant contains stilbene glycosides, rhaponticin, and rhapontigenin, which is a metabolite (Tayade et al., 2012). In terms of different types of phytochemicals present in the roots, *R. acuminatum, R. officinale*, and *R. australe* are relatively similar and can be used as substitute for each other in traditional medicines. Significant amounts of resveratrol is also present in the rheum species (Rokaya et al., 2012) and oxalic and malic acid are also the major organic acids present in rhubarb (Rumpunen & Henriksen, 1999).

9.6 STRUCTURAL CHARACTERIZATION OF BIOACTIVE COMPONENTS

9.6.1 Anthraquinones

Anthraquinones are a class of secondary metabolites that are structurally linked to 9,10-dioxoanthracene and their glycosides (Chien et al., 2015).

9.6.1.1 Rhein

Rhein is a lipophilic anthraquinone with the presence of carboxylic acid in its structure, as shown in Figure 9.2 (Zargar et al., 2011). It is an aromatic polyketide with International Union of Pure and Applied Chemistry (IUPAC) name 4,5-dihydroxyanthraquinone-2-carboxylic acid, chemical formula $C_{15}H_8O_6$ and molecular weight 284.22 g/mol (Zhou et al., 2017; HMDB, 2022d). The common occurrence form of rhein is rhein-8-glucoside or glucorhein (Tabin et al., 2016). Rhein has potential pharmacological applications because of its ability to modify its structure to produce several rhein derivatives by various methods, including carboxyl substitution, hydroxyl substitution, amidation, esterification, etc. One such example is of 1,8-dihydroxy-3-[2'-(4″-methylpiperazin-1″-yl)] ethyl-9,10-anthraquinone-3-carboxylate, which is a carboxyl-substituted rhein derivative which exhibits anti-tumor properties. Another rhein derivative with anti-MRSA (methicillin-resistant *Staphylococcus aureus*) activity is synthesized by substituting the carboxyl group of α-hydroxyphosphate at the third position (Cheng et al., 2021).

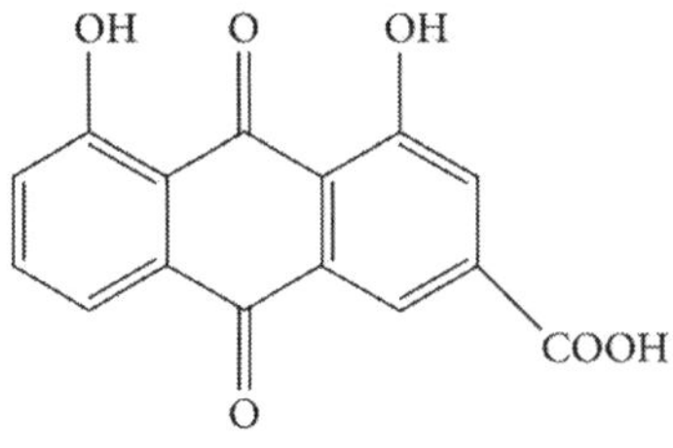

FIGURE 9.2 Chemical structure of rhein.

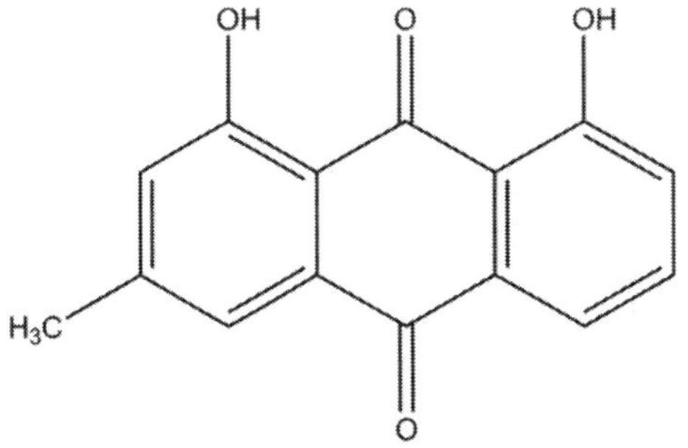

FIGURE 9.3 Chemical structure of chrysophanol.

9.6.1.2 Chrysophanol

Chrysophanol, also known as chrysophanic acid or 1,8-dihydroxy-3-methyl-anthraquinone, is a natural anthraquinone obtained from the rhubarb plant (Xie et al., 2019). It is a mono-nuclear compound having methyl substitution at C-3 and functions as building block for synthesis of different natural products (Su et al., 2020) (Figure 9.3).

9.6.1.3 Emodin

An anthranoid derivative, emodin (1,3,8-trihydroxy-6-methylanthraquinone), has antibacterial, antidiuretic, anti-cancerous, and powerful antioxidant properties (Singh et al., 2012). The findings reveal that the middle quinonoid moiety effectively isolates the two terminal rings from one another. As a result, the placements of the H-2 and H-4 signals in chrysophanol match those of H-7 and H-5 in physcion and emodin. C-1, C-2, C-3, and C-4 signals in chrysophanol are readily distinguishable from C-8, C-7, C-6, and C-5 signals in physcion and emodin. In chrysophanol, aloe emodin, and rhein, neither the H-5, H-6, and H-7 signals nor the C-5, C-6, C-7, and C-8 signals differ greatly in location. Tables 9.3 & 9.4 describe the chemical shifts and coupling of hydrogen and carbon atoms in different anthraquinones, respectively (Danielsen et al., 1992).

9.6.1.4 Aloe Emodin

A natural active component of rhubarb is aloe emodin (1, 8-Dihydroxy-3-(hydroxymethyl)-9, 10-anthraquinone) possessing antiviral, antimicrobial, and anticancer properties (Tabin et al., 2016). Also known as rhabarberone, aloe emodin is an aromatic polyketide and a secondary metabolite. The chemical formula of aloe emodin is $C_{15}H_{10}O_5$ with a molecular weight of 270.24 g/mol and a melting point of 221–223 °C (HMDB, 2022a) (Figure 9.4).

9.6.1.5 Physcion

Physcion, with the chemical formula $C_{16}H_{12}O_5$ and the molecular weight 284.267 g/mol, was identified in the methanol extracts of *R. emodi* by using GC-MS. It is present in both the rhizome and the

TABLE 9.3
^{1}H NMR Chemical Shifts and Coupling for Anthraquinones

Proton	Chrysophanol	Physcion	Emodin	Aloe-emodin	Rhein
H-2	7.22	6.68	6.59	7.29	7.77
H-4	7.55	7.36	7.12	7.69	8.14
H-5	7.71	7.62	7.49	7.72	7.75
H-6	7.80	_	_	7.80	7.84
H-7	7.38	7.07	7.16	7.38	7.41
OH-1	11.87	12.30	12.08	11.91	
OH-8	11.97	12.11	12,01	11.97	
OH-3	_	_	11.37	5.62	
CH_3	2.44	2.44	2.41	_	_
OCH_3	_	3.93	_	_	_
CH_2	_	_	_	4.63	_
H-2 (*m*)	0.82, 1.15	2.57	2.46, 0.35	1.66	1.47
H-4 (*m*)	0.49, 1.15	2.57	2.28, 0.35	1.66	1.47
H-5 (*m*)	1.31	0.55, 0.92	0.35, 0.53	1.16	1.10
H-5 (*o*)	7.55	_	_	7.63	7.54
H-6 (*o*)	7.55, 8.37	_	_	7.63, 8.29	7.54, 8.22
H-7 (*m*)	1.31	0.74, 0.92	0.35, 0.70	1.16	1.10
H-7 (*o*)	8.37	_	_	8.29	8.22
CH_3 (*m*)	0.82, 0.49	0.55, 0.74	0.53, 0.70	_	_

TABLE 9.4
^{13}C NMR Chemical Shifts for Anthraquinones*

Carbon	Chrysophanol	Physcion	Emodin	Aloe-emodin	Rhein
C-1	161.67	165.20	164.54	161.72	161.51
C-2	124.16	106.78	108.02	120.78	124.21
C-3	149.26	166.56	165.64	153.80	138.20
C-4	120.64	108.22	108.84	117.18	119.05
C-5	119.41	121.29	120.58	119.43	119.48
C-6	137.41	148.44	148.35	137.42	137.63
C-7	124.49	124.51	124.23	124.49	124.64
C-8	161.41	162.51	161.51	161.43	161.27
C-9	191.72	190.82	189.83	191.74	191.49
C-10	181.57	182.05	181.48	181.59	181.25
C-11	133.40	133.23	132.91	133.54	133.41
C-12	115.94	113.69	113.46	116.01	116.33
C-13	113.85	19.27	109.06	114.57	118.48
C-14	133.10	135.27	135.21	133.22	133.61
CH_3	21.71	22.15	21.60	_	_
OCH_3	_	56.07	_	_	_
CH_2	_	_	_	62.15	_
COOH	_	_	_	_	165.52

* Coupling constants (one bond) for 1,8-dihydroxyanthraquinone are as follows: C-2 (164.4 Hz), C-4 (167.9 Hz), C-5 (167.9 Hz), C-6 (164.6 Hz), C-7 (164.4 Hz)

Source: Danielsen et al., 1992.

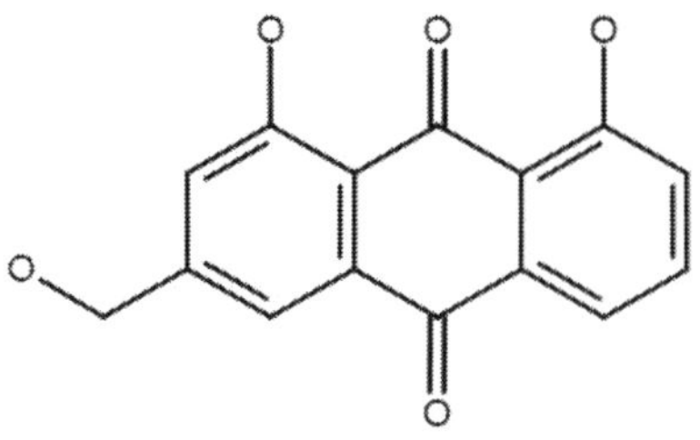

FIGURE 9.4 Chemical structure of aloe emodin.

Source: **Meier et al., 2017.**

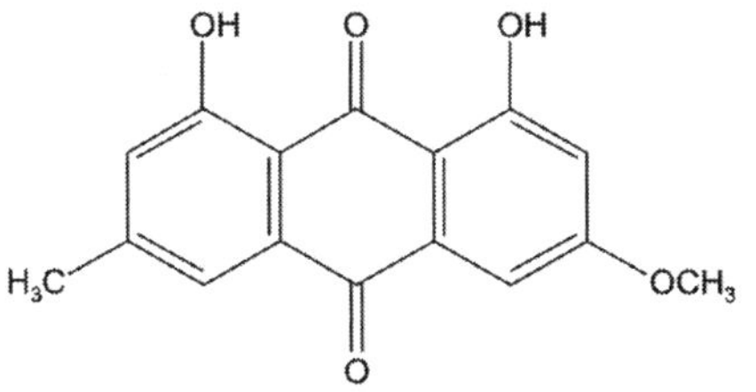

FIGURE 9.5 Chemical structure of Physcion.

fruit parts of the plant. Physcion can exists in three tautomeric forms with each tautomeric form forming two anions. Hence, depending on the pH, it can be present in 12 different forms, i.e. six mononanions, three dianions, and three non-ionized forms (Fain et al., 2005) (Figure 9.5).

9.6.2 Stilbene Glycosides

When the carbohydrate moiety is present in the stilbene skeleton by glycosidic linkage, this produces stilbene glycoside (HMDB, 2022b; HMDB, 2022c). Stilbene compounds have anti-bacterial, anti-fungal, anti-cancer, and anti-malarial properties (He et al., 2019). Rhaponticin and rhapontigenin are the two principal stilbene compounds present in rhubarb plant.

9.6.2.1 Rhaponticin

Rhaponticin, also known as rhapontin or trans-rhaponticin, is a stilbene compound with chemical name 3,3′,5-trihydroxy-4′-methoxystilbene 3-*O*-β-D-glucoside and chemical formula $C_{21}H_{24}O_9$ (Kolodziejczyk-Czepas, & Czepas, 2019; HMDB, 2022b). The molecular weight of compound is 420.4 g/mol (Kim et al., 2010). It is an aromatic heteromonocyclic compound and a secondary metabolite (HMDB, 2022b) (Figure 9.6).

9.6.2.2 Rhapontigenin

The biological active form of rhapontin is rhapontigenin (Figure 9.7), which is synthesized when rhapontin is hydrolyzed by glucosidase enzyme. It is an aromatic homomonocyclic compound having the molecular weight 258 g/mol (HMDB, 2022c; Kim et al., 2010). The IUPAC name of rhein is 5-[(Z)-2-(3-hydroxy-4-methoxyphenyl) ethenyl] benzene-1,3-diol having the chemical formula $C_{15}H_{14}O_4$. The compound has a melting point of 186–187 °C and contain 1,2-diphenylethylene moiety in its structure (HMDB, 2022c).

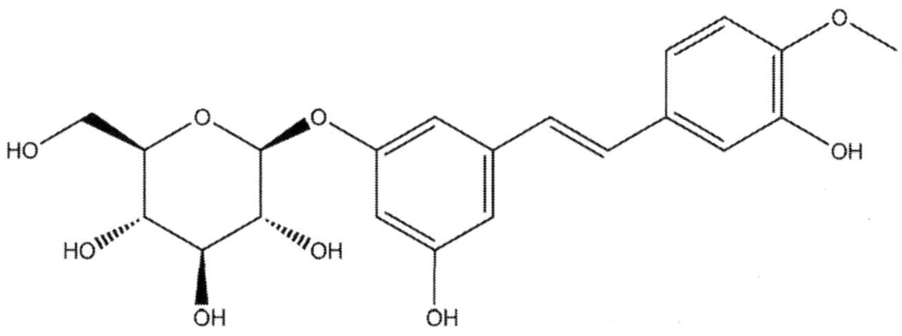

FIGURE 9.6 Chemical structure of Rhaponticin.

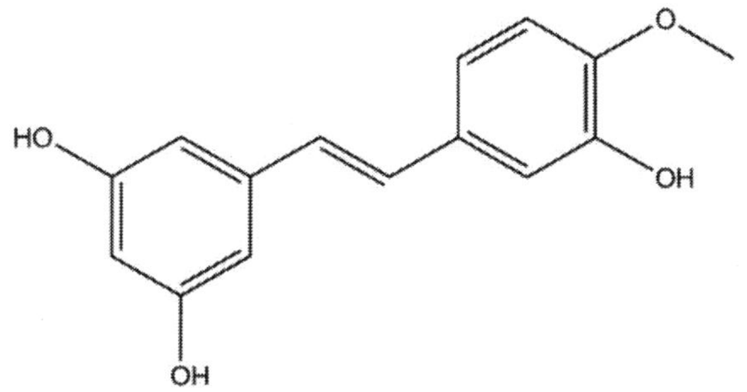

FIGURE 9.7 Chemical structure of rhapontigenin.

9.6.3 Other Classes of Bioactive Components Present in Rhubarb

Apart from anthraquinones, their glycosides and stilbene glycosides, anthrones, flavonoids and acylglucosides are also present in different rhubarb species. Cascaroside, 10-hydroxycascaroside, palmidin (A, B, and C), rendin (A, B, and C), revandchinone (1, 2, and 4), rhenoside (A, B, C, and D), sennoside (A, B, C, D, E, and F), etc. are among the major anthrones present in rhubarb possessing pharmacological properties In the case of flavonoids, the ring structure contains phenolic hydroxyl groups which are attached to it. Catechin, epicatechin-3-O-gallate, gallic acid, kaempferol-3-O-rhamnoside, etc. are among the major flavonoids present in the rhubarb species. Another bioactive component of importance is acylglucosides. These include gallic acid-3-*O*-β-*D*-glucopyranoside, lindleyin, coumaroyl-*O*-galloyl-glucose, 2-*O*-galloylsucrose, and 1-*O*-galloyl-6-*O*-cinnamoyl-β-*D*-glucose (He et al., 2019).

9.7 EXTRACTION AND ISOLATION TECHNIQUES

9.7.1 Conventional Extraction Techniques (Soxhlet Apparatus)

The solvent of choice is heated to reflux in the Soxhlet flask. The combination of solvent molecules and extracted bioactive components migrate into the bulk solvent from the matrix. The extraction chamber is continuously loaded with fresh solvent since the solvent is boiling away. This method is suitable for extraction in batches. It is repeated for a considerable time until all of the plant material has been used up. The last extract in the extractor flask is concentrated and augmented in target bioactive components.

Using a Soxhlet apparatus, rhubarb root (100 g) was incessantly extracted for 3 days using methanol (1:1) in the form of a fine powder. The methanol was vaporized at a low pressure, and the resultant concentration was hydrolyzed (250 ml of 7.5% HCl, 100 °C, 3h). Repeated extractions with diethyl ether were used to recover the anthraquinones from the cooled solution. The anthraquinones were isolated by preparative high-performance liquid chromatography using silica gel 60H, 15 μm (Merck) as the stationary phase in a laboratory-packed column (Jobin-Yvon Miniprep LC) (Danielsen et al., 1992).

Malik et al. (2010) used Soxhlet extractions to extract the bioactive components from dried and powdered rhizomes using different types of solvents (hexane, chloroform, and ethyl acetate). Under reduced pressure, the extracts were sifted and the solvents were extracted. The gradient elution method was used to apply concentrated extracts derived from various solvents to column chromatography over silica gel. After column purification with methanol–chloroform in the ratio 20:80 (v/v), chrysophanol glycoside and emodin glycoside were obtained from ethyl acetate extracts. Emodin was obtained from chloroform extract after column purification using ethyl acetate-hexane (30:70, v/v), whereas physcion and chrysophanol were obtained from hexane extract after column purification using ethyl acetate-hexane (5:95, v/v).

9.7.2 Ultrasonic Assisted Extraction

The use of ultrasonography is a cutting-edge 21st-century method. It is now widely researched in a variety of sectors, including cosmetics, medicines, chemistry, and food processing. The quantity of publications has expanded dramatically over the previous two decades (Jambrak et al., 2013). The ultrasonic frequencies utilized in chemistry range from 20 kHz to 10 MHz and the waves produce cavitation bubbles which implode, resulting in high-velocity inter-particle collisions, macro-turbulence and disturbance in the sample's micro-porous particles. Furthermore, the cavitation phenomena increase pressures and temperatures, enabling even more mass transfer. Microfractures and cell wall disintegration appear to promote extraction by enabling the entry of solvent into the tissue. In contrast to traditional maceration or extraction, this cell disruption releases the bioactive components like emodin, aloe-emodin and rhein more efficiently (Wang et al., 2008).

9.7.3 Microwave-assisted Extraction (MAE)

Microwave-assisted extraction (MAE) is a newer method, which combines microwaves with classic solvent extraction method. MAE is now a popular technique for the extraction of natural compounds from plants. Microwaves are electromagnetic fields that generate energy using frequencies that range from 300 MHz to 300 GHz to heat the solution. The instrument's design differs and may have an impact on the extraction process (Destandau & Michel, 2013). The degree of heating is determined by nature, specificity and polarity of solvent. Polar solvents such as water, having a high dielectric constant, are influenced by microwave energy, whereas non-polar solvents, such as heptane, lack the physical qualities to heat up under microwave energy. As a result, non-polar solvents are known as microwave-transparent solvents.

Arvindekar and Laddha (2016) established the ability of MAE to produce significant amounts of anthraquinones, namely five 1,8-dihydroxyanthraquinones (rhein, emodin, aloe emodin, physcion and chrysophanol) isolated from rhubarb. The optimal solvent, as assessed by an experimental setup, was a 0.255 M NaOH solution. Based on the result, various solvent options for MAE extraction can be provided, depending on the target molecules.

9.7.4 Pressurized Liquid Extraction

While using atmospheric pressure, it is impossible to raise the temperature over the boiling point of the solvent due to solvent evaporation. Enhanced Solvent Extraction (ESE)/Accelerated Solvent Extraction (ASE)/Pressurized Liquid Extraction (PLE)/High-Pressure Solvent Extraction (HSPE) uses the aqueous and organic solvents in the same way as other standard liquid extraction methods, but at a higher pressure. This allows the temperature to rise above their atmospheric pressure boiling point. Solvents such as methanol, water, heptane, or acetone are often employed in PLE at the temperature range of 75 to 150 °C. This increase in temperature improves the process of extraction by enhancing the rate kinetics by affecting viscosity and surface tension of the solvent thereby increasing solubility and mass transfer rate resulting in a faster extraction (Luthria et al., 2019). Rhein, physcion, emodin, aloe emodin, and chrysophanol from rhubarb can be extracted by pressurized liquid extraction using methanol as a solvent (Gong et al., 2005).

9.7.5 Supercritical Fluid Extraction

In recent times, there has been an expansion in the number of reports relating to supercritical fluid extraction (SFE) for active compounds derived from food products, vegetable substrates, and plants. Supercritical carbon dioxide (Sc-CO_2) is the most commonly used solvent in SFE. This is an appealing candidate because of its high diffusivity mixed with its readily adjustable solvent strength via temperature and pressure variations. To enhance extraction yield and selectivity, a tiny amount of an organic solvent (typically referred to as a "modifier") can be added (Duval et al., 2016). One of the chief factors for the interest in SFE applications is the possibility to use lower temperatures in comparison to PFE, which allows for the extraction of chemicals that may be oxidized or damaged when high temperatures are applied. Lo et al. (2012) used the SFE method for the extraction of chrysophanol from *Rheum palmatum*. The results show that a higher pressure extracted more chrysophanol and that the extraction yield was better when the pressure exceeded 210 atm. The data demonstrate that as the extraction time surpassed 30 minutes, the extraction rate fell down. Thus, the best extraction conditions in this experiment were 30 minutes, 85 °C and 210 atm.

9.8 APPLICATION OF ACTIVE INGREDIENTS

Compounds isolated from *Rheum emodi* have been shown to have antioxidant, anti-diabetic, antibacterial, antifungal, cytotoxic, hepatoprotective, and nephroprotective properties.

9.8.1 Antibacterial and Antifungal Action

Because many micro-organisms are immune to a range of antibiotics, herbal medicines have been researched for their antibacterial capabilities. Commercial rhubarb aloe emodin, rhein, and emodin all have a considerable antimicrobial effect on four strains of *S. aureus*, which are resistant to methicillin, and one strain of *S. aureus*, which is methicillin-sensitive. Aloe emodin has antimicrobial properties against these pathogens, with a minimum inhibitory concentration (MIC) of 2 µg/ml. Rhein has a MIC value of 128 µg/ml against *E. coli* K12 (Zargar et al., 2011). Physcion, chrysophanol, rhein, and aloe-emodin, all isolated from the underground stem of *Rheum emodi*, were shown to be antifungal against *Trichophyton mentagrophytes*, *Aspergillus fumigatus*, *Candida albicans*, and *Cryptococcus neoformans* with a MIC of 25–250 µg/ml, which causes skin diseases (Agarwal et al., 2000). Revandchinone-1 and revandchinone-2 (oxanthrone esters), revandchinone-3 (anthraquinone ether), and revandchinone-4 (oxanthrone ether) were identified from Himalayan rhubarb rhizomes. Revandchinone-1 and 3 demonstrated lower levels of antibacterial activity than revandchinone-4. Revandchinone-4 was discovered to have good antiseptic activity against Gram-positive bacteria (*Staphylococcus aureus*, *Bacillus sphaericus*, and *Bacillus subtilis*) using penicillin G as a control and Gram-negative bacteria (*Pseudomonas aeruginosa*, *Klebsiella aerogenes*, and *Chromobacterium violaceum*) using streptomycin as a control. When clotrimazole is employed as a control, revandchinone-1, 3, and 4 exhibit mild antifungal efficacy against *Rhizopus oryzae* and *Aspergillus niger* (Babu et al., 2003). The ethanolic extract of *Rheum emodi* was discovered to have promising action against several strains of Helicobacter pylori (Ibrahim et al., 2006).

It was reported that the five hydroxyanthraquinones extracted from rhubarb have antibacterial action in the following order: rhein>emodin>aloe-emodin>chrysophanol>physicion. The -COOH, -OH, and -CH_2OH functional groups on the phenyl ring in hydroxyanthraquinones may boost antibacterial action (Wang et al., 2010). Research demonstrated that rhubarb (*Rheum palmatum*) root extract has a high level of antibacterial action, inhibiting the development of *Porphyromonas gingivalis*, an oral pathogen, with a MIC of 0.32 mg/m (Nanasombat et al., 2018). It is important to note that rhubarb methanol extract inhibits *E. coli*, *L. monocytogenes*, *K. pneumoniae*, and *B. subtilis* more effectively than typical antibiotics. Furthermore, rhubarb inhibits the growth of *S. aureus* by causing damage to the microbial cellular structure and affecting the cell membrane permeability, as shown in Figure 9.8 (Xiang et al., 2020). Bacterial biofilm production is a major cause of bacterial multidrug resistance. *S. suis* is a major swine infectious agent capable of causing chronic sickness in humans. Rhubarb has been proven to inhibit biofilm production by decreasing the expression of transduction channels and boosting the levels of DNA-binding protein and transcriptional regulatory components (Ding et al., 2017). Furthermore, emodin from rhubarb can prevent *Pseudomonas aeruginosa* and *Stenotrophomonas maltophilia* biofilm production, while aloe emodin can cause bacterial membrane damage through the coupling of phosphatidylethanolamine and phosphatidylglycerol (Xiang et al., 2020). In addition, investigations have also shown that rhubarb peony beverage can re-establish the Th17/Treg (T helper 17/regulatory T cells) balance in the human intestine by increasing the amount of *Firmicutes* and *Actinobacteria* while decreasing the number of *Proteobacteria* and *Bacteroidetes* (Luo et al., 2019). Furthermore, by regulating gut flora, rhubarb can minimize hepatotoxicity resulting from severe alcohol use as well as influencing the content of branched-chain amino acids and phenolics in faeces (Neyrinck et al., 2017).

9.8.2 Regulation of Gastrointestinal Activity

For thousands of years, Chinese doctors have utilized rhubarb as a first-line therapy for constipation as the plant has definite purgative properties. The strength and efficacy of Chinese herbs can be altered during processing. The purgative action of raw rhubarb differs from that of processed rhubarb. The most biologically important components of the cathartic effect, stimulating the stomach and inducing diarrhea, are assumed to be combination anthraquinones, comprising "sennosides,"

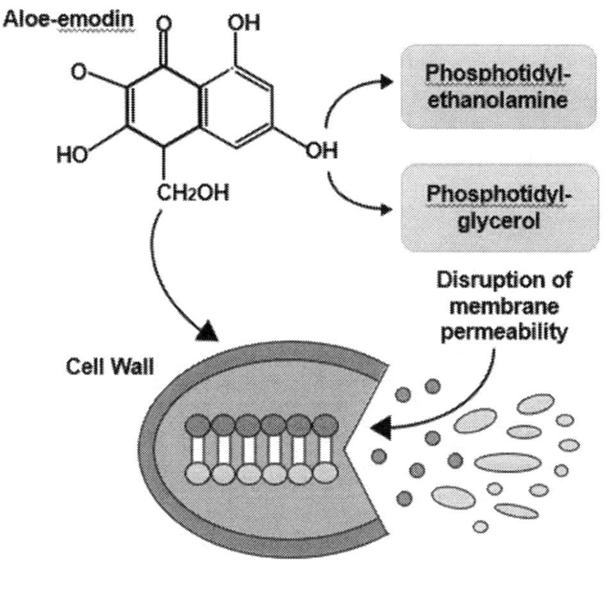

FIGURE 9.8 Antibacterial activity of rhubarb.

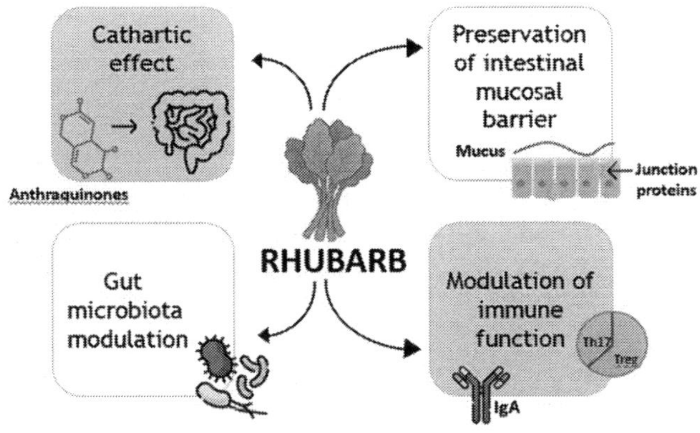

FIGURE 9.9 Gastrointestinal function regulation by rhubarb.

"rheinosides," and "anthraquinone aglyconesa. In the intestinal canal, combined anthraquinone is converted into a free metabolite, which has a laxative effect (Cao et al., 2017). Anthraquinone derivatives can activate the "submucosal nerve plexus," inhibit potassium ion channel conductivity on the cytoplasmic membrane, and increase the electrical reactivity of gastrointestinal smooth muscles. Furthermore, rhubarb tannins have been demonstrated to have antidiarrheal qualities, with the actual mechanism perhaps being an enhancement in protein coagulation (Xiang et al., 2020).

The epidermal barrier, immune barrier, gastrointestinal microbiota barrier, and chemical barrier comprise the intestinal mucosal barrier, which serves to separate the contents of the intestinal cavity and prevent dangerous antigens from entering. Many human and animal studies have shown that rhubarb can preserve the intestinal mucosal barrier, and the method by which it does so is complicated. Xiang et al. (2020) proposed numerous possible methods (Figure 9.9). They are:

1. Rhubarb can help to maintain the gut microbiota balance, restore disrupted flora, and prevent gut microbiota from shifting;
2. Rhubarb reduces mucosal injury by raising the production of junction proteins, which modulates intestinal permeability;
3. Rhubarb can regulate the immune function of the gastrointestinal tract. Rhubarb, in brief, lowers excessive intrinsic immune-mediated inflammatory reactions and intestinal injury by suppressing the activity of the inflammasome nucleotide-binding oligomerization domain-like receptor protein 3". Furthermore, anthraquinones from rhubarb can increase the production of "secretory immunoglobulin A"(SIgA) and maintain the T1/T2 (transitional B cells) and Th17/Treg balance. Furthermore, rhubarb has the capacity to alter intestinal lymphocyte homing, improve intestinal mucosal immunological function, and relieve ulcerative colitis;
4. Rhubarb promotes the growth of "goblet cells" in the intestinal mucosa, which generate a lot of mucus and assist in reinforcing the intestinal mucosal barrier by forming a mucosal layer (Xiang et al., 2020).

9.8.3 Anti-inflammatory Activity

Rhubarb has anti-inflammatory properties including heat clearance and detoxification. It is used to cure inflammations produced by a range of factors. The present focus of rhubarb anti-inflammatory mechanism research is on its constituents such as "emodin, "rhein, and "aloe emodin" (Cao et al., 2017). Rhaponticin, and its aglycone rhapontigenin, had anti-inflammatory effects by inhibiting the activity of cyclooxygenase, lipoxygenase, and hyaluronoglucosaminidase, as well as altering various pro-inflammatory responses (Kolodziejczyk-Czepas & Czepas, 2019). Other anti-inflammatory rhubarb components include "emodin, "rhein, "chrysophanol, and "aloe-emodin, in addition to "rhaponticin" (Xiang et al., 2020).

An inflammatory model was created in one investigation by the activation of macrophage cells with lipopolysaccharide and then addressing these using rhein. The findings indicate that rhein has an anti-inflammatory effect (Wen et al., 2020). Jiang et al. (2013) found that the intervention impact of "emodin lipid nano-microbubble" dramatically reduced the protein expressions of phosphorylated P38, phosphorylated ERK, phosphorylated JNK, and the secretion amounts of inflammatory cytokines such as tumor necrosis factor-α, interleukin-1, and interleukin-6. Hu et al. (2014) revealed that "aloe emodin" was also a therapeutic rhubarb constituent with anti-inflammatory properties. By blocking the nuclear factor kappa-B cells, mitogen-activated protein kinase, and phosphoinositide 3-kinase pathways, "aloe emodin" at 0.01 mM or 0.02 mM levels was able to reduce the production of pro-inflammatory cytokines in the LPS-induced macrophages. Feng et al. (2012) administered rhubarb decoction (150 mg/kg) orally to rats suffering from acute pancreatitis (AP) and discovered that rhubarb likely alleviated AP by blocking MAPK activation.

9.8.4 Anti-fibrotic Activity

Fibrosis is a common side effect of acute hepatic disease, chronic renal illness, interstitial lung disorders, and other severe illnesses. Fibrosis is caused by parenchymal cell degeneration and necrosis, which can stimulate macrophages to produce pro-inflammatory mediators. In the presence of these active factors, "static extracellular matrix" (ECM)-producing cells differentiate into "myofibroblasts, which can accelerate ECM synthesis and eventually lead to fibrosis. Rhubarb reduces fibrosis via blocking or reversing parenchymal cell necrosis, lowering monocyte activation and migration to injured tissues, suppressing fibroblast activation, relieving collagen deposition, and increasing collagen breakdown. The immunological response has been linked to the development and recurrence of fibrosis. On the one hand, inflammation causes harm to parenchymal cells while also activating matched macrophages and releasing active substances that promote fibrosis development and progression as shown in Figure 9.10 (Xiang et al., 2020).

Idiopathic pulmonary fibrosis (IPF) can significantly impair lung function and lead to mortality. Emodin, a component of rhubarb, has been shown to protect against lung disturbances. Tian et al. (2018) discovered that emodin (20 mg/kg) protected rats from lung fibrosis by decreasing NF-κB activation and reducing the production of pro-inflammatory factors such as tumor necrosis factor-α, interleukin-6, and interleukin-1β.

9.8.5 Antitumor Activity

Many researchers have discovered that the anthraquinones found in rhubarb can prevent the development and multiplication of some cancer cells. Emodin, aloe emodin, and rhein are rhubarb compounds with anti-cancer properties in a variety of human cancer cells, comprising cell cycle arrest and death (Chen et al., 2010). The unbound anthraquinones of rhubarb were studied to develop and characterize the multiple molecular pathways underpinning rhubarb's anti-cancer action, with emodin being the most often-studied ingredient among the active components. First, "emodin" reduced tumor cell growth and promoted apoptosis. Second, it has the potential to block the formation of tumor blood vessels. Moreover, rhubarb reduced the spread of cancer cells (Cao et al., 2017). Modulating the Rac1/ROS/MAPK/AP-1 pathway and suppressing matrix metalloproteinases might offer a method for the treatment of ovarian cancer. Rhein suppresses matrix metalloproteinase synthesis in human ovarian cancer cells by altering the Rac1/ROS/MAPK/AP-1 pathway. Its inhibitory effects were connected to a rise in the production of metalloproteinase tissue inhibitors NM23-H1, TIMP-1, and TIMP-2. Mechanistic investigations demonstrated that rhein suppresses the production of reactive oxygen species and lowers NADPH oxidase activity (Zhou et al., 2017). The invasion of human tongue cancer SCC-4 cells was prevented by emodin, aloe emodin, and rhein by inhibiting

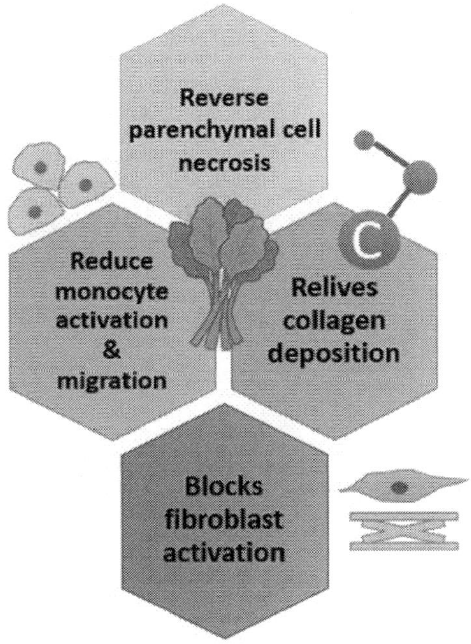

FIGURE 9.10 Anti-fibrosis activity of rhubarb.

Source: **Xiang et al., 2020.**

the activity of matrix metalloproteinase-2 (MMP-2), and by regulating the protein levels and the gene expression of MMP-9. MMP-9 (gelatinase-B) plays a key role in tumor migration, invasion, and metastasis in a variety of human malignant tumors (Chen et al., 2010).

According to Liu et al. (2011), emodin suppresses the growth of a human pancreatic cancer cell line (Panc-1), and this inhibitory activity is connected to the concentration of emodin administered. He et al. (2009) used a quantitative endogenous alkaline phosphatase test and staining assay to analyze the 95% ethanol extract and four subsequent fractions of rhubarb root and five anthraquinones extract on zebra fish. They discovered that anthraquinones with acidic or polar, hydrophilic alterations at C-6 or C-3 sites strongly reduced angiogenesis. These findings imply that rhubarb chemical monomers have a wide range of effects on various biological systems connected to cancer. In ovarian cancer, lung cancer, nephroblastoma, and stomach cancer, the Wnt/β-catenin signaling pathway is required for cell proliferation and invasion. Rhubarb has been found in previous studies to decrease tumor growth by interfering with the Wnt signaling pathway and increasing β-catenin protein degradation. It is worth mentioning that rhubarb can help prevent stomach cancer, ovarian cancer, and lung cancer from spreading. It is unknown if this is connected to rhubarb's influence on the Wnt/β-catenin signaling pathway (Hong et al., 2014; Xiang et al., 2020). The anti-tumor activity of rhubarb is shown in Figure 9.11.

9.8.6 Hepatoprotective

Rhubarb research has recently centered on its hepatoprotective effects. Rhubarb tannins and anthraquinones have different effects on liver protection and damage. Anthraquinones decrease liver fibrosis and cell damage more than tannins, whereas in large doses tannins may damage the liver. Rhubarb and its free anthraquinones are also being researched for their potential use in the treatment of liver fibrosis, intrahepatic cholestasis, hepatic encephalopathy, and other disorders (Cao et al., 2017).

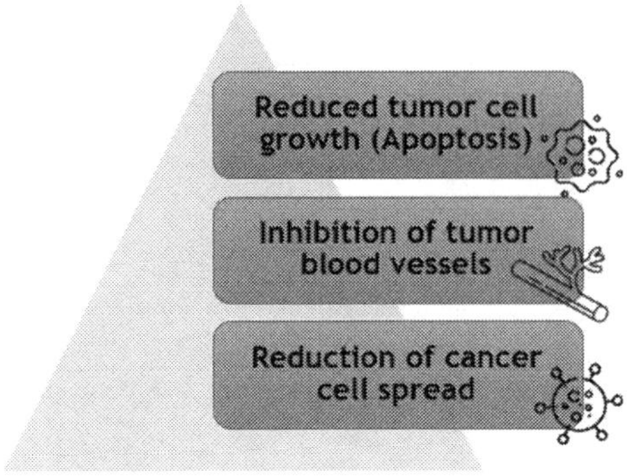

Figure 9.11 Anti-tumor activity of rhubarb.

Source: Cao et al., 2017.

9.8.7 Cardiovascular System

Rhubarb not only improves hemorheology but also processes the impact of hemostasis, reducing bleeding without causing any obstruction. It has been established that chrysophanol and tannin have a hemostasis effect. In mice, chrysophanol can shorten clotting time, raise platelet count, and accelerate blood coagulation, while tannin induces localized vascular constriction through a convergent effect. It is also believed that inhibiting Na+-K+-ATPase causes an increase in plasma osmotic pressure, driving water into the blood arteries, increasing blood volume, and assisting in the treatment of microcirculation disorders (Cao et al., 2017). Rhubarb's effects on blood lipids and antioxidation in hyperlipidemic rats were investigated. For 28 days, *Rheum tanguticum* was administered intragastrically in two doses (3.0 g/kg and 1.0 g/kg). *Rheum tanguticum* therapy was shown to diminish blood levels of low-density lipoprotein, total cholesterol, triglycerides, and malonaldehyde. They may also increase high-density lipoprotein levels, increase the activity of superoxide dismutase and glutathione peroxidase, and reduce blood viscosity (Wang et al., 2015).

Du et al. (2015) explored the pathway of rhubarb's anti-hyperlipidemia and revealed that it can impede clotting factors and increase blood circulation, hinder inflammatory cytokine production and preserve fat metabolism balance, enhance cholesterol excretion, prevent cholesterol, and triglyceride synthesis, and have other lipid-lowering impacts. Chen et al. (2017) observed that chrysophanol can diminish the blood cholesterol levels in zebra fish after a high-fat diet, implying that the mechanism may interfere with lipid absorption in the intestinal tract.

9.8.8 Urinary System

The kidney-protective effect of rhubarb is attributable to a combination of many pharmacological effects. Rhubarb, which possesses anti-inflammatory, diuretic, immune-boosting, metabolism-regulating, anti-bacterial, and other qualities, among other things, speeds up waste removal, stimulates protein synthesis, and maintains damaged kidney tissue. Rhubarb is often used in clinical research to address kidney disorders such as diabetic nephropathy and chronic functional failure. In recent years, the renal protective effect of rhubarb and its active components, notably emodin and rhein, has received a lot of attention. However, consensus on the treatment mechanism has yet to be reached. Rhein lowers anti-oxidative stress, reduces urine protein excretion, and improves glycolipid metabolism, which might be one of the techniques for avoiding and treating diabetic nephropathy. To treat renal problems, emodin primarily includes cells and cytokines such as connective tissue growth factor, transforming growth factor-β, thrombospondin-1, and monocyte chemoattractant protein-1 (Cao et al., 2017).

9.9 CLINICAL APPLICATIONS

The most prevalent causes of cell damage are inflammation, infection, and oxidative stress. Rhubarb, due to its vast pharmacological actions, can interfere with the development of a variety of disorders. The bulk of research have focused on the effectiveness of rhubarb in the treatment of chronic renal failure (CRF), sepsis, severe acute pancreatitis (SAP), and other conditions. Most of the clinical trials performed with rhubarb and its active constituents use randomized, double-blind, placebo-controlled trials as the study design, and the results revealed oxygenation function in acute respiratory distress syndrome (ARDS) patients, intestinal permeability in sepsis patients, feeding tolerance in critically ill patients etc. have been improved remarkably through the use of rhubarb. This plant, in particular, has had a considerable impact on SAP sufferers. It has been effective in relieving distension, abdominal pain and other symptoms, and also shortens the time period of the SAP patients in the ICU and hospital, by improving gastrointestinal function, promoting pancreatic exudate absorption, mitigating SAP-related kidney and liver damage, and reducing systemic inflammation (Xiang et al., 2020).

9.9.1 Constipation

Rhubarb has a laxative effect and is utilized for the treatment of constipation caused by several factors (Cao et al., 2017). A constipation model developed by Chen et al. (2019) with dyspepsia-induced gastrointestinal accumulated heat for investigating rhubarb's efficacy at six doses ranging from 0.135 to 8.1 g $kg^{-1}d^{-1}$ to elucidate two major effects of rhubarb, "Removing accumulation with purgation" (E1) and "Clearing heat and purging fire" (E2), both of which were consistent with traditional Chinese medicine. The study showed that low dosages of rhubarb had little or no impact, and that while rhubarb liquid at large concentrations boosted therapeutic efficiency, it also had specific negative effects on the body. In another study, by Gao et al. (2021), it was found that rhubarb extract, by attracting mast cells to the colon, can stimulate colonic mucus secretion and release acetylcholine through the stimulation of the submucosae neural system. Rhubarb extract generates endoplasmic reticulum stress in colonic mucosal goblet cells and introduces changes in intestinal flora and short-chain fatty acids, ultimately alleviating constipation.

In a similar research, pathological sections and UPLC-Q-TOF/MSE were used to investigate the fire-purging and laxative effects of six rhubarb dosages (0.135, 0.27, 0.81, 1.35, 4.05, and 8.1 g/kg). The findings revealed that significant lesions existed in the colon and stomach of the mice model. Conditions improved to some extent in rhubarb-treated groups. The threshold for the effective dosage of rhubarb (EC_{20}–EC_{80} range) in rat serum was found to be 0.31–4.5 g/kg and 0.29–2.1 g/kg, corresponding to 3.44–50.00 g and 3.22–23.33 g in the clinic, respectively, in the faeces. According to faecal metabolomics and integrated serum, rhubarb doses ranging from 0.31 to 2.1 g/kg are safe and can be effectively used in constipation therapy (Chen et al., 2020a). Likewise, using loperamide, the laxative effects of rhubarb in constipation in rats were studied. According to the findings, rhubarb can greatly increase faecal water content and the frequency of defecation, decrease gastrointestinal transit time, preserve the colon mucus layer and ameliorate gastrointestinal hormone disruption. According to 16S rRNA gene sequencing, rhubarb might help constipated rats in the relief of the problem of intestinal microbiota. Following rhubarb therapy, beneficial bacteria, such as *Limosilalactobacillus*, *Ligilactobacillus*, and *Prevotellaceae* UCG-001, for example, increased greatly, whereas a significant decrease was observed in pathogens such as *Escherichia Shigella*. After rhubarb treatment, metabolites such as cholic acid, chenodeoxycholic acid, α-linolenic acid, and prostaglandin F2α increased significantly in rats suffering from constipation; however, metabolites such as 10-hydroxystearic acid, calcidiol, and lithocholic acid were significantly reduced in such animals (Yang et al., 2022). In order to assess the safety and effectiveness of rhubarb navel plasters for chronic constipation patients, Wei et al. (2021) performed a prospective multi-centric randomized controlled experiment to determine if rhubarb navel plaster administration throughout an 8-day treatment period using the Bristol Stool Scale (BSS), 24 h defecating frequency and Cleveland Constipation Score (CCS). Results showed significantly improved bowel function. The findings imply that rhubarb navel plasters are a realistic, safe, and effective therapy option for people suffering from persistent constipation. In the clinic, rhubarb is frequently used to treat patients by acupoint treatment rather than oral ingestion (Cao et al., 2017).

9.9.2 Severe Acute Pancreatitis

SAP is a prevalent clinical acute abdominal condition having a fatality rate ranging from 20% to 30%. Numerous clinical investigations have demonstrated that rhubarb enema can lower high endotoxin levels and sensory C-reactive protein (CRP), serum inflammatory cytokines, as well as restore intestinal mucosal barrier integrity and relieve systemic inflammatory stress in SAP patients. Furthermore, in SAP patients, the symptoms of pancreatic leakage can be reduced by giving hot compresses with rhubarb. Meanwhile, gastrointestinal retention and paralysis in SAP patients can be treated through nasal injections of rhubarb. Furthermore, early stages of SAP include disorders such as microthrombus formation, blood stasis, capillary ischemia, and other microcirculation disorders and rhubarb can help in reversing the decline in pancreatic blood flow and minimizing pancreatic bleeding, possibly because of its blood circulation promotion, anti-inflammatory, and blood stasis removal effects (Xiang et al., 2020).

In a study, two intraperitoneal injections of cerulein with a 5-hour constraint water-immersion stress generated severe acute pancreatitis in rats. Before the first cerulein injection, oral dosage of 75–150 mg/kg of rhubarb was given. The researchers observed the serum amylase levels, degree of pancreatic edema, local pancreatic blood flow (PBF) and other histological changes. According to the results, control group showed diffuse hemorrhage and severe edema in the pancreas, as well as an increase in pancreatic wet weight and serum amylase, whereas rats treated with 150 mg/kg rhubarb showed a significant reduction in the polymorphonuclear neutrophil (PMN) infiltration in the pancreas and necrosis. The study indicated that rhubarb can protect against SAP by suppressing pancreatic inflammation, increasing pancreatic microcirculation, and changing exocrine secretion (Zhao et al., 2004). The combination of early enteral nutrition and rhubarb may improve enteral nutrition efficacy, reduce systemic inflammation and disease severity, and improve gastrointestinal function, resulting in improved treatment success in SAP patients (Chen et al., 2020b; Wan et al., 2014).

Another study looked at how rhubarb affected the toll-like receptors (TLRs) and intestinal flora in rats with SAP. For the study, seven male Sprague-Dawley rats were assigned randomly to each group which received a sham-operated surgical model of SAP without or with postoperative rhubarb therapy. During the postoperative 24 hours, rats in the rhubarb group were tube-fed with a 10% rhubarb decoction (1 m/200 g) every 8 hours. In the case of SAP patients without rhubarb groups, levels of TLR2 and TLR4 messenger RNA expression in intestinal mucosa were considerably greater than in the SAP patients with rhubarb groups. The number of intestinal bifidobacteria and lactobacilli was considerably higher in the control group than in the SAP without rhubarb group, although the intestinal *Escherichia coli* amount was comparatively larger in the control group than in the rhubarb treatment group. Rhubarb may preserve the intestinal mucosal barrier in rats with SAP, by controlling intestinal flora and reducing the intestinal inflammatory response (Yao et al., 2015).

A study was carried out to look into the effects of free total rhubarb anthraquinones (FTRAs) on the prevention of intestinal damage and the alteration of intestinal barrier function in rats with SAP through regulation of the intestinal immune function. A solution of 3.5% sodium taurocholate was injected to the rat model of SAP. They were then treated immediately and, 12 hours later, with FTRAs (36 or 72 mg/kg) or normal saline in the case of the control. FTRAs dramatically reduced the levels of endotoxin, TNF-α, interleukin 1, and nitric oxide in the blood, as well as TNF-α, IL-1 and intestinal damage and protein extravasation in the intestinal tissues of SAP rats. Furthermore,

FTRAs showed significant decrease in the apoptosis-associated speck-like protein containing a CARD domain, expressions of nucleotide-binding oligomerization domain-like receptors 3 and casepase-1, as well as the Th1/Th2 ratio and the number of Tregs in the SAP rats' intestinal tissues and/or mesenteric lymph node cells, while significantly increasing the expression of secretary immunoglobulin A (Xiong et al., 2018).

9.9.3 Sepsis

Sepsis is a systemic inflammatory response syndrome (SIRS) induced by infection, specifically acting on the gastrointestinal tract in the promotion of SIRS following infection. Patients with sepsis frequently have immunological dysfunction and impaired coagulation function (Xiang et al., 2020). Rhubarb may be useful in minimizing sepsis-induced damage to the gastrointestinal tract, lung, and liver, and its potential mechanisms include lowering oxidative stress and inflammation, improving microcirculatory disruption, and preserving immunological balance (Lai et al., 2015). The combination of ulinastatin (UTI), a protease inhibitor, and rhubarb have both been shown to be beneficial in the treatment of sepsis (Meng et al., 2020). A study by Wang et al. (2017) showed that rhubarb monomer has been found effective in reducing mucosal damage in the treatment of sepsis by increasing the translation, transcription, and expression of junction proteins. These findings imply that rhubarb, due to its anticoagulation, anti-inflammation, prevention of bacterial gastrointestinal protection and toxin translocation, and gastrointestinal protection effects may be employed in the therapeutic treatment of sepsis. Moreover, rhubarb can be said to boost T cell subsets, which indicates that the modulation of immune function has the potential to contribute to the mechanism of rhubarb in treatment of sepsis patients (Xiang et al., 2020).

9.9.4 Diabetic Nephropathy and Chronic Renal Failure

Rhubarb is a prominent traditional Chinese herb found to be useful in the treatment of chronic renal failure (CRF) and diabetic nephropathy (DN). According to a significant number of linked literatures, rhubarb was frequently utilized to treat diabetic nephropathy when used in combination with other medications. Rhubarb could improve the condition of patients suffering from diabetic nephropathy by lowering blood lipids, reducing the excretion of urinary protein, regulating the abnormal expression of TGF-β1, MMP-2, MMP-9 and MCP-1 in the blood, improving the renal function of diabetic nephropathy patients, and inhibiting fibrosis process and renal inflammation (Xiong, 2012). Renal fibrosis is a frequent disease of CRF, which affects the glomerulus and renal interstitium. The early symptoms of CRF can be alleviated by rhubarb and slow down the onset of renal failure. Potential processes consist of renal fibrosis inhibition, toxin excretion promotion, metabolic disorder recovery, oxidative stress damage, and renal cell protection from excessive inflammation. In the treatment of CRF, rhubarb is mostly supplied through retention enema, and the dose depends on the number of times the patient defecates each day (Xiang et al., 2020).

9.10 CONCLUSION

The chapter discussed the active ingredients, bioactive components, and nutritional and medicinal values of *Rheum emodi*. Due to its ethnomedicinal properties, it has been found helpful in treatment of gastrointestinal and digestive ailments, diarrhea, blood purification, stomach problems, menstrual problems etc. and a considerable part of global population continues its use as a form of traditional medicine and reason being the presence of phytochemicals and antioxidants . In addition to from medicinal uses, rhubarb is also used for culinary and industrial application such as dye making, syrup, and liquor manufacture etc. Different methods for the extraction of active components of rhubarb were also discussed, such as Soxhlet extraction, microwave-assisted extraction, supercritical fluid extraction, and pressurized liquid extraction. The active ingredients of rhubarb exhibit anti-tumor, anti-fibrotic, anti-inflammatory, and anti-bacterial activities. Clinical studies reveal that the phytochemicals found in rhubarb has been found helpful in dealing with a range of medical conditions: preventing chronic renal failure, constipation, and severe acute pancreatitis, maintaining a healthy cardiovascular system, and much more. With all the studies carried out to date, it can be said that rhubarb is undoubtedly a wondrous plant, which has many potential benefits for human health. Apart from medicinal use, its utilization in the form of health supplements or functional foods can be explored further.

REFERENCES

Agarwal, S. K., Singh, S. S., Verma, S., & Kumar, S. (2000). Antifungal activity of anthraquinone derivatives from *Rheum emodi*. *Journal of Ethnopharmacology*, *72*(1–2), 43–46.

Arvindekar, A. U., & Laddha, K. S. (2016). An efficient microwave-assisted extraction of anthraquinones from *Rheum emodi*: Optimisation using RSM, UV and HPLC analysis and antioxidant studies. *Industrial Crops and Products*, *83*, 587–595.

Babu, K. S., Srinivas, P. V., Praveen, B., Kishaore, K. H., Murthy, U. S., & Rao, J. M. (2003). Antimicrobial constituents from the rhizomes of *Rheum emodi*. *Phytochemistry*, *62*(2), 203–207.

Bezzi, A. (1987). Proceedings of the conference on officinal plants. *Trento*, 9(10), 423–502.

Cao, Y. J., Pu, Z. J., Tang, Y. P., Shen, J., Chen, Y. Y., Kang, A., Zhou, G. S., & Duan, J. A. (2017). Advances in bio-active constituents, pharmacology and clinical applications of rhubarb. *Chinese Medicine*, *1236*.

Chen, J. Q., Chen, Y. Y., Tao, H. J., Pu, Z. J., Shi, X. Q., Zhang, J., Tan, Y. J., Yue, S. J., Zhou, G. S., Shang, E. X., Tang, Y. P., & Duan, J. A. (2020a). An integrated metabolomics strategy to reveal dose-effect relationship and therapeutic mechanisms of different efficacy of rhubarb in constipation rats. *Journal of Pharmaceutical and Biomedical Analysis*, *177*, 112837.

Chen, J. Q., Li, D. W., Chen, Y. Y., Tao, H. J., Pu, Z. J., Zhang, J., Tan, Y. J., Shi, X. Q., Yue, S. J., Zhou, G. S., Tang, Y. P., & Duan, J. A. (2019). Elucidating dosage-effect relationship of different efficacy of rhubarb in constipation model rats by factor analysis. *Journal of Ethnopharmacology*, *238*, 111868.

Chen, K., Wang, C. Q., Fan, Y. Q., Han, Z. H., Wang, Y., Gao, L., & Zeng, H.-S. (2017). Lipid-lowering effect of seven traditional Chinese medicine monomers in zebrafish system. *Sheng Li Xue Bao:[Acta Physiologica Sinica]*, *69*(1), 55–60.

Chen, X., Yang, K., Jing, G., Yang, J., & Li, K. (2020b). Meta-analysis of efficacy of rhubarb combined with early enteral nutrition for the treatment of severe acute pancreatitis. *Journal of Parenteral and Enteral Nutrition*, *44*(6), 1066–1078.

Chen, Y. Y., Chiang, S. Y., Lin, J. G., Ma, Y. S., Liao, C. L., Weng, S. W., Lai, T. Y., & Chung, J. G. (2010). Emodin, aloe-emodin and rhein inhibit migration and invasion in human tongue cancer SCC-4 cells through the inhibition of gene expression of matrix metalloproteinase-9. *International Journal of Oncology*, *36*(5), 1113–1120.

Cheng, L., Chen, Q., Pi, R., & Chen, J. (2021). A research update on the therapeutic potential of rhein and its derivatives. *European Journal of Pharmacology*, *899*, 173908.

Chien, S. C., Wu, Y. C., Chen, Z. W., & Yang, W. C. (2015). Naturally occurring anthraquinones: Chemistry and therapeutic potential in autoimmune diabetes. *Evidence-Based Complementary and Alternative Medicine*, *18*, 357357.

Danielsen, K., Aksnes, D. W., & Francis, G. W. (1992). NMR study of some anthraquinones from rhubarb. *Magnetic Resonance in Chemistry*, *30*(4), 359–360.

Daniyal, M., Tahir, I. M., Akram, M., Zahid, R., Zainab, R., Riaz, Z., & Bin, L. (2019). Pharmacological effects of *Rheum emodi*: a multiple purpose plant in health and disease. *Pakistan Journal of Medical & Biological Sciences*, *2*(1), 68–73.

Destandau, E., & Michel, T. (2013). Microwave assisted extraction. In M. A. Rostagno, & J. M. Prado (Eds.), *Natural Product Extraction: Principles and Applications* (pp. 113–156). RSC Publishing.

Ding, W. Y., Li, Y. H., Lian, H., Ai, X. Y., Zhao, Y. L., Yang, Y. B., Han, Q., Liu, X., Chen, X. Y., & He, Z. (2017). Sub-minimum inhibitory concentrations of rhubarb water extracts inhibit *Streptococcus suis* biofilm formation. *Frontiers in Pharmacology*, *8*(425), 1–9.

Du, L., Yuan, B., Zhang, B., Zhang, Y. L., Gao, X. Y., & Wang, Y. (2015). Study on mechanism for anti-hyperlipidemia efficacy of rhubarb through assistant analysis systems for acting mechanisms of traditional Chinese medicine. *China Journal of Chinese Material Medica*, *40*(19), 3703–3708.

Duval, J., Destandau, E., Pecher, V., Poujol, M., Tranchant, J. F., & Lesellier, E. (2016). Selective enrichment in bioactive compound from *Kniphofia uvaria* by super/subcritical fluid extraction and centrifugal partition chromatography. *Journal of Chromatography A*, *1447*, 26–38.

Fain, V. Y., Zaitsev, B. E., & Ryabov, M. A. (2005). Tautomerism of the natural anthraquinones physcion and emodin and their analogs. *Chemistry of Natural Compounds*, *41*(5), 501–507.

Feng, Z., Fei, J., Wenjian, X., Jiachen, J., Beina, J., Zhonghua, C., Xiangyi, Y., & Shaoying, W. (2012). Rhubarb attenuates the severity of acute necrotizing pancreatitis by inhibiting MAPKs in rats. *Immunotherapy*, *4*(12), 1817–1821.

Gao, C. C., Li, G. W., Wang, T. T., Gao, L., Wang, F. F., Shang, H. W., Yang, Z. J., Guo, Y. X., Wang, B. Y., & Xu, J. D. (2021). Rhubarb extract relieves constipation by stimulating mucus production in the colon and altering the intestinal flora. *Biomedicine and Pharmacotherapy*, *138*, 111479.

Goel, V., Ooraikul, B., & Basu, T. K. (1997). Cholesterol lowering effects of rhubarb stalk fiber in hypercholesterolemic men. *Journal of the American College of Nutrition*, *16*(6), 600–604.

Gong, Y. X., Li, S. P., Wang, Y. T., Li, P., & Yang, F. Q. (2005). Simultaneous determination of anthraquinones in Rhubarb by pressurized liquid extraction and capillary zone electrophoresis. *Electrophoresis*, *26*(9), 1778–1782.

He, J., Wang, L., Guo, H., Zhao, H., & Sun, J. (2019). Chemistry, pharmacology and processing method of rhubarb (*Rheum* species): A review. *Journal of Food Bioactives*, *8*, 42–50.

He, Z. H., He, M. F., Ma, S. C., & But, P. P. H. (2009). Anti-angiogenic effects of rhubarb and its anthraquinone derivatives. *Journal of Ethnopharmacology*, *121*(2), 313–317.

HMDB (2022a). *Aloe emodin* [Dataset]. The Metabolomics Innovation Centre. https://hmdb.ca/metabolites/HMDB0030829

HMDB (2022b). *Rhaponticin* [Dataset]. The Metabolomics Innovation Centre. https://hmdb.ca/metabolites/HMDB0304731

HMDB (2022c). *Rhapontigenin* [Dataset]. The Metabolomics Innovation Centre. https://hmdb.ca/metabolites/HMDB0031842

HMDB (2022d). *Rhein* [Dataset]. The Metabolomics Innovation Centre. https://hmdb.ca/metabolites/HMDB0032876

Hong, J. Y., Chung, H. J., Bae, S. Y., Trung, T. N., Bae, K., & Lee, S. K. (2014). Induction of cell cycle arrest and apoptosis by physcion, an anthraquinone isolated from rhubarb (rhizomes of *Rheum tanguticum*), in MDA-MB-231 human breast cancer cells. *Journal of Cancer Prevention*, *19*(4), 273.

Hu, B., Zhang, H., Meng, X., Wang, F., & Wang, P. (2014). Aloe-emodin from rhubarb (*Rheum rhabarbarum*) inhibits lipopolysaccharide-induced inflammatory responses in RAW 264.7 macrophages. *Journal of Ethnopharmacology*, *153*(3), 846–853.

Ibrahim, M., Khan, A. A., Tiwari, S. K., Habeeb, M. A., Khaja, M. N., & Habibullah, C. M. (2006). Antimicrobial activity of *Sapindus mukorossi* and *Rheum emodi* extracts against *H pylori*: *In vitro* and *in vivo* studies. *World Journal of Gastroenterology*, *12*(44), 7136–7142.

Integrated Taxonomic Information System-Report. (2011). *Rheum* L [Dataset]. ITIS-report. https://www.itis.gov/servlet/SingleRpt/SingleRpt?search_topic=TSNandsearch_value=21318#null

Jambrak, A. R., Nutrizio, M., Pingret, D., Fabiano-Tixier, A. S., & Chemat, F. (2013). Ultrasound assisted extraction. In M. A. Rostagno and J. M. Prado (Eds.), *Natural Product Extraction: Principles and Applications* (pp. 117–139). RSC Publishing.

Jeelani, M. A. A. S. (2015). Production technology for endangered medicinal plants of Kashmir Himalayas-I. Cultivation profile of *Rheum Australe* D. Don. *Life Sciences Leaflets*, *63*, 205–216.

Jiang, Y. N., Mo, H. Y., & Ren, H. (2013). Effect of emodin lipid nano-microbubble on MAPK signal pathway and inflammation cytokine in AT-II cells by mechanical stretch. *Journal of Chinese Medicinal Materials*, *36*(6), 967–971.

Khattak, A. K., Syeda, M. H., & Shahzad, S. M. (2020). General Overview of phytochemistry and pharmacological potential of *Rheum palmatum* (Chinese rhubarb). *Innovare Journal of Ayurvedic Sciences*, *8*(6), 1–5.

Kim, J. K., Kim, N., & Lim, Y. H. (2010). Evaluation of the antibacterial activity of rhapontigenin produced from rhapontin by biotransformation against *Propionibacterium acnes*. *Journal of Microbiology and Biotechnology*, *20*(1), 82–87.

Kolodziejczyk-Czepas, J., & Czepas, J. (2019). Rhaponticin as an anti-inflammatory component of rhubarb: A minireview of the current state of the art and prospects for future research. *Phytochemistry Reviews*, *18*(5), 1375–1386.

Kolodziejczyk-Czepas, J., & Liudvytska, O. (2021). *Rheum rhaponticum* and *Rheum rhabarbarum*: A review of phytochemistry, biological activities and therapeutic potential. *Phytochemistry Reviews*, *20*(3), 589–607.

Kounsar, F., Rather, M. A., Ganai, B. A., & Zargar, M. A. (2011). Immuno-enhancing effects of the herbal extract from Himalayan rhubarb *Rheum emodi* Wall. Ex Meissn. *Food Chemistry*, *126*(3), 967–971.

Lai, F., Zhang, Y., Xie, D. P., Mai, S. T., Weng, Y. N., Du, J. D., Wu, G. P., Zheng, J. X., & Han, Y. (2015). A systematic review of rhubarb (a traditional Chinese medicine) used for the treatment of experimental sepsis. *Evidence-Based Complementary and Alternative Medicine*, *2015*, 1–12.

Liu, A., Deng, Z. F., Hu, J. X., Han, S. W., Huang, L. L., Ke, P. Y., & Lin, S. Z. (2011). Effect of emodin on cell proliferation and apoptosis of human pancreatic cancer cell line Panc-1. *Chinese Traditional and Herbal Drugs*, *42*(4), 756–759.

Lo, T. C. T., Nian, H. C., Chiu, K. H., Wang, A. Y., & Wu, B. Z. (2012). Rapid and efficient purification of chrysophanol in *Rheum palmatum* LINN by supercritical fluid extraction coupled with preparative liquid chromatography in tandem. *Journal of Chromatography B*, *893*, 101–106.

Luo, S., Wen, R., Wang, Q., Zhao, Z., Nong, F., Fu, Y., Huang, S., Chen, J., Zhou, L., & Luo, X. (2019). Rhubarb Peony Decoction ameliorates ulcerative colitis in mice by regulating gut microbiota to restoring Th17/Treg balance. *Journal of Ethnoparmacology*, *231*, 39–49.

Luthria, D., Vinjamoori, D., Noel, K., & Ezzell, J. (2019). Accelerated solvent extraction. In D. L. Luthria (Ed.), *Oil extraction and analysis* (pp. 25–38). AOCS Publishing.

Malik, S., Sharma, N., Sharma, U. K., Singh, N. P., Bhushan, S., Sharma, M., Sinha, A. K., & Ahuja, P. S. (2010). Qualitative and quantitative analysis of anthraquinone derivatives in rhizomes of tissue culture-raised *Rheum emodi* Wall. plants. *Journal of Plant Physiology*, *167*(9), 749–756.

Mehra, P., & Jain, S. (2020). Rhubarb: An acidic vegetable. *Krishi Science – eMagazine For Agricultural Sciences*, *1*(2), 77–80.

Meier, N., Meier, B., Peter, S., & Wolfram, E. (2017). In-silico UHPLC method optimization for aglycones in the herbal laxatives *Aloe barbadensis* Mill., *Cassia angustifolia* vahl pods, *Rhamnus frangula* L. Bark, *Rhamnus purshianus* DC. bark, and *Rheum palmatum* L. roots. *Molecules*, *22*(11), 1838.

Meng, F., Du, C., Zhang, Y., Wang, S., Zhou, Q., Wu, L., Wang, Y., & Yang, X. (2020). Protective effect of rhubarb combined with ulinastatin for patients with sepsis. *Medicine*, *99*(7).

Nanasombat, S., Kuncharoen, N., Ritcharoon, B., & Sukcharoen, P. (2018). Antibacterial activity of thai medicinal plant extracts against oral and gastrointestinal pathogenic bacteria and prebiotic effect on the growth of *Lactobacillus acidophilus*. *Chiang Mai Journal of Science*, *45*(1), 33–44.

Nazir, S., Sharma, M., Saxena, M, Abrar, M., & Ajaz, M. (2013). *Rheum emodi*: Phytochemistry, bioactive compounds and their biological activity. *International Journal of Phytopharmacology*, *4*(4), 272–276.

Neyrinck, A. M., Etxeberria, U., Taminiau, B., Daube, G., van Hul, M., Everard, A., Cani, P. D., Bindels, L. B., & Delzenne, N. M. (2017). Rhubarb extract prevents hepatic inflammation induced by acute alcohol intake, an effect related to the modulation of the gut microbiota. *Molecular Nutrition and Food Research*, *61*(1), 1–12.

Pandith, S. A., Dar, R. A., Lattoo, S. K., Shah, M. A., & Reshi, Z. A. (2018). *Rheum australe*, an endangered high-value medicinal herb of North Western Himalayas: A review of its botany, ethnomedical uses, phytochemistry and pharmacology. *Phytochemistry Reviews*, *17*(3), 573–609.

Pandith, S. A., Hussain, A., Bhat, W. W., Dhar, N., Qazi, A. K., Rana, S., Razdana, S., Wani, T. A., Shah, M. A., Bedi, Y. S., Hamid, A., & Lattoo, S. K. (2014). Evaluation of anthraquinones from Himalayan rhubarb (*Rheum emodi* Wall. ex Meissn.) as antiproliferative agents. *South African Journal of Botany*, *95*, 1–8.

Park, S. K. and Lee, Y. K. (2021). Antioxidant Activity in *Rheum emodi* Wall (Himalayan Rhubarb). *Molecules*, *26*(9), 1–9.

Radhika, R., Krishnakumari, S., Umashankar, V., & Sudarsanam, D. (2010). Effect of enzymatic antioxidants of *Rheum emodi* in alloxan induced diabetic rats. *International Journal of Biological and Chemical Sciences*, *4*(6), 1905–1913.

Rana, M. K., & Duhan, D. S. (2018). Rhubarb. In M. K. Rana (Ed.), *Vegetable Crops Science* (pp. 875–884). CRC Press.

Rehman, H., Begum, W., Anjum, F., & Tabasum, H. (2014). *Rheum emodi* (Rhubarb): A fascinating herb. *Journal of Pharmacognosy and Phytochemistry*, *3*(2), 89–94.

Rokaya, M. B., Maršík, P., & Münzbergová, Z. (2012). Active constituents in *Rheum acuminatum* and *Rheum australe* (Polygonaceae) roots: A variation between cultivated and naturally growing plants. *Biochemical Systematics and Ecology*, *41*, 83–90.

Rumpunen, K., & Henriksen, K. (1999). Phytochemical and morphological characterization of seventy-one cultivars and selections of culinary rhubarb (*Rheum* spp.). *The Journal of Horticultural Science and Biotechnology*, *74*(1), 13–18.

Singh, D., Rawat, M. S. M., Semalty, A., & Semalty, M. (2012). Emodin–phospholipid complex: A potential of herbal drug in the novel drug delivery system. *Journal of Thermal Analysis and Calorimetry*, *108*(1), 289–298.

Singh, R., & Chaturvedi, P. (2018). Phytochemical screening and determination of antioxidant activity in callus and different parts of *Rheum emodi* Wall ex. Messin. *Journal of Pharmacognosy and Phytochemistry*, *7*, 2541–2547.

Singh, R., & Chaturvedi, P. (2019). Phytochemical Characterization of rhizome, fruit, leaf and callus of *Rheum emodi w*all. using GC-MS. *Pharmacognosy Journal*, *11*(3), 617–623.

Singh, R., Tiwari, T., & Chaturvedi, P. (2017). *Rheum emodi* Wall ex. meissn (Indian Rhubarb): Highly endangered medicinal herb. *Journal of Medicinal Plants Studies*, *5*(4), 13–16.

Su, S., Wu, J., Gao, Y., Luo, Y., Yang, D., & Wang, P. (2020). The pharmacological properties of chrysophanol, the recent advances. *Biomedicine and Pharmacotherapy*, *125*, 110002.

Tabin, S., Gupta, R. C., Bansal, G., & Kamili, A. N. (2016). Comparative HPLC analysis of emodin, aloe emodin and rhein in *Rheum emodi* of wild and in vitro raised plants. *Journal of Pharmacognosy and Phytochemistry*, *5*(2), 121.

Tayade, A., Dhar, P., Ballabh, B., Kumar, R., Chaurasia, O. P., Bhatt, R. P., & Srivastava, R. B. (2012). *Rheum webbianum* royle: A potential medicinal plant from trans-Himalayan cold deserts of Ladakh, India. *Plant Archives*, *12*(2), 603–606.

Tian, S. L., Yang, Y., Liu, X. L., & Xu, Q. B. (2018). Emodin attenuates bleomycin-induced pulmonary fibrosis via anti-inflammatory and anti-oxidative activities in rats. *Medical Science Monitor: International Medical Journal of Experimental and Clinical Research*, *24*, 1–10.

United States Department of Agriculture. (2018). *Rhubarb, raw* [Dataset]. USDA-ARS. https://fdc.nal.usda.gov/fdc-app.html#/food-details/167758/nutrients

Voltolina, G. (1987). Prospects for the development of officinal plant cultivation in mountainous districts and the results of various cultural trials conducted in Cansiglio. AttiConvegno Sulla Coltivazionedelle PianteOfficinali. *Trento*, *9*(10), 2.

Wan, B., Fu, H., Yin, J., & Xu, F. (2014). Efficacy of rhubarb combined with early enteral nutrition for the treatment of severe acute pancreatitis: A randomized controlled trial. *Scandinavian Journal of Gastroenterology*, *49*(11), 1375–1384.

Wang, J., Zhao, H., Kong, W., Jin, C., Zhao, Y., Qu, Y., & Xiao, X. (2010). Microcalorimetric assay on the antimicrobial property of five hydroxyanthraquinone derivatives in rhubarb (*Rheum palmatum* L.) to *Bifidobacterium adolescentis*. *Phytomedicine*, *17*(8–9), 684–689.

Wang, L., Cui, Y. L., Zhang, Z., Lin, Z. F., & Chen, D. C. (2017). Rhubarb monomers protect intestinal mucosal barrier in sepsis via junction proteins. *Chinese Medical Journal*, *130*(10), 1218–1225.

Wang, L., Li, D., Bao, C., You, J., Wang, Z., Shi, Y., & Zhang, H. (2008). Ultrasonic extraction and separation of anthraquinones from *Rheum palmatum* L. *Ultrasonics Sonochemistry*, *15*(5), 738–746.

Wang, Z., Guo, M., Ma, D., & Wang, R. Q. (2015). Effects of rhubarbs from different regions on blood lipid and antioxidation of hyperlipidemia rats. *Chinese Journal of Applied Physiology*, *31*(3), 278–281.

Wani, I. A., Verma, S., Ahmad, P., El-Serehy, H. A., & Hashim, M. J. (2022). Reproductive Biology of *Rheum webbianum* Royle, a vulnerable medicinal herb from alpines of North-Western Himalaya. *Frontiers in Plant Science*, *13*, 1–19.

Wasim, A., Mohd, M., Sayeed, A., & Sma, Z. (2013). Current strategy for research on quality identification of *Rheum emodi* Wall. rhizome. *International Journal of Drug Development and Research*, *5*(1), 297–304.

Wei, L., Luo, Y., Zhang, X., Liu, Y., Gasser, M., Tang, F., Ouyang, W. W., Wei, H., Lu, S., Yang, Z., Waaga-Gasser, A. M., Deng, C., & Lin, M. (2021). Topical therapy with rhubarb navel plasters in patients with chronic constipation: Results from a prospective randomized multicenter study. *Journal of Ethnopharmacology*, *264*, 113096.

Wen, Q., Miao, J., Lau, N., Zhang, C., Ye, P., Du, S., Mei, L., Weng, H., Xu, Q., Liu, X., Chen, D., Zhang, F., Li, C., & Li, H. (2020). Rhein attenuates lipopolysaccharide-primed inflammation through NF-κB inhibition in RAW264.7 cells: targeting the PPAR-γ signal pathway. *Canadian Journal of Physiology and Pharmacology*, *98*(6), 357–365.

Xiang, H., Zuo, J., Guo, F., & Dong, D. (2020). What we already know about rhubarb: A comprehensive review. *Chinese Medicine*, *15*(88), 1–22.

Xie, L., Tang, H., Song, J., Long, J., Zhang, L., & Li, X. (2019). Chrysophanol: A review of its pharmacology, toxicity and pharmacokinetics. *Journal of Pharmacy and Pharmacology*, *71*(10), 1475–1487.

Xiong, Y., Chen, L., Fan, L., Wang, L., Zhou, Y., Qin, D., Sun, Q., Wu, J., & Cao, S. (2018). Free total rhubarb anthraquinones protect intestinal injury via regulation of the intestinal immune response in a rat model of severe acute pancreatitis. *Frontiers in Pharmacology*, *9*(75), 1–13.

Xiong, Z. H. (2012). *The study of intervention effect and molecular mechanism of rhubarb used in diabetic nephropathy* [Unpublished doctoral thesis]. Guangzhou University of Chinese Medicine.

Yang, L., Wan, Y., Li, W., Liu, C., Li, H., Dong, Z., Zhu, K., Jiang, S., Shang, E., Qian, D., & Duan, J. (2022). Targeting intestinal flora and its metabolism to explore the laxative effects of rhubarb. *Applied Microbiology and Biotechnology*, *106*(4), 1615–1631.

Yao, P., Cui, M., Li, Y., Deng, Y., & Wu, H. (2015). Effects of rhubarb on intestinal flora and toll-like receptors of intestinal mucosa in rats with severe acute pancreatitis. *Pancreas*, *44*(5), 799–804.

Zargar, B. A., Masoodi, M. H., Ahmed, B., & Ganie, S. A. (2011). Phytoconstituents and therapeutic uses of *Rheum emodi* wall. ex Meissn. *Food Chemistry*, *128*(3), 585–589.

Zhao, Y. Q., Liu, X. H., Ito, T., & Qian, J. M. (2004). Protective effects of rhubarb on experimental severe acute pancreatitis. *World Journal of Gastroenterology*, *10*(7), 1005–1009.

Zhou, G., Peng, F., Zhong, Y., Chen, Y., Tang, M., & Li, D. (2017). Rhein suppresses matrix metalloproteinase production by regulating the Rac1/ROS/MAPK/AP-1 pathway in human ovarian carcinoma cells. *International Journal of Oncology*, *50*(3), 933–941.

10 Costus (*Saussurea costus*)

Kawaljeet Kour, Ilknur Ucak, and Raees ul Haq

10.1 INTRODUCTION

Plants are used as major natural resources for traditional as well as modern medicinal systems across the world. Plants and their parts are used in pharmaceuticals to provide natural cures and remedies, making them a significant part of the world's medical markets. The Indian subcontinent contains a very broad range of vegetation because of its range of different geographic, climatic, and biological characteristics. It is home to approximately 15,000 flowering plant species, accounting for approximately 6% of all the plant varieties found around the globe. In India, where they have been utilized for therapeutic purposes since antiquity, some 3000 plants have been scientifically proven to offer health advantages (Ali and Venkatesalu, 2022). There are over 30,000 species in the Asteraceae family, which is distributed worldwide. Within this, the *Saussurea* genus contains approximately 300 species, with about 61 of them being identified in India. *Saussurea costus*, the most renowned of the *Saussurea* species, has a number of applications in a range of conventional medical care practices (Vijayalakshmi et al., 2022). "Costus" is the usual name for *Saussurea costus* in English, but the plant is also known by a variety of other terms in India, including "kuth" in Hindi, "postkhai" in Kashmiri, "kostum" in Tamil, "kot" in Punjabi, "kut" in Gujrati, "kur" in Bengali, "sepuddy" in Malayalam, "kushta" in Sanskrit and Marathi, "Kustam" in Telugu, "Koshta" in Kannada, "Kustullnally" in Arabic and "Muk heung" in Canton (Madhuri et al., 2012; El-Far et al., 2018).

Conventionally, *Saussurea costus* has been utilized for its anti-cancer, hepatoprotective, anti-arthritic, anti-convulsant, anti-microbial, and anti-viral activities (Anantharam & Chittibabu, 2021). This plant's root has historically been used to cure lung inflammation, paralysis, and pain. In the Ayurveda tradition, it is used to improve the complexion, treat leucoderma, itching, ringworm, vomiting, scabies, headaches, and epilepsy (Madhuri et al., 2012). The numerous terpenes contained within it have anti-inflammatory and anti-tumor effects (Lunz & Stappen, 2021). The sesquiterpene lactones costunolide and dehydrocostus lactone, which have undergone significant pharmacological testing, are the two key components of the roots (Alhoot, 2021).

In *Saussurea costus*, the main active components are terpenes, but it also includes flavonoids, anthraquinones, and alkaloids. *Saussurea costus* has been linked to a large number of sesquiterpene lactones, phytosterols, triterpenes, flavonoides, lignans, alkaloids, and phenyl propanoids (Kumar & Pundir, 2021). Flavonoids are antibacterial (Abdallah et al., 2017), and chlorogenic acid inhibits oxidation and reverses the damage caused by free radicals (Gheraibia et al., 2020). Three anthraquinone compounds found in *Saussurea costus* prevent the synthesis of protein tyrosine phosphatase (Dubey et al., 2022). This plant yielded seven non-cytotoxic compounds, including novel sesquiterpenes (Choodej et al., 2019). Other substances were also obtained from *S. costus*, including chlorogenic acid, cynaropicrin, saussurine, dehydrocostus lactone, *Saussurea*mines A–C, and reynosin (Semwal et al., 2020).

10.2 OCCURRENCE

Saussurea costus is a perennial or biennial herb that grows in the wild at higher elevations in North America, Europe, and Asia, particularly in some areas of the Himalaya region (Nadda et al., 2020). *Saussurea costus* is reported to be geographically restricted to some of these areas, i.e., Himachal

DOI: 10.1201/9781003292340-10

Pradesh, Jammu and Kashmir, and Uttrakhand, at elevations between 3200 and 3800 metres above mean sea level (msl) (Ali & Venkatesalu, 2022). *Saussurea costus* has been found to grow naturally in India's highest elevations in the Kishenganga and Chenab valleys of the Jammu and Kashmir region, while it also grows sporadically in the neighboring regions of Kashmir, Uttarakhand, and Himachal Pradesh. However, the industrial exploitation of *Saussurea costus* started at the beginning of the 20th century in the native growing regions of Kashmir, Himachal Pradesh (Lahul), and Uttar Pradesh (Garhwal) as a result of a lack of wild plant supply to meet customer needs (Ali & Venkatesalu, 2022). *Saussurea costus* is a Himalayan endemic, but, because of its very narrow geographic range, it is one that is highly vulnerable to extinction (Rathore et al., 2021). The majority of the naturally growing *S. costus* are either subjected to devastating exploitation or have been destroyed because of the enormous market for *S. costus* roots in the pharmacy sector and also because of its ethnomedicinal value (Vijayalakshmi et al., 2022).

10.3 SYSTEMATIC CLASSIFICATION

Saussurea costus is a perennial herb with an upward stem of approximately 12 m. The stout root measures up to 60 cm in length and has a potent aroma. Transverse sections of the periderm clearly show the plant's phloem and xylem. The stem of the costus is thick and woody and the lobes and stalked leaves are roughly 1 m in length. The blooms are arranged into axillary and terminal clusters, ranging in color from dark bluish purple to black. The flower heads are 3–5 cm in diameter, stalkless, stiff, and spherical in shape. The fruit is compressed, cupped, and curled, and measures around 3 mm in length (Zahara et al., 2014). The dried root is 8–12 cm in length, 1–3 cm in width, and has a somewhat pungent flavor. The exterior of the plant ranges in color from dark grey to yellowish. It has creases and wrinkles, and its secondary roots are mostly cylindrical in nature (Ali & Venkatesalu, 2022). The taxonomic classification of *Saussurea costus* is presented in Table 10.1

10.4 NUTRITIONAL ATTRIBUTES OF THE PLANT

Saussurea costus contains a number of volatile components. The amounts of volatile elements vary significantly between cultivars as the result of a wide range of climatic factors, including temperature, irradiation, and humidity. Additionally, a plant's genetic heritage has an impact on the chemistry of its volatile compounds. The roots also include the alkaloid saussurin (a bitter resin) and tannins. A, B1, B2, B12, and C are just a few of the vital vitamins to be found in plants in trace

TABLE 10.1
Taxonomic Classification of *Saussurea costus*

Kingdom	Plantae
Sub-kingdom	Viridaeplantae
Infrakingdom	Streptophyta
Superdivision	Embryophyta
Division	Tracheophyta
Subdivision	Spermatophytina
Infradivision	Angiospermae
Class	Magnoliopsida
Superorder	Asteranae
Order	Asterales
Family	Asteraceae
Genus	*Saussurea*
Species	*costus*

levels. According to analysis, the following minerals are also present: calcium (Ca), magnesium (Mg), iron (Fe), manganese (Mn), zinc (Zn), lead (Pb), strontium (Sr), copper (Cu), nickel (Ni), and chromium (Cr) (Ali et al., 2021).

Saussurea costus has an overtly aromatic odor due to the presence of volatile or essential oils, which are primarily restricted to roots. The principal constituents in this fragrant volatile oil are hydrocarbons, oxides, ketones, esters, alcohols, aldehydes and acids (Chang and Kim, 2008). In one analysis, a total of 39 components have been identified in the essential oil of the *Saussurea costus* root. Dehydrocostus lactone (44–47%) was the main component, followed by 8-cedren-13-ol with concentrations ranging from 3 to 5%, β-costol (12–14%), δ-elemene (11–13%), α-curcumene (3–5%), costunolide (6–9%), α-selinene (3–5%), 4-terpinol (2–4%), α-costol (2–4%), elemol (1–3%), and α-ionone (1–3%). There are also minute amounts of (+)-selina-4, 2-pinene, (-)-caryophyllene, 11-diene, (-)-(E)-trans-bergamota-2,12-dien14-al, (+)--costal, and 12-methoxy-dihydro-dehydro costus lactone (Ali et al., 2021).

10.5 PHYTOCHEMICAL OVERVIEW OF THE PLANT

As outlined above, *Saussurea costus* has a wide range of secondary metabolites, including terpenes, alkaloids, tannins, flavonoids, and sesquiterpenes (Ambavade et al., 2009). In 1914, Semmler and Feldstein were the first to published a phytochemical report on *Saussurea costus*; the same year, they also published a second supplementary study (Semmler and Feldstein, 1914). From the oil of the roots, they separated two lactones with the chemical formulae $C_{15}H_{22}O_2$ and $C_{15}H_{20}O_2$. The plant is abundant in terpenes and substances similar to sesquiterpenoid lactones, such as dehydrocostus lactone and costunolide (Robinson et al., 2008; Wang et al., 2008; Kim & Choi, 2019). *S. lappa* (also known as *Saussurea costus*) contains an abundance of alkaloids, steroids, flavonoids, and resins (Abdallah et al., 2017). In addition, numerous bioactive compounds, including sesquiterpene lactones, costunolide, isodehydrocostus, isozaluzanin-C, guiainolide, cynaropicrin, reynosin, santamarine, and many others, were discovered and extracted from *S. costus* (Pandey et al., 2007). The abundance of phytochemical substances in *S. costus* is unquestionably responsible for the variety of biological activities of this plant.

A number of monoterpenes have been identified in the *S. costus* root, including anethole, phellandrene, citronellyl propionate, thymol, α-pinene, β-pinene, estragole, α-terpineol, α-thujene, camphene, camphor, myrcene, sabinene, menthone, linalool, terpinen-4-ol, cryptone, ocimene, citronellal, α-terpinolene, p-cymene, γ-terpinene, limonene, 1,8 cineol, etc. (Chang and Kim 2008; Gwari et al. 2013). Rutin, apigenin-7-O-β-D-glucoside and luteolin-7-O-β-D-glucoside were flavonoids discovered from roots of *S. costus* with one glucoside substituent (Alaagib and Ayoub 2015); however, flavonoids, containing multiple glucoside substituents, were less abundant.

Sesquiterpene lactones are the main compounds derived from numerous medicinal plants and they have historically been used to treat inflammatory illnesses. Lappadialactone, germacrenes, and 4β-methoxy-dehydrocostus lactone are all present in the petroleum extract taken from the root extracts of *Saussurea costus* (de Kraker et al., 2001). The methanolic root extract yielded mokko lactone, lignan glycoside (-)-massoniresinol 4″-*O*-β-D-glucopyranoside, betulinic acid, methyl ester, and saussureamines A to E (Matsuda et al., 2000). The volatile components were separated from the root of *S costus* using hydro distillation and evaluated by gas chromatography-mass spectroscopy. They were found to have a total of 35 aromatic compounds, accounting for around 92.81% of the entire composition. In contrast to additional ketones such as dehydrocostus lactone (16.7%), γ-costol (1.80%), terpinen-4-ol (1.60%), alcohols like elemol (5.84%), valerenol (4.20%), and vulgarol B (3.14%). The aldehyde-like (10Z, 7Z, 13Z)-7, 10, 13-hexadecaterinal was discovered to be a prominent component (Gwari et al., 2013). Another component found in roots, santamarin, lowered COX-2 protein, decreased murine peritoneal macrophages, and inhibited nitric oxide synthase protein (iNOS) (Choi et al., 2012). From the hexane root extract, β-cyclocostunolide and dihydrocostunolide have been extracted (Robinson et al., 2008).

Lappasterol and 3-epilappasterol, two phytosterols, were also isolated from the roots (Singh et al., 2017; Idriss et al., 2022). From the roots, arbuscalin A, β-costic acid, lactone, costunolide, syringin, reynosin, 4-allyl-2,6- dimethoxyphenyl glucoside, costunolid15-β-D-glucopyranoside and chlorogenic acid were also isolated (Cao et al., 2016). Geracrenes are present in newly harvested costus roots (de Kraker et al., 2001). Gas chromatography was used to analyze 63 aroma components, comprising 26 hydrocarbons, 13 alcohols, 11 ketones, 4 aldehydes, 1 carboxylic acid and 1 oxide. Oligomeric glucofructans and fructofuranosese were also extracted from the roots (Olennikov et al., 2011).

10.6 STRUCTURAL CHARACTERIZATION OF BIOACTIVE COMPONENTS

The main active ingredients in *S. costus* are terpenes, although it also contains flavonoids, anthraquinones, and alkaloids. The different terpenes found in *S. costus*, including costunolide and dihydrocostunolide, have anti-inflammatory and anticancer properties (Peng et al., 2017). 10α-methoxyartemisinic acid (Cao et al., 2016) and dehydrocostuslactone, costunolide, were identified from *S. costus*. In addition, linoleic acids, costic, palmitic, α-cyclocostunolide, β-sitosterol, isoalantolactone, β-cyclocostunolide, alantolactone, and isodihydrocostunolide have also been identified (Guan et al., 2015; Lim et al., 2015). Researchers have also found a wide range of additional substances: santamarin (Choi et al., 2012), δ-elemene, β-castol (Negi et al., 2013), luteolin-7-O-β-D-glucoside, apigenin-7-O-β-D – glucoside, umbelliferon glucoside and rutin (Alaagib and Ayoub, 2015), 1β-hydroxy arbusculin A (Choi et al., 2008), betulinic acid, mokko lactone, betulinic acid methyl ester (Kour et al., 2015), arbusculin B (Julianti et al., 2011), dihydroneoclovene, alloaromadendrene, hexadecanoic acid, dehydrocostus lactone, glaucyl alcohol, dehydrocostus, lactone, and caryophyllene oxide (Benedetto et al., 2019). From *S. costus*, seven non-cytotoxic compounds were isolated, comprising two novel sesquiterpenes, a guaianolide compound with a C_{17} structure, 1β, 6α-dihydroxycostic acid ethyl ester, and lappalone (Sun et al., 2003). Several additional components that were discovered from *S. costus*, including chlorogenic acid (Pandey et al., 2004), cynaropicrin (Choi et al., 2012), saussurine (Semwal et al., 2020), dehydrocostus lactone, 12 methoxydihydrocostunolide, and *Saussure* amines A–C (Matsuda et al., 2000). Some of the compound structures found in *Saussurea costus* are presented in Figure 10.1.

10.7 EXTRACTION AND ISOLATION TECHNIQUES

One of the crucial steps in creating high-quality herbal products is the extraction of ingredients for the development of such medicines. The important phases in obtaining top-notch bioactive compounds are the selection of an appropriate solvent, extracting methods, phytoconstituent testing techniques, fractionation methods, and detection methods. Polar, intermediate-polar, and nonpolar solvents are commonly employed to extract medicinal herbs. Examples of such solvents include water and alcohols, acetone, and dichloromethane. Maceration, digestion, decoction, infusion, percolation, Soxhlet extraction, superficial extraction, ultrasound-assisted extraction, and microwave-assisted extraction are some of the methods of extraction (Yadav et al., 2020). Phytoconstituents molecules can be segregated and filtered using a variety of chromatographic methods, including paper chromatography, gas chromatography, thin-layer chromatography, and high-performance liquid chromatography (Raafat et al., 2019). A range of identification techniques, such as infrared spectroscopy, ultraviolet spectroscopy, nuclear magnetic resonance spectroscopy, and mass spectroscopy, are used to identify substances (Heftmann, 2004). Using healthy *Saussurea costus* roots, leaves, seeds, and flowers, a plant extract is initially made for the extraction process. For this purpose, various plant components were carefully washed, dried in the shade, and turned into a refined powder. Three types of extract are prepared from *Saussurea costus* viz. ethanolic, methanolic, and aqueous forms.

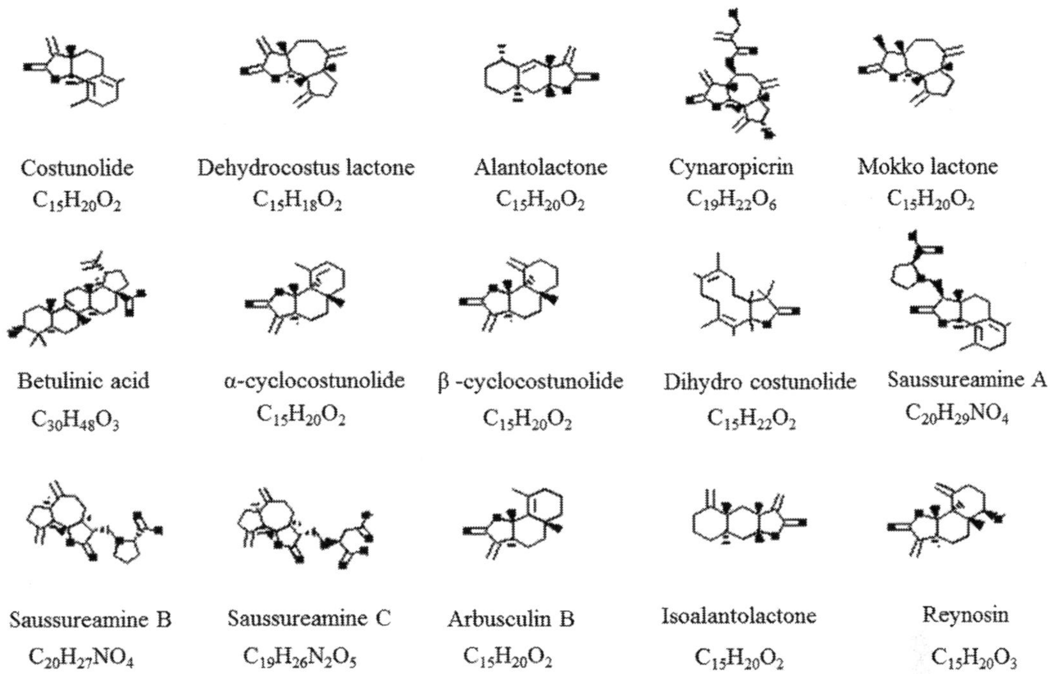

FIGURE 10.1 Bioactive compounds with their structures in *Saussurea costus*. (Compiled from: https://pubchem.ncbi.nlm.nih.gov)

10.7.1 Preparation of Extracts from *Saussurea costus*

10.7.1.1 Preparation of Ethanol Extract from *Saussurea costus* (SCEE)

In the extraction procedure, the roots were left to dry, ground into a little coarse powder, and then placed in airtight equipment (Elshaer et al., 2022). In order to obtain a crude *S. costus* extract, the dry herbal roots (200g) were crushed into a powder, separated with 70% aqueous ethanol at ambient temperature using a basic maceration for 72 hours, screened three times, and then vaporized using a rotary evaporator (Tag et al., 2016). The obtained extract was kept at 4 °C for further use.

10.7.1.2 Preparation of Methanol Extract from *Saussurea costus* (SCME)

In this method, *Saussurea costus*' root was pulverized and then filtered with the help of a 70% methanol solution to create *S. costus* methanol extract (SCME). The clear liquid was then separated using filter paper and then concentrated into a viscous semi-solid slurry with the help of a rotary evaporator; It was finally diluted in bi-distilled water (Alnahdi, 2017). The obtained extract was stored at 4 °C for further usage.

10.7.1.3 Preparation of Aqueous Extract from *Saussurea costus* (SCME)

The aqueous extract of *Saussurea costus* (SCAE) was made by cleaning the plant material and then crushing it. The crushed material (in powdered form) was then boiled with distilled water for 30 min, strained via filter paper, and condensed under decreased pressure into a viscous, semi-solid slurry before being lyophilized. The freeze-dried material was measured, dispersed in water, and then saved for later action (Saleem et al., 2013). Until it was used, the resulting extract was kept at 4 °C.

10.7.2 Typically used Techniques in the Extraction Procedure

10.7.2.1 Maceration

In this method, the menstruum is layered over coarsely powdered drug material, which can be leaves or bark, and thoroughly wrapped. After that, the vessel is sealed and stored for at least three days (Ingle et al., 2017). To facilitate complete separation, the mixture should be blended occasionally and, if it is in a bottle, occasionally agitated. The micelle and marc have been isolated from each other using filtration or decantation at the completion of separation. After that, the micelle and menstruum are segregated by evaporation in an oven or water bath (Doughari, 2012). This technique is quite practical in the real world and is being used in different fields.

10.7.2.2 Infusion

This is a method of extraction similar to maceration. A clean container is filled with the medication material after it has been pounded into a fine powder. The material is then covered with the extracting agent, either hot or cold, and allowed to soak for a brief period (Azwanida, 2015). This technique works well for the isolation of easily soluble active compounds. It is an appropriate approach for the preparation of fresh extracts. The ratio of the solvent to the sample is commonly 4:1 or 16:1, varied according to the intended use (Majekodunmi, 2015).

10.7.2.3 Digestion

This extraction technique uses a moderate amount of heat to carry out the extraction. The powdered drug ingredient is added to the solvent in a clean glass container. The extraction mixture is placed in an oven or over a water bath at a temperature of roughly 50 °C (Pandey & Tripathi, 2014). During the whole extraction process, heat was used to decrease the solvent's viscosity and improve the removal of secondary metabolites. This approach works well for easily soluble plant components (Ingle et al., 2017).

10.7.2.4 Decoction

This procedure entails continuous hot extraction utilizing a predetermined amount of water as a solvent. Plant matter which has been dehydrated, pulverized, and powdered is put into a sterile container. Water is then added and the two ingredients are mixed together. Heat is then applied in order to speed up the process (Azwanida, 2015). It takes just a few minutes, typically around 15, to complete the process,. The ratio of the solvent to the crude medication is typically 4:1 or 16:1. This method is employed to extract heat- and water-resistant plant components (Pandey & Tripathi, 2014).

10.7.2.5 Percolation

The device employed in this process is known as a percolator. This resembles a small, cone-shaped glassware container with openings across sides. The plant matter is dehydrated, pulverized, and thoroughly powdered, and then steeped with the extractant in a tidy vase. The mixture is held for four hours after a larger quantity of solvent is poured in. The material is then placed into a percolator, with the lower half sealed, and allowed to stand for about 24 hours (Majekodunmi, 2015). The active substance is subsequently saturated thoroughly by the extraction solvent, which is poured from above. The bottom half of the percolator is unlocked, and the liquid is then permitted to flow gradually. The extraction process, governed by gravitation, drives a constant volume of solvent through the sample (Azwanida, 2015). When amount of solvent introduced is equal to 75% of the whole amount anticipated for the preparation, the solvent infusion ends. The extract is isolated via filtration, and then decanted. The final quantity of solvent is then added to the expressed marc to create the appropriate volume (Pandey & Tripathi, 2014).

10.7.2.6 Soxhlet Extraction

Another name for this procedure is "continuous hot extraction," with the glass instrument used in this technique referred to as a Soxhlet extractor. The apparatus involved comprises a round-bottomed flask, an extraction compartment, a syphon tube, and a condensing unit at the top. A perforated pouch (thimble) made from durable filter paper or cleaned fabric is filled with dehydrated and finely grounded matter, and the pouch is then tightly sealed (Doughari, 2012). The extraction solvent is then added to the bottom flask before the thimble is placed inside the extraction column. The solvent is warmed in the bottom flask, vaporized via the condensing unit, descends to the extraction column, and, by coming into contact with the drug, extracts it. When the proportion of solvent in the extracting column approaches top of the siphon, the solvent and the extracted material thereafter find their way back to the flask (Hossain et al., 2014). The procedure continues until all of the medication has been extracted. The benefit of this extraction method is that a substantial amount of medication can be extracted using only a small volume of solvent. It can also be employed on plant materials that can withstand heat. There is no need for filtering, and a lot of heat can be applied (Majekodunmi, 2015). Its drawback includes the method's inability to handle thermolabile materials as heat is mandatory for evaporation of the solvent (Azwanida, 2015).

10.7.2.7 Extraction with Microwave Assistance

This is the most effective technique for the extraction of therapeutic plant compounds. The method employs the relocation of ionic species found in the solvent and medicinal material, as well as the dipole rotation and ionic transfer mechanisms. This technique works well for the extraction of flavonoids. Electromagnetic radiation with frequencies ranging from 300 MHz and 300 GHz and possessing wavelengths from 1 cm to 1 m are used in this process (Altemimi et al., 2017; Xiong et al., 2016). Microwaves with a frequency of 2450 Hz produced energy between 600 and 700 W. The method blasts an object with microwave radiation which has the ability to soak up electromagnetic radiation and turn it into thermal energy. The resultant heat facilitates the entry of the solvent into the medicinal complex (Bhan et al., 2017). Utilizing a polar solvent increases solvent penetration and helps with extraction by causing ion migration and dipole rotation. It is inadvisable to use non-polar solvents in this procedure because they reduce the amount of heat produced by the microwave radiation (Doughari, 2012). The special benefits of microwave-assisted extraction include a reduction in the amount of solvent used, cutting down on extraction time, and an improvement in results (Pandey & Tripathi, 2014). However, only flavonoids and phenolic chemicals can be used in this approach. Due to the high temperature involved, substances like tannins and anthocyanins might be destroyed (Altemimi et al., 2017).

10.7.2.8 Extraction with Ultrasound Assistance

The use ultrasonic waves causes an increase in the surface area by breaking the tissues and the creation of new surface enhances overall mass transfer rate. The process results in release of bioactive compounds in greater proportion. In this technique, plant material must be dried before being pounded into a fine powder and properly sieved. After mixing the prepared sample with the necessary extraction solvent, the prepared sample is placed into the ultrasonic separator (Altemimi et al., 2017). The extraction procedure was accelerated by the high sound waves used; this reduced the need for heating. Small samples can be used for ultrasound-assisted extraction, which increases yield while reducing both extraction time and solvent usage (Azwanida, 2015). It is difficult to reproduce this technique. Additionally, the increased energy input could result in a degradation of the phytochemical through the production of free radicals (Majekodunmi, 2015).

10.7.3 Isolation Techniques

Fractionation is the technique by which plant extracts are separated into various fragments. The fractions are then split up into chemically diverse categories. The process is then repeated until the isolation of a pure chemical (Rimando et al., 2001). When more than one solvent is needed for the fractionation process, the additions should be made in the order of increasing polarity. Fundamentally, there are two types of fractionation techniques: physical and chemical. These are outlined below.

10.7.3.1 Chemical Methods

The basis for this extraction method is the types of functional groups that each constituent in the blend contains. If the right correct are utilized, materials can be separated or purified through the use of chemical processes (Doughari, 2012).

10.7.3.2 Physical Methods

Physical techniques for segregating compounds from mix include fractional distillation, fractional crystallization, fractional liberation, sublimation, and the separation funnel method (Doughari, 2012).

10.7.3.2.1 Separation funnel technique

The fractionation process begins by hydrating and completely combining the extract with 250 ml of water after four distinct solvents (n-hexane, chloroform, acetone, and n-butanol) have been chosen. The material is then transferred, shaken, and allowed to settle into a separating funnel. In addition, the less polar solvent was mixed with 250 ml of hexane. After the matter subsides, the aqueous layer can be eliminated by releasing the bottom of the funnel. In order to acquire the n-hexane component, the leftover material in the separatory funnel was put into a fresh vessel (Sasidharan et al., 2011). Once more, add an equal volume of n-hexane, shake it, and then separate it. After pouring n-hexane and shaking, the process proceeded until no discernible amount of extract was observed to flow into the n-hexane portion (Heftmann, 2004). A similar cycle was carried out for the purpose of obtaining fractions of chloroform, acetone, and n-butanol. Since the crude extract was initially dispersed in water, the proportion that is left after fractionation is known as the residual aqueous fraction (RAF) (Rimando et al., 2001).

10.7.3.2.2 Fractional distillation

Compounds in a mixture are separated or purified using this procedure. Based on the differences in boiling points, purification is accomplished. Every apparatus used in fractional distillation is built in such a way that, whenever heated, it causes each chemical to vaporize and separate at its boiling temperature. As a result, every fractionated component will condensate and be extracted as a distinct substance using a number of syphons connected to the fractional distillation device (Doughari, 2012).

10.7.3.2.3 Fractional crystallization

In nature, a large majority of the chemicals found in plant extracts naturally crystallize. The development of crystals during the heating or cooling of an extract concentrate results in their separation into individual compounds (Doughari, 2012).

10.7.3.2.4 Fractional liberation

This technique works well for the extraction of chemicals from mixtures that readily form precipitates. In most cases, the precipitate is created by salting the chemicals (Doughari, 2012).

10.7.3.2.5 Sublimation

This process includes transitioning directly from a solid to a gaseous form, bypassing the liquid state. When heat is applied, certain compounds, including volatile oils, dissociate and turn straight into gaseous form (Doughari, 2012).

10.7.3.2.6 Chromatography

These specialized methods are employed to isolate compounds from mixtures based on their size, form, and polarity. The concept of chromatography entails the use of an extraction solvent as the mobile phase and a stationary phase, like silica gel or sephadex, with calcium sulphate as a binding agent (Heftmann, 2004). Silica gel can be used to extract alkaloids, amino acids, carbohydrates, fatty acids, lipids, and other substances. Sephadex is employed to separate proteins and amino acids. In order to separate phenols, steroids, alkaloids, vitamins, and carotenes, aluminum is useful. Alkaloids, food colors, and amino acids are separated using cellulose powder. By contrast, steroids and organic cations are separated using celite (Sasidharan et al., 2011). When substances are separated using chromatographic methods, a number of mechanisms may be at play, including adsorption, segregation, attraction, ion exchange, and physical degradation (Bhan et al., 2017). Paper chromatography, column chromatography (CC), thin layer chromatography (TLC), liquid chromatography (LC), gas chromatography (GC), and high-performance liquid chromatography (HPLC) are among the examples of chromatographic procedures (Ingle et al., 2017).

The chromatographic separation can occur in a number of different ways: partition chromatography or absorption chromatography or ion exchange chromatography or affinity chromatography. These are briefly outlined below:

a. Adsorption chromatography: The basis for separation in this technique is the interplay of the components to be segregated and the stationary phase. In this scenario, molecules will be attracted to, and removed from, the stationary phase by hydrophobic, non-covalent Van der Waals forces. The mobile phase will initially elute the loosely bound compound (Ingle et al., 2017).
b. Partition chromatography: When two or more immiscible solvents are added to an extract mixture, compounds can be separated. Each of the chemicals will separate from the mixture by absorbing the solvent in the area where it is soluble (Sasidharan et al., 2011). The individual compounds will next be obtained by the separation of the immiscible liquids using a separating funnel. An alternative name for this chromatography is liquid/liquid separation (Heftmann, 2004).
c. Affinity chromatography: This technique divides substances based on how a protein interacts with a certain ligand. The stationary phase is a ligand placed in a segregating section. The chemicals that lacked affinity for the stationary phase were removed by the mobile phase. As a result, substances with a strong attraction to the stationary phase are drawn to it and separated (Ingle et al., 2017).
d. Ion exchange chromatography: In order to separate polar molecules according to the type of charge they have, ion exchange chromatography is used. As we know, like charges attract whereas unlike charges repel. Because of this, components with the same charge are drawn to one another and separated from the blend (Sasidharan et al., 2011).
e. Size exclusion chromatography: This technique uses various widths of mesh to separate substances according to their molecular size. This technique is often referred to as molecular sieving or gel filtration (Heftmann, 2004). It involves the use of first a lower size mesh, then a medium size mesh, and, eventually, a large pore size mesh.

10.8 APPLICATION OF ACTIVE INGREDIENTS

Plant secondary metabolites are becoming more and more important today because of their numerous biological and medicinal uses. The identification and comprehension of cellular pathways, which are crucial phases in the drug-development procedure, is greatly aided by these natural compounds, which tend to be a provenance of advancement in the pharmaceutical industry. One sesquiterpene lactone with anti-tumor properties is costunolide. The ability of costunolide to suppress multiplication, penetration, and metastasis in addition to induce apoptosis suggests that it has the prospect of being a robust, comprehensive anti-tumor treatment (Rasul et al., 2012). It has antioxidant, anti-inflammatory, anti-allergic, bone-remodeling, neuro-protective, hair growth-promoting, anti-cancer, and anti-diabetic properties. Mechanistic information on several of these bioactivities, including the modification of different cellular pathways implicated in triggering tissue inflammation, tumor progression, bone depletion, and neurodegeneration, is used to support them. The main molecular targets of costunolide are redox-regulated transcription factors such as nuclear factor-kappaB, activator protein-1, signal transducer and activator of transcription and intracellular kinases like mitogen-activated protein kinases, cyclin-dependent kinases, Akt kinase, cyclins, and telomerase. Pro-inflammatory mediators, including inducible nitric oxide synthase, cyclooxygenase-2, prostaglandins, nitric oxide, and cytokines, were similarly reduced in production and/or expression by the substance (Kim & Choi, 2019).

Dehydrocostus lactone (DHE), a naturally occurring sesquiterpene lactone, has the capacity to significantly inhibit the biological features of gliomas. The anti-apoptotic Bcl-2/BAX ratio was decreased by DHE, and it also promoted the liberation of cytochrome from the mitochondria into the cytosol, which stimulates the caspase-signaling pathway and causes mitochondria-mediated apoptosis. DHE has the ability to effectively reduce the biological features of glioma by diminishing the NF-κB/COX-2-signaling pathway. DHE can pass quickly through the blood–brain boundaries. These results offer preclinical proof that DHE could be developed as a possible treatment for glioma (Wang et al., 2021).

Natural substance alantolactone (ALT) has the ability to cure several diseases. Recent *in vitro* and *in vivo* research has shown that ALT has cytotoxic effects on a number of malignancies, including breast, colorectal, and liver cancers. The inhibitory effects of ALT are dependent on a number of signaling pathways linked to cancer as well as aberrant regulatory elements in cancer cells. It is a prospective therapeutic option for cancer therapy since it has exceptional anti-tumor capabilities in addition to exhibiting anti-inflammatory and antibacterial activities. A few studies have shown that ALT can work in concert with chemical cancer treatments such as quercetin and oxaliplatin to increase their anticancer benefits. ALT was also said to be able to improve the therapeutic sensitivity of cancer treatment (Cai et al., 2021).

Another compound, cynaropicrin, demonstrated viable action towards all strains of hepatitis C virus (HCV). It has also demonstrated a variety of additional pharmacological attributes, including antispasmodic, anti-trypanosomal, anti-malarial, anti-hyperlipidemic, antifeedant, anti-photoaging, and anti-tumor intervention, in addition to the stimulation of bitter sensory receptors and anti-inflammatory characteristics. The excellent impact of this drug in dealing with HCV is greatly helped by these pharmacological effects. Cynaropicrin can be formulated into therapeutic injections due to its solubility in water; it can be thought of as a viable medication candidate. This tiny chemical is also simple to produce (Elsebai et al., 2016).

Familiar for its antiviral, antimalarial, and anti-inflammatory properties, betulinic acid is a pentacyclic triterpene that occurs in nature (Chowdhury et al., 2002). More recently, topoisomerase inhibition research has shown that betulinic acid may also have anticancer effects. Caspase activation, changes to the mitochondrial membrane, and DNA fragmentation all occur in neuroectodermal tumor cells as a result of betulinic acid-induced apoptosis. The inactive proenzymes which are created as caspases are converted into active versions by the proteolytic process. When these proteases work together, caspases can activate one another in a process in a process known as a proteolytic cascade. When the caspase cascade begins, endonucleases such as caspase-activated DNAase (CAD) may become active. CAD helps with DNA deterioration after activation. By having an immediate impact on the mitochondria, betulinic acid causes apoptosis (Thurnher et al., 2003).

10.9 PHARMACOLOGICAL APPLICATIONS

A large number of bioactive components were discovered and extracted from various sections of *Saussurea costus* using a range of *in vitro* and *in vivo* techniques. These substances were found to have properties which were effective against fungal infections, diabetes, tumors, ulcers, inflammation, pesticides, and many more toxins.

10.9.1 Anticancer/Antitumor

S. costus root extract has been shown to have anti-cancerous properties on a number of tumors, particularly oral cancer (Moon et al., 2013), gastric cancer (Ko et al., 2005), prostate cancer (Kim et al., 2008), and lung cancer (Hung et al., 2010). Hexane extracts of *S. costus* roots caused apoptosis in human prostate cells. This means that prostate cancer cells cannot spread (Kim et al., 2008). This prevents prostate cancer cells from migrating (Kim et al., 2012). The investigation considered how autophagy and apoptosis interacted in systemic lupus erythematosus mediated prostate cells. The findings of this study showed that SLE regulates prostate cancer cells' ability to perish and to undergo autophagy, which thereby mediates the anticancer action of SLE (Tian et al., 2017). In the case of mouth cancer in humans, the roots of the plant have been proven to stop the spread of malignant cells. Western blotting and the DNA fragmentation test were used to conduct the research (Moon et al., 2013). The gastric malignant cells are affected by the ethanolic extract of the plant's roots, which has therapeutic powers for the disease, whether used conventionally or in conjunction with contemporary chemical therapy. This is a dosage and time-dependent approach, meaning that a specific dosage and a specific amount of time are needed for the desired results. For example, the treatment of gastric cancer cells requires 80 µg/ml and a period of 48 h (Ko et al., 2005). Similarly, breast cancer cell metastasis was inhibited by costunolide. After a few days of observation, the trials were completed on mice, with the results revealing a considerable reduction of breast cancer cell proliferation and metastasis. Therefore, it can be said that the costunolide obtained from *Saussurea* prevent the spreading, proliferation, and metastasis of breast cancer cells (Choi et al., 2013).

Dehydrocostus lactone from *Saussurea costus* displayed the highest level of cytotoxicity against a number of drug-resistant cell lines, leukemia cell lines, and acute myeloid leukemia (AML) stem cells taken from a patient's samples (Ding et al., 2016). However, normal hematopoietic cells showed no toxic effects. Enhancing the expression of the apoptosis-related protein Bax, which was preceded by caspase-3, caspase-9, and poly ADP-ribose polymerase cleavage (PARP), was the main mechanism by which dehydrocostus lactone induced apoptosis (Kumar et al., 2016). Dehydrocostus lactone and costunolide derived from *S. costus* were investigated for their ability to induce apoptosis and prevent metastasis against neuroblastoma cell lines using 3-(4,5-dimethylthiazol-2-yl)-2,5-diphenyltetrazolium bromide (MTT) assay (Tabata et al., 2015). According to the analysis, both substances displayed anti-metastatic action by preventing migration and invasion as well as cytotoxic effects on neuroblastoma cells through the induction of apoptosis as a result of caspase-7 activation. They are therefore potential resources for the creation of novel neuroblastoma-targeted anticancer medications. Combining costunolide and dehydrocostuslactone, both derived from *S. costus*, had a synergic anti-cancer effect that primarily worked through six distinct routes: inducing apoptosis, inhibiting expansion, controlling cell cycle, preventing invasion, and preventing both angiogenesis and metastasis (Lin et al., 2015).

10.9.2 Anti-inflammatory

The different bioactive substances which were extracted from *Saussurea costus* have an anti-inflammatory effect on both people and other living creatures. These compounds work in a dose-dependent pattern to reduce inflammation. The characteristics of acute inflammation include redness, heat, discomfort, and swelling. Inflammation is brought about by an increase in the flow of plasma and leukocytes from the circulation. *Costus* has a number of chemicals, including alantolactone, costic acid, and costunolide, which can help the healthcare sector in the development of medications to treat skin inflammation, which is a fairly frequent dermatological condition. Experiments using these substances on infection cells serve to demonstrate this effect (Lim et al., 2015).

10.9.3 Hepatoprotective

The active substances which have been found to exhibit the substantial inhibition of the hepatitis B surface antigen include costunolide and dehydrocostus. Results indicated that dehydrocostus lactone and costunolide could both be utilized in future anti-HBV medications (Chen et al., 1995). According to Alnahdi, male albino rats' plasma levels of liver biomarkers were decreased by a 300 mg/kg root aqueous extract of *S. costus* (Alnahdi et al., 2017). Similarly, *S. costus* aqueous and methanol extracts prevented mice from developing liver damage caused by D-galactosamine (D-GalN) and lipopolysaccharide (LPS) (Yaeesh et al., 2010). The hepatitis B surface antigen (HBsAg) was expressed in human hepatoma Hep3B cells in a dose-dependent pattern through the application of constunolide and dehydrocostus lactone, with IC50 values of 1.0 and 2.0 M, respectively. This inhibition occurred predominantly at the mRNA level. According to the results, costunolide and dehydrocostus lactone have the ability to be used as particular anti-HBV medication option (Chen et al., 1995).

10.9.4 Anti-ulcer

Dehydrocostuslactone and costunolide prevented the development of ethanol-induced stomach ulcers through the reduction of oxidative stress, suppressing inflammatory responses, and promoting cell growth (Zheng et al., 2016). As a result, *Saussurea costus* considerably accelerated the formation of a duodenal ulcer, improved intestinal electromyographic function, and accelerated intestinal thrust. It also raised cholecystokinin and motilin expression levels in the serum and duodenum. This provides a pharmacological foundation for the treatment of chronic duodenal ulcers (Zhang et al., 2016). The cytoprotective action of the medicine, through strengthening the stomach or duodenal mucosa and enhancing mucosal defense, caused the ethyl acetate extract of the *S. costus* root to inhibit gastric acid, free acid, and total acid at a dose of 400 mg/kg (Sutar et al., 2011).

The development of stomach ulcers induced by the withdrawal of water in mice was inhibited by acetone and costunolide from SLE. Wistar rats were used in the trials to test this characteristic, which involved administering the test substance orally in a dose-dependent manner (Madhavi et al., 2012). UL-409, a herbal formulation activity (Kaur et al., 2019) renowned for its antiulcer capability, is in *Saussurea costus*. To test this, male pigs and Wistar rats were each given 600mg/kg orally. The findings demonstrated a healing impact on stomach ulcers brought on by alcohol, aspirin, and resistance to the cold. It also increases stomach mucus secretion in all varieties of ulcers. The investigation demonstrated that UL-409 is an effective antiulcer drug (Kaur et al., 2019).

10.9.5 Immunomodulator

The hydroalcoholic extract of the roots of *S. costus* displayed the exceptional immunostimulatory action of the immune system, suggesting that it may be used as a long-term treatment for people with immunosuppression (Pandey, 2012). Other guaianolides, including the mokko lactone and dehydrocostus lactone from *S. costus*, also inhibited the cytotoxic T lymphocytes (CTLs) ability to kill and increased ICAM-1 expression (Pandey et al., 2007). Similar phagocytic action and DNA disintegration were observed at ID50 values between 3.6 and 10 μM for costunolide and dehydrocostus lactone, which were isolated from *S. costus*. At 13.6 M, costunolide considerably reduced granule exocytosis; at other concentrations, including 43.0, 13.6, 4.3, and 1.36 M, it even prevented an increase in tyrosine phosphorylation. According to the study, costunolide operates in a predictable way by suppressing CTL lethal actions and blocking the rise in tyrosine phosphorylation brought on by T-cell receptor cross-linking (Ali and Venkatesalu, 2022).

10.9.6 Antioxidant Activity

An effective antioxidant activity was demonstrated by a methanolic extract of *S. costus* root in a 2,2-diphenyl-1-picrylhydrazyl (DPPH) radical scavenging experiment (Mathew and Subramanian, 2014). The total phenolic and flavonoid composition of *S. costus* roots in n-butanol (1,000 ppm) showed the highest ability to inhibit DPPH radicals and reduce power and may be able to stop the return of antioxidant stress (Chang et al., 2012). The methanol fraction of the *S. costus* root had the best DPPH scavenging activity when compared to the other fractions, which included the chloroform, water, and ethanol fractions (Thara and Zuhra, 2012). In comparison to other fractions, chloroform fractions also demonstrated a considerable amount of DPPH scavenging activity. The ability of *S. costus* to absorb DPPH, nitric oxide, and superoxide radicals, as well as to prevent lipid peroxidation and GSH oxidation, was examined in order to determine its antioxidant activity. As a result, the plant's medicinal effects can be explained by its antioxidant properties that are most likely caused by the presence of chlorogenic acid (Pandey et al., 2005).

10.9.7 Antimicrobial Activity

Saussurea lappa's methanolic extracts show potent antibacterial activity (Parekh & Sumitra, 2007). This plant has been determined to help in the development of new medications to treat bacterial infections and has a broad range of potent antibacterial activities against different food-borne germs (Khalid et al., 2011). Gastric cancer, gastritis, and dyspepsia are only a few of the functional problems of the digestive tract caused by *Helicobacter pylori*. It can also induce endocrine issues (Kaur et al., 2019). Five distinct strains of *H. pylori* were examined, and in all of the tests the bacteria were found to be significantly inhibited by the ethanol extracts of *Saussurea lappa* (Li et al., 2005). *Streptococcus mutans'* water-insoluble glucan is prevented from growing, producing acid, or adhering by the ethanolic extracts from *Saussurea costus*. It operates in terms of dosage, which is about 0.5 mg/ml to 4 mg/ml (Yu et al., 2007). The roots of *Saussurea costus* shows activities against fungus, particularly against fungus that causes diseases. It was examined with 105 kinds of Indian plants, and significant antifungal properties were discovered (Kaur et al., 2019). *Aspergillus flavus, Aspergillus niger, Aspergillus ochraceus, Aspergillus flavus, Aspergillus versicolor, Penicilium funiculosum, Penicilium ochrochloron, Cladosporium cladosporioides, Trichodermaviride*, and *Alternaria* were the ten fungal strains against which different components were separated from *Saussurea costus* and examined. The substance had moderate-to-strong antifungal properties (Al Otibi et al., 2020).

10.9.8 Antiparasitic Activity

It was determined that *Saussurea costus* has antiparasitic properties against *Trypanosoma cruzi, Clonorchis sinensis*, and various nematodal diseases. The proportion decrease in the number of faecal eggs per gram (EPG) in infants who had acquired the worms naturally, *Saussurea costus*' antinematodal efficacy was examined (Onifade et al., 2013). Axeni cultures were used to evaluate *Saussurea lappa* extracts via *in vitro* experimental procedure against the epimastigote form of *Trypanosoma cruzi* (Lirussi et al., 2004). Ethyl acetate extract from the roots of *S. costus* and isolated compounds (viz., dehydrocostuslactone, zaluzanin D, α-cyclocostunolide, arbusculin B, costunolide, parthenolide, and eupatoriopicrin) were examined for *in vitro* antitrypanosomal activities and cytotoxicity against *Trypanosoma brucei rhodesiense*. With 96% inhibition at 4.8 μg/ml, the extract was particularly efficient against *T. b. rhodesiense*. However, the IC50s of isolated compounds were between 0.8 and 22 μM, and those for cytotoxic compounds were between 1.6 and 19 μM. The effectiveness of *Saussurea costus* as a novel medication to treat trypanosome infections was thus proven (Julianti et al., 2011).

10.9.9 Pesticidal Activity

The *Tribolium castaneum* (Herbst) (red flour beetle) was evaluated for resistance to the acetone extracts of *S. costus* leaves. The study found that plant extract might effectively replace synthetic insecticides in the management of large numbers of *T. castaneum* in store goods by having an effect on pupation and the proportion which reached adulthood (Abdallah et al., 2017). The larvae of the Culicidae mosquito, *Aedes albopictus*, were used as test subjects for the essential oil extracted from the roots of *S. costus* and its separated components (dehydrocostus lactone and costunolide). Dehydrocostus lactone and costunolide offered excellent larvicidal efficacy against *A. albopictus* (Liu et al., 2012). Additionally, the essential oil of *S. costus* was examined for its ability to deter adults from eating and preventing the growth of *Tribolium castaneum* larvae. It demonstrated outstanding repellency at 0.1 and 0.5% concentrations after 24 hours; after 48 hours, it was shown to be exceptionally effective (Madhavi et al., 2012). Investigation was also made into the effectiveness of *S. costus* and seven other plant species as repellents against *Tribolium castaneum* (herbst) and as effective antifeedants against *Rhyzopertha dominica*. Rhizomes from *S. costus* exhibited the strongest repulsive effects on *T. castaneum*. However, they lacked any antifeedant activity against *R. dominica* (Malik and Naqvi, 1984).

10.9.10 Angiogenesis Effect

The anti-angiogenic substance costunolid, which was isolated from the roots of *S. costus*, prevented endothelial cells from proliferating when vascular endothelial growth factor (VEGF) was present. Human umbilical vein endothelial cells (HUVECs) experienced a significant suppression of VEGF-induced chemotaxis at an IC50 of 3.4 M. Additionally, it was shown to block VEGFR KDR/Flk-1 via a signaling route (Rathore et al., 2021).

10.9.11 Cardiovascular Activity

The cardiovascular activity of an aqueous *S. costus* root material was examined in rats with myocardial damage brought on by isoproterenol. The extract considerably improved the levels of aspartate aminotransferase (AST), creatine kinase (CK), lactate dehydrogenase (LDH), thiobarbituric acid-reactive substances (TBARS) and glutathione (GSH)in the myocardium at three different doses (Saleem et al., 2013). The first three doses (0.5/g, 2.5/g, and 5.0/g) of the methanolic extract of *S. costus* roots displayed a prominent positive inotropic impact, but all dosages tested in the separated highly vascularized rabbit heart showed a prominent negative chronotropic influence on the velocity of cardiac fluid. These side effects resembled those of diltiazem and digoxin (Akhtar et al., 2013).

10.9.12 Spasmolysis and Analgesia

Constipation and spasms have both previously been treated with *Saussurea costus*. Several studies were conducted to verify the scientific validity of this characteristic, and the results showed that the aqueous-methanol raw matter of *Saussurea costus* has an anti-spasmodic action and can treat cramps. It has a diminishing effect on spasms at concentration-dependent doses (Gilani et al., 2007). The findings suggested that it has spasmolytic components. The research provides evidence that *Saussurea costus* root can be utilized to treat constipation by indicating that its crude extract may include cholinergic-type gut stimulatory components (Gilani et al., 2007).

10.9.13 Antiepileptic Activity

It was found that the ethanol extract of *Saussurea costus* root significantly reduces seizures brought on by both pentylenetetrazol (PTZ) as well as maximal electroshock (MES). PTZ-induced clonic seizures caused by *S. costus* at doses of 50, 100, and 200 mg/kg were greatly reduced in the tonic extensor phase of the convulsion and can significantly reduce this form of muscle spasm (Gupta Pushparaj et al., 2009).

10.9.14 Anticonvulsant

The roots of *Saussurea costus* were utilized in the Ayurveda tradition to cure epilepsy. Experiments utilizing various costus root extracts were conducted on mice to ascertain the scientific foundation of the anticonvulsant effect (Ambavade et al., 2009; Gupta Pushparaj et al., 2009). The outcomes showed the strong anticonvulsant properties of petroleum ether extract. As a result, the seizure threshold was raised through a GABAergic mechanism (Ambavade et al., 2009). The ethanolic extract prevents the convulsions brought on by pentylenetetrazol. Different techniques were used for the experiments (Gupta Pushparaj et al., 2009). The investigations stated above demonstrated that *Saussurea costus* had anticonvulsant properties.

10.9.15 Bronchitis

The experiments were conducted to learn more about the effects of several *S. costus* extracts on asthma and chronic bronchitis (Gautam & Asrani, 2018). In the treatment of bronchitis, asthma, and cough, it has proved very beneficial. The plant's essential oil helps in bronchial muscle relaxation (Qureshi et al., 2007). The alkaloidal portion in the extract was discovered to be not harmful as it had a lower impact on the blood pressure and breathing of the cats and rabbits (Vashisht, 2021). Preliminary findings thus seemed to corroborate the reputation of the medicine as a helpful treatment for asthma and chronic bronchitis.

10.9.16 Diabetes-fighting Activity

The methanol extract of *S. costus* was separated into mokko lactone, betulinic acid, dehydrocostus lactone, betulinic acid methyl ester, and nine additional substances. In albino rats, the ethanol extract of *S. costus* roots increased hepatic glycogen and elicited a hypoglycemic response. After the administration of medication, plasma insulin was seen to significantly decrease before returning to normal levels. These results support the Ayurvedic system's use of *S. costus* for the treatment of diabetes (Abdel-Rahman et al., 2020).

10.9.17 Antihyperlipidemic Activity

In a study, the ethanol extract of *Saussurea costus* was utilized to lower lipid profiles. Treatment with *S. costus* ethanol extract (200 and 400 mg/kg) prevented significant weight gain; however, it did not reduce food intake when compared to rats consuming a high-cholesterol diet (Gautam & Asrani, 2018).

10.9.18 Antidiarrheal Activity

At doses of 100, 300, and 500 mg/kg body mass, methanol extract of *S. costus* roots caused Wistar rats to exhibit diarrhea (Negi et al., 2013). Additionally, a dosage of 500 mg per kg of body weight generated outcomes that were remarkably similar to those of the common medication loperamide (68.02% at 5 mg/kg). The research backs up the conventional usage of *Saussurea costus* as an anti-diarrheal medication.

10.9.19 Miscellaneous

Cough relief is obtained through taking root powder of *Saussurea costus* with warm water (Kumar et al., 2009). Similarly, toothaches can also be treated with root powder and root juice combined with sweetener is used to treat rheumatism (Mahmood et al., 2011). The plant's natural essential oil also helps in the relaxation of bronchial muscles (Qureshi et al., 2007). In another application, typhoid fever can be treated with an extraction from the root (Rahmat et al. 2019). Root paste is taken with milk to cure scabies (Mala et al., 2012) and is used to treat dysentery (Kala, 2006). Due to its appealing scent, *Saussurea* oil is frequently used in perfumes, being blended with various fragrances. The above-ground parts of the plant are used both as firewood and as feed for animals (Butola & Samant, 2010).

10.10 CLINICAL TRIALS

S. costus has a numerous important conventional and medical usages. Chemicals taken from the plant's roots shows a variety of pharmacological activities that are employed to cure a variety of illnesses and ailments. In order to evaluate the LD50 and safety margin of several extracts of *S. costus* roots, acute intraperitoneal toxicity tests on Swiss albino mice were undertaken. The extract was administered to the mice at doses of 250, 500, 750, 1000, 1500, and 2000 mg/kg body weight. The mice were then monitored for 24 h to look for any toxic signs. The LD50 of more than 1000 mg/kg body mass can be considered appropriate because there was no fatality and no discernible impact on mass or soundness measures (Abou-Elella et al., 2019). Phikud Navakot, a blend of nine herbal plant species, including *S. costus*, is widely used in Thai traditional medicine to treat hyperlipidemia and cardiovascular disease. The ingested early onset and late onset (90-day) lethality of this mixture were studied (Kengkoom et al., 2012). Sprague Dawley rats received single and repeated doses of the crude extract in accordance with the OECD code 420 guideline for acute tests and the OECD code 408 guideline for sub-chronic tests, respectively, to measure various parameters. Acute toxicity tests conducted throughout the observation period revealed that none of the rats experienced any fatalities or toxicological symptoms after receiving the limit dose of 2000 mg/kg. No treatment-related harmful effects were found in the results of the sub-chronic tests. According to the study, Phikud Navakot was not involved in any fatalities and nor did it produce any discernible behaviorally, hematologically, or morphologically unfavorable consequences in acute and sub-chronic lethality trials in rats (Kengkoom et al., 2012).

The efficacy of *S. costus* alcoholic root extract as a subgingival irrigant as a supplement to scaling and root planing (SRP) was investigated in an experiment. In this study, 30 individuals with chronic periodontitis, aged 25 to 45, were split into three groups: the first group received SRP alone; the second group received SRP together with *S. costus* extract; and the third group received SRP combined with chlorhexidine as an irrigant. Over the course of the trial, periodontal metrics improved in all three groups. On the other hand, the *S. costus*-treated group distinguished itself significantly from the chlorhexidine-treated group, resulting in a decrease in periodontitis accompanied by fluid release and no negative effects. The findings suggested that *S. costus* subgingival irrigation may be a promising addition to conventional therapy methods for the early detection of periodontal infections (Trivedi and Dixit, 2015). Similar to this, when given orally with Brahmi ghrita at doses of 1, 2.5, and 5 g/kg, *S. costus* did not result in any fatalities or behavioral disorder in either an acute toxicity study or a sub-chronic toxicity study in Swiss strain albino mice (Ali and Venkatesalu, 2022).

The efficacy, stability, and affordability of *S. costus* roots as a migraine treatment were investigated in a clinical trial. Patients' calcium and magnesium levels drastically decreased during two weeks of therapy with *Saussurea costus*, although their serotonin levels increased considerably. While *S. costus* had no effect, the medications significantly reduced alkaline phosphatase levels. It was also discovered to be a reliable and effective supplementary therapeutic agent for migraine (Moujir et al., 2020). *Saussurea costus* was shown as being most effective for overweight hyperglycemia in detailed research on powerful hypoglycemic plants from different parts of India, which was done to determine the antidiabetic herbs utilized in Indian folklore (Singh & Pandey, 2016). Additionally, while giving birth those who smelled the essential oil of *S. costus* reported minor indications of labor discomfort. The medicine developed moderate drowsiness, eased anxiety, and had no harmful effects on either the mother or the fetus (Kour et al., 2015). The antinematodal effectiveness of *S. costus* roots was compromised by mixed nematodes, the majority of which carried *Ascaris lumbricoides*. The percent decrease in the number of fecal eggs per gram (EPG) counts was used to gauge the efficiency of interaction. On days 7 and 15, an oral dose of 40 and 50 mg/kg of powdered *S. costus* roots significantly reduced EPG count (Pc 0.05 or 0.001), with proportion declines equivalent to those found with 10 mg/kg pyrantel pamoate. The exact same results were obtained with a 50 mg/kg dosage of a methanolic extract of *S. costus* with no side effects (Onifade et al., 2013).

A research was done to determine the sensitive-inducing potential of recently found scent sensors, including methyl heptine carbonate (MHC) and *S. costus* root oil (SRO), compared to Dinitrochlorbenzene (DNCB). 150 patients with conditions such as tumors underwent sensitization using 20% SRO + 3% LAS in petrolatum, 30% MHC + 3% LAS in hydrophilic ointment, and 5% DNCB + 3% LAS in petrolatum at the same time. Following this, the patients underwent patch testing using 3% MHC, 3% LAS in petrolatum, and 0.1% DNCB in ethanol. The results showed that all DNCB-sensitized patients also had CRO and MHC sensitivities. Compared to DNCB, MHC and SRO were substantially less harmful. In the study, the use of these compounds had no consequences for people with persistent dermatitis (Ali and Venkatesalu, 2022).

A study was made of the combined effects of dehydrocostuslactone and costunolide, both extracted from the volatile oil of *Saussurea costus* root, on breast cancer cells (Peng et al. 2017). The Ingenuity Pathways Analysis (IPA) results indicated that both affect protein kinase A (PKA) signaling and 14-3-3-mediated and c-Myc-mediated cell death based on the differentially expressed proteins. Both p53 and p-14-3-3 levels were significantly upregulated, whilst c-Myc, p-AKT, and p-BID expressions were significantly downregulated. Cell cycle arrest and apoptosis were generated in a dose-dependent way. The findings imply that they trigger cell cycle arrest and death in breast cancer cells via the c-Myc/p53 and AKT/14-3-3 signaling pathways. With no obvious side effects, the volatile oil made from *Saussurea costus* root (VOSL) demonstrated a powerful therapeutic effect on SMMC-7721 xenografts. By stopping the cell cycle at the S and G2/M phases and triggering apoptosis in HCC cells by upregulating Caspase 3 mechanisms, VOSL reduced the development of HCC cells. Additionally, VOSL's MMP-9 suppression decreased the ability of HCC cells to penetrate and migrate. Additionally, a pharmacologic study discovered that VOSL inhibits the activation of the epithelial growth factor receptor, inhibiting the downstream effects of the MEK/P38 and PI3-K/Akt pathways. These findings suggest that VOSL might be a suitable therapeutic goal for the treatment of cancer (Lin et al., 2016). Dehydrocostus lactone, a newly discovered thioredoxin reductase inhibitor (TrxR1) isolated from the methanol extract, destroyed human cervical cancer HeLa cells by autophagy and suppressing TrxR1 function through causing an accumulation of reactive oxygen species (ROS) and a breakdown of the intracellular redox balance, which led to apoptosis (Hu et al., 2022).

The consequences of costunolide on cell viability, cell cycle, and apoptosis in human SK-MES-1 lung squamous carcinoma cells were examined using the MTT technique (Hua et al. 2016). In SK-MES-1 cells, costunolide led to morphological changes, decreased cell viability, induced apoptosis, and generated cell cycle arrest at the G1/S phase in a dosage-dependent pattern by up-regulating the expression of p53 and Bax and down-regulating the expression of Bcl2 and activating caspase 3. The outcomes demonstrated the potential anticancer effects of costunolide. A series of 1,4-disubstituted-1,2,3-triazoles conjugates synthesized through diastereo selective Michael addition followed by regioselective Huisgen 1,3-dipolar cycloaddition reactions were examined for their anticancer properties against the cancer cell lines DU-145, SIHA, MDA-MB-231, IMR-32, PANC 1, and A549 using the sulforodamine B assay. The results show that these counterparts were highly cytotoxic and numerous synthesized derivatives outperformed costunolide and dehydrocostus lactone in terms of anticancer efficacy (Ali and Venkatesalu, 2022).

To diagnose the MCF-7 xenograft mice, costunolide and dehydrocostus lactone, were examined in order to gauge their effectiveness and investigate their potential therapeutic framework of cancer development suppression. Owing to probable changes in serum and urine metabolic profiles, both were efficient in reducing physiologic fluctuations in MCF-7 xenograft mice and effectively prevented the formation of MCF-7 xenograft tumors (Peng et al., 2015). Similarly, a number of sesquiterpene mustards with mustard nitrogen and GSH-reactive-methylene-gamma-butyro-location groups were synthesized and developed as potential dysfunctional anticancer drugs (Xu et al., 2019).

An increase in liver glutathione (GSH), a reduction in hepatic malonaldehyde (MDA), and an inhibition of the activation of hepatic stellate cells (HSC) were all demonstrated to hasten the recovery and transition of liver fibrosis in rats (Elgharabawy et al., 2021). The cytotoxic effects of costunolide, which was synthesized from an extract of *S. costus* roots, were examined *in vitro* against breast cancer (MCF-7), mouse melanoma (B-16), lung cancer (A-549), cervical cancer (Hela), and prostate cancer (DU-145) cell lines. Several of the counterparts were more effective than costunolide against the DU-145 and MCF-7 cell lines (Vadaparthi et al., 2015).

10.11 CONCLUSION

Saussurea costus has various positive conventional and medical applications. In ethnobotany, it is employed to cure various ailments and conditions, such as headaches, nausea, leprosy, epilepsy, and typhoid. Chemicals extracted from the root of the plant also demonstrate a number of pharmacological activities, including anticancer, anti-inflammation, and antibacterial effects. *Saussurea costus* has seen an increase in interest, but there are still many unanswered questions regarding how it might be used. The first difference is that, despite *Saussurea costus* being widely recognized for its effectiveness in curing a variety of abnormalities, the potential pathway of action is still not known, and the active ingredients attached to therapeutic actions (such as those that are anti-diabetic, antioxidant, anti-diarrheal, anti-epileptic, and anticonvulsant) have not yet been identified. More studies should undoubtedly be prioritized in order to identify particular chemical components that are biolavailable and to better comprehend the mechanism involved of *Saussurea costus'* medicinal characteristics. A variety of pharmacological characteristics have been identified for cyclocostunolide, dehydrocostus lactone, alantolactone, costunolide, isodihydrocostunolide, isoalantolactone, and other active substances. However, some active ingredients could be turned into new medicines that might have the potential to cure a range of disorders. Examples of substances that could be utilized to diagnose inflammatory illnesses include costunolide, dehydrocostus lactone, and cynaropicrin.

It is deduced that *Saussurea costus* could indeed help in potential therapeutic and chemical research in future because of its term effectiveness and ethnobotanical purposes, as well as the existence of numerous substantial active ingredients that can lead to the retrieval and development of new chemicals. *Saussurea costus* is a critically endangered species that is limited to a small area of the Himalayas. The majority of the species' natural populations are either subject to destructive harvesting or have vanished entirely as a result of the enormous demands from the pharmaceutical sector. Therefore, a multi-branched strategy that incorporates *ex situ* and *in situ* preservation as well as the screening of superior genotypes and the subsequent amplification of those genotypes could be a long-term fix for the issue.

REFERENCES

Abdallah, E. M., Qureshi, K. A., Ali, A. M., & Elhassan, G. O. (2017). Evaluation of some biological properties of *Saussurea costus* crude root extract. *Bioscience Biotechnology Research Communications*, *10*(4), 601–611.

Abdel-Rahman, M., Rezk, M. M., Ahmed-Farid, O. A., Essam, S., & Abdel Moneim, A. E. (2020). *Saussurea lappa* root extract ameliorates the hazards effect of thorium induced oxidative stress and neuroendocrine alterations in adult male rats. *Environmental Science and Pollution Research*, *27*(12), 13237–13246.

Abou-Elella, F. M., Hanafy, E. A., Maoud, M. T., Kassem, H. A., & Kassem, N. M. (2019). C/EBPA as a biomarker gene in cancer cell and its relation to natural anticancer compounds. *Plant Archives*, *19*, 2693–2700.

Akhtar, M. S., Bashir, S., Malik, M. N. H., & Manzoor, R. (2013). Cardiotonic activity of methanolic extract of *Saussurea lappa* Linn roots. *Pakistan Journal of Pharmaceutical Sciences*, *26*(6), 1197–1201.

Alaagib, R. M. O., & Ayoub, S. M. H. (2015). On the chemical composition and antibacterial activity of *Saussurea lappa* (Asteraceae). *The Pharma Innovation*, *4*(2, Part C), 73.

Alhoot, M. A. (2021). Study of antiphlogistic effect of *Saussurea lappa* roots' ethanol extract in comparison to paracetamol effect. *Malaysian Journal of Microscopy*, *17*(1).

Ali, A., Alqaseer, K., Fatlawi, D., Shehab, S., Falah, M., Hassan, M., & Radhi, O. (2021). *Saussurea lappa*: An important medicinal plant for treatment different diseases: a review. *Kufa Journal for Nursing Sciences*, *11*(1), 1–19.

Ali, S. I., & Venkatesalu, V. (2022). Botany, traditional uses, phytochemistry and pharmacological properties of *Saussurea costus*: An endangered plant from Himalaya—A review. *Phytochemistry Letters*, *47*, 140–155.

Alnahdi, H. S. (2017). Injury in metabolic gland induced by pyrethroid insecticide could be reduced by aqueous extract of *Sassura lappa*. *International Journal of Pharmaceutical Research & Allied Sciences*, *6*(2).

Altemimi, A., Lakhssassi, N., Baharlouei, A., Watson, D. G., & Lightfoot, D. A. (2017). Phytochemicals: extraction, isolation, and identification of bioactive compounds from plant extracts. *Plants*, *6*(4), 42.

Al Otibi, F., Rizwana, H., Alharbi, R. I., Alshaikh, N., & Albasher, G. (2020). Antifungal effect of *Saussurea lappa* roots against phytopathogenic fungi and resulting morphological and ultrastructural changes. *Gesunde Pflanzen*, *72*(1), 57–67.

Ambavade, S. D., Mhetre, N. A., Muthal, A. P., & Bodhankar, S. L. (2009). Pharmacological evaluation of anticonvulsant activity of root extract of *Saussurea lappa* in mice. *European Journal of Integrative Medicine*, *1*(3), 131–137.

Anantharam, G., & Chittibabu, C. V. (2021). Herbal gardens of educational institutions and entrepreneurship development: a case study from south India. *Center for Development Economics*, *8*(10), 62–84.

Azwanida, N. N. (2015). A review on the extraction methods use in medicinal plants, principle, strength and limitation. *Medicinal & Aromatic Plants*, *4*(196), 2167–2412.

Bhan, M., Satija, S., Garg, C., Dureja, H., & Garg, M. (2017). Optimization of ionic liquid-based microwave assisted extraction of a diterpenoid lactone-andrographolide from Andrographis paniculata by response surface methodology. *Journal of Molecular Liquids*, *229*, 161–166.

Benedetto, C., D'Auria, M., Mecca, M., Prasad, P., Singh, P., Singh, S., & Milella, L. (2019). Chemical and biological evaluation of essential oil from *Saussurea costus* (Falc.) Lipsch. from Garhwal Himalaya collected at different harvesting periods. *Natural Product Research*, *33*(16), 2355–2358.

Butola, J. S., & Samant, S. S. (2010). *Saussurea* species in Indian Himalayan Region: diversity, distribution and indigenous uses. *International Journal of Plant Biology*, *1*(1), e9.

Cai, Y., Gao, K., Peng, B., Xu, Z., Peng, J., Li, J., & Yan, Y. (2021). Alantolactone: A natural plant extract as a potential therapeutic agent for cancer. *Frontiers in Pharmacology*, *12*, 781033.

Cao, K., Qian, W., Xu, Y., Zhou, Z., Zhang, Q., & Zhang, X. (2016). A new sesquiterpenoid from *Saussurea lappa* roots. *Natural Product Research*, *30*(19), 2160–2163.

Chang, K. M., Choi, S. I., & Kim, G. H. (2012). Anti-oxidant activity of *Saussurea lappa* C.B. clarke roots. *Preventive Nutrition and Food Science*, *17*(4), 306–309.

Chang, K. M., & Kim, G. H. (2008). Comparison of volatile aroma components from *Saussurea lappa* CB Clarke root oils. *Preventive Nutrition and Food Science*, *13*(2), 128–133.

Chen, H. C., Chou, C. K., Lee, S. D., Wang, J. C., & Yeh, S. F. (1995). Active compounds from *Saussurea lappa* Clarks that suppress hepatitis B virus surface antigen gene expression in human hepatoma cells. *Antiviral Research*, *27*(1–2), 99–109.

Choodej, S., Pudhom, K., Yamauchi, K., & Mitsunaga, T. (2019). Inhibition of melanin production by sesquiterpene lactones from *Saussurea lappa* and their analogues. *Medicinal Chemistry Research*, *28*(6), 857–862.

Choi, H. G., Lee, D. S., Li, B., Choi, Y. H., Lee, S. H., & Kim, Y. C. (2012). Santamarin, a sesquiterpene lactone isolated from *Saussurea lappa*, represses LPS-induced inflammatory responses via expression of heme oxygenase-1 in murine macrophage cells. *International Immunopharmacology*, *13*(3), 271–279.

Choi, J. Y., Na, M., Hyun Hwang, I., Ho Lee, S., Young Bae, E., Yeon Kim, B., & Seog Ahn, J. (2008). Isolation of betulinic acid, its methyl ester and guaiane sesquiterpenoids with protein tyrosine phosphatase 1B inhibitory activity from the roots of *Saussurea lappa* CB Clarke. *Molecules*, *14*(1), 266–272.

Choi, Y. K., Cho, S. G., Woo, S. M., Yun, Y. J., Jo, J., Kim, W., & Ko, S. G. (2013). *Saussurea lappa* Clarke-derived costunolide prevents TNFα-induced breast cancer cell migration and invasion by inhibiting NF-κB activity. *Evidence-Based Complementary and Alternative Medicine*, *2013*, 936257.

Chowdhury, A. R., Mandal, S., Mittra, B., Sharma, S., Mukhopadhyay, S., & Majumder, H. K. (2002). Betulinic acid, a potent inhibitor of eukaryotic topoisomerase I: identification of the inhibitory step, the major functional group responsible and development of more potent derivatives. *Medical Science Monitor, 8*(7), 254–260

de Kraker, J. W., Franssen, M. C., de Groot, A., Shibata, T., & Bouwmeester, H. J. (2001). Germacrenes from fresh costus roots. *Phytochemistry, 58*(3), 481–487.

Ding, Y. H., Gao, X., Long, J., Kuang, B. J., Chen, Y., & Zhang, Q. (2016). Di spiro-cyclo-propyldehydrocostus lactone selectively inhibits acute myelogenous leukemia cells. *Bioorganic & Medicinal Chemistry Letters, 26*(4), 1165–1168.

Doughari, J. H. (2012). *Phytochemicals: extraction methods, basic structures and mode of action as potential chemotherapeutic agents* (pp. 1–33). Intech Open.

Dubey, A., Gupta, S., Khatoon, M., & Gupta, A. K. (2022). *Saussurea costus* (Kust) and *Senna alexandrina* (Senna). In A. Husen (Ed.), *Herbs, Shrubs, and Trees of Potential Medicinal Benefits* (pp 261–290). CRC Press.

El-Far, A., Shaheen, H., Alsenosy, A., El-Sayed, Y., Al Jaouni, S., & Mousa, S. (2018). Costus speciosus: traditional uses, phytochemistry, and therapeutic potentials. *Pharmacognosy Reviews, 12*(23).

Elsebai, M. F., Mocan, A., & Atanasov, A. G. (2016). Cynaropicrin: A comprehensive research review and therapeutic potential as an anti-hepatitis C virus agent. *Frontiers in Pharmacology, 7*, 472.

Elgharabawy, R. M., El Sayed, I. E. T., Rezk, N. A. A., & Tousson, E. (2021). Therapeutic impact of Costus (*Saussurea lappa*) against ehrlich solid tumor-induced cardiac toxicity and DNA damage in female mice. *Frontiers in Pharmacology, 12*.

Elshaer, S. E., Hamad, G. M., Hafez, E. E., Baghdadi, H. H., El-Demerdash, F. M., & Simal-Gandara, J. (2022). Root extracts of *Saussurea costus* as prospective detoxifying food additive against sodium nitrite toxicity in male rats. *Food and Chemical Toxicology, 166*, 113225.

Gautam, H., & Asrani, R. K. (2018). Phytochemical and pharmacological review of an ethno medicinal plant: *Saussurea lappa. Veterinary Research International, 6*, 1–9.

Gheraibia, S., Belattar, N., & Abdel-Wahhab, M. A. (2020). HPLC analysis, antioxidant and cytotoxic activity of different extracts of *Costus speciosus* against HePG-2 cell lines. *South African Journal of Botany, 131*, 222–228.

Gilani, A. H., Shah, A. J., & Yaeesh, S. (2007). Presence of cholinergic and calcium antagonist constituents in *Saussurea lappa* explains its use in constipation and spasm. *Phytotherapy Research, 21*(6), 541–544.

Gupta Pushpraj, S., Jadhav, S. S., Ghaisas, M. M., & Deshpande, A. D. (2009). Anticonvulsant activity of *Saussurea lappa. Pharmacologyonline, 3*, 809–814.

Guan, Z., Prokop, J., Roques, P., Lapeyrie, J., & Nel, A. (2015). Revision of the enigmatic insect family *Anthracoptilidae* enlightens the evolution of Palaeozoic stem-dictyopterans. *Acta Palaeontologica Polonica, 61*(1), 71–87.

Gwari, G., Bhandari, U., Andola, H. C., Lohani, H., & Chauhan, N. (2013). Volatile constituents of *Saussurea costus* roots cultivated in Uttarakhand Himalayas, India. *Pharmacognosy Research, 5*(3), 179.

Heftmann, E. (2004). *Chromatography: Fundamentals and Applications of Chromatography and Related Differential Migration Methods-Part B: Applications*. Elsevier.

Hossain, M. A., Al-Hdhrami, S. S., Weli, A. M., Al-Riyami, Q., & Al-Sabahi, J. N. (2014). Isolation, fractionation and identification of chemical constituents from the leaves crude extracts of *Mentha piperita* L grown in sultanate of Oman. *Asian Pacific Journal of Tropical Biomedicine, 4*, 368–372.

Hua, P., Zhang, G., Zhang, Y., Sun, M., Cui, R., Li, X., & Zhang, X. (2016). Costunolide induces G1/S phase arrest and activates mitochondrial-mediated apoptotic pathways in SK-MES 1 human lung squamous carcinoma cells. *Oncology Letters, 11*(4), 2780–2786.

Hung, J. Y., Hsu, Y. L., Ni, W. C., Tsai, Y. M., Yang, C. J., Kuo, P. L., & Huang, M. S. (2010). Oxidative and endoplasmic reticulum stress signaling are involved in dehydrocostuslactone-mediated apoptosis in human non-small cell lung cancer cells. *Lung Cancer, 68*(3), 355–365.

Hu, X. F., Liu, W. X., Zhang, R., Zhang, W., Wang, C., Chen, M., & Wang, Q. (2022). Essential oil from s inhibits proliferation and migration of Eca109 cells via mitochondrial apoptosis and STAT3 signaling. *Asian Pacific Journal of Tropical Biomedicine, 12*(6), 253.

Idriss, H., Siddig, B., Maldonado, P. G., Elkhair, H. M., Alakhras, A. I., Abdallah, E. M., & Elzupir, A. O. (2022). Phytochemical discrimination, biological activity and molecular docking of water-soluble inhibitors from *Saussurea costus* herb against main protease of SARS-CoV-2. *Molecules, 27*(15), 4908.

Ingle, K. P., Deshmukh, A. G., Padole, D. A., Dudhare, M. S., Moharil, M. P., & Khelurkar, V. C. (2017). Phytochemicals: Extraction methods, identification and detection of bioactive compounds from plant extracts. *Journal of Pharmacognosy and Phytochemistry, 6*(1), 32–36.

Julianti, T., Hata, Y., Zimmermann, S., Kaiser, M., Hamburger, M., & Adams, M. (2011). Antitrypanosomal sesquiterpene lactones from *Saussurea costus. Fitoterapia, 82*(7), 955–959.

Kaur, L., Singh, A., Kaur, J., & Kaur, N. (2019). A brief review of remedial uses of *Saussurea lappa*. *Journal of Pharmacognosy and Phytochemistry*, *8*(3), 4423–4430.

Kala, C. P. (2006). Medicinal plants of the high altitude cold desert in India: diversity, distribution and traditional uses. *The International Journal of Biodiversity Science and Management*, *2*(1), 43–56.

Kengkoom, K., Chaimongkolnukul, K., Cherdyu, S., Inpunkaew, R., & Ampawong, S. (2012). Acute and sub-chronic oral toxicity studies of the extracts from herbs in Phikud Navakot. *African Journal of Biotechnology*, *11*(48), 10903–10911.

Khalid, A., Rehman, U. U., Sethi, A., Khilji, S., Fatima, U., Khan, M. I., & Murtaza, G. (2011). Antimicrobial activity analysis of extracts of *Acacia modesta, Artemisia absinthium, Nigella sativa* and *Saussurea lappa* against Gram positive and Gram negative microorganisms. *African Journal of Biotechnology*, *10*(22), 4574–4580.

Kim, D. Y., & Choi, B. Y. (2019). Costunolide—A bioactive sesquiterpene lactone with diverse therapeutic potential. *International Journal of Molecular Sciences*, *20*(12), 2926.

Kim, E. J., Hong, J. E., Lim, S. S., Kwon, G. T., Kim, J., Kim, J. S., & Park, J. H. Y. (2012). The hexane extract of *Saussurea lappa* and its active principle, dehydrocostus lactone, inhibit prostate cancer cell migration. *Journal of Medicinal Food*, *15*(1), 24–32.

Kim, E. J., Lim, S. S., Park, S. Y., Shin, H. K., Kim, J. S., & Park, J. H. Y. (2008). Apoptosis of DU145 human prostate cancer cells induced by dehydrocostus lactone isolated from the root of *Saussurea lappa*. *Food and Chemical Toxicology*, *46*(12), 3651–3658.

Ko, S. G., Kim, H. P., Jin, D. H., Bae, H. S., Kim, S. H., Park, C. H., & Lee, J. W. (2005). *Saussurea lappa* induces G2-growth arrest and apoptosis in AGS gastric cancer cells. *Cancer Letters*, *220*(1), 11–19.

Kour, S., Singh, S., & Kaloo, Z. A. (2015). Conservation Strategies of *Saussurea costus*, Critically Endangered Medicinal Herb Growing in Kashmir Himalaya—A Review. *International Journal of Science and Research (IJSR)*, *4*(7), 257–260.

Kumar, J., & Pundir, M. (2021). Phytochemistry and pharmacology of *Saussurea* genus (*Saussurea lappa, Saussurea costus, Saussurea obvallata, Saussurea involucrata*). *Materials Today: Proceedings*, *56*(3), 1173–1181.

Kumar, M., Kaur, V., Kumar, S., & Kaur, S. (2016). Phytoconstituents as apoptosis inducing agents: strategy to combat cancer. *Cytotechnology*, *68*(4), 531–563.

Kumar, M., Paul, Y., & Anand, V. K. (2009). An ethnobotanical study of medicinal plants used by the locals in Kishtwar, Jammu and Kashmir, India. *Ethnobotanical Leaflets*, *2009*(10), 5.

Li, Y., Xu, C., Zhang, Q., Liu, J. Y., & Tan, R. X. (2005). In vitro anti-Helicobacter pylori action of 30 Chinese herbal medicines used to treat ulcer diseases. *Journal of Ethnopharmacology*, *98*(3), 329–333.

Lim, H. S., Jin, S. E., Kim, O. S., Shin, H. K., & Jeong, S. J. (2015). Alantolactone from *Saussurea lappa* exerts antiinflammatory effects by inhibiting chemokine production and STAT1 phosphorylation in TNF-α and IFN-γ-induced in HaCaT cells. *Phytotherapy Research*, *29*(7), 1088–1096.

Lin, X., Peng, Z., Fu, X., Liu, C., Xu, Y., Ji, W., & Su, C. (2016). Volatile oil *from Saussurea lappa* exerts antitumor efficacy by inhibiting epithelial growth factor receptor tyrosine kinase-mediated signaling pathway in hepatocellular carcinoma. *Oncotarget*, *7*(48), 79761.

Lin, X., Peng, Z., & Su, C. (2015). Potential anti-cancer activities and mechanisms of costunolide and dehydrocostuslactone. *International Journal of Molecular Sciences*, *16*(5), 10888–10906.

Lirussi, D., Li, J., Prieto, J. M., Gennari, M., Buschiazzo, H., Rios, J. L., & Zaidenberg, A. (2004). Inhibition of Trypanosoma cruzi by plant extracts used in Chinese medicine. *Fitoterapia*, *75*(7–8), 718–723.

Liu, Z. L., He, Q., Chu, S. S., Wang, C. F., Du, S. S., & Deng, Z. W. (2012). Essential oil composition and larvicidal activity of *Saussurea lappa* roots against the mosquito *Aedes albopictus (Diptera: Culicidae)*. *Parasitology Research*, *110*(6), 2125–2130.

Lunz, K., & Stappen, I. (2021). Back to the roots—an overview of the chemical composition and bioactivity of selected root-essential oils. *Molecules*, *26*(11), 3155.

Madhavi, M., Mallika, G., Lokanath, N., Vishnu, M. N., Chetty, C. M., & Saleem, T. M. (2012). A review on phytochemical and pharmacological aspects of *Saussurea lappa*. *International Journal of Life Science and Medical Research*, *2*, 24–31.

Madhuri, K., Elango, K., & Ponnusankar, S. (2012). *Saussurea lappa* (Kuth root): review of its traditional uses, phytochemistry and pharmacology. *Oriental Pharmacy and Experimental Medicine*, *12*(1), 1–9.

Mahmood, A., Malik, R. N., Shinwari, Z. K., & Mahmood, A. (2011). Ethnobotanical survey of plants from Neelum, Azad Jammu and Kashmir, Pakistan. *Pakistan Journal of Botany*, *43*(1), 105–110.

Majekodunmi, S. O. (2015). Review of extraction of medicinal plants for pharmaceutical research. *Merit Research Journal of Medicine and Medical Sciences*, *3*, 521–527.

Mala, F. A., Lone, M. A., Lone, F. A., & Arya, N. (2012). Ethno-medicinal survey of Kajinaag range of Kashmir Himalaya, India. *International Journal of Pharma and Bio Sciences*, *3*(2), 442–449.

Malik, M. M., & Naqvi, S. H. M. (1984). Isolation and identification of insect repellent compound from *Saussurea lappa* CB Clarke. *Pakistan Journal of Scientific and Industrial Research*, *27*(3), 144–145.

Mathew, M., & Subramanian, S. (2014). In vitro screening for anti-cholinesterase and antioxidant activity of methanolic extracts of ayurvedic medicinal plants used for cognitive disorders. *PloS One*, *9*(1), e86804.

Matsuda, H., Kageura, T., Inoue, Y., Morikawa, T., & Yoshikawa, M. (2000). Absolute stereostructures and syntheses of *Saussurea*mines A, B, C, D and E, amino acid–sesquiterpene conjugates with gastroprotective effect, from the roots of *Saussurea lappa*. *Tetrahedron*, *56*(39), 7763–7777.

Moon, S. M., Yun, S. J., Kook, J. K., Kim, H. J., Choi, M. S., Park, B. R., & Kim, C. S. (2013). Anticancer activity of *Saussurea lappa* extract by apoptotic pathway in KB human oral cancer cells. *Pharmaceutical Biology*, *51*(11), 1372–1377.

Moujir, L., Callies, O., Sousa, P. M., Sharopov, F., & Seca, A. M. (2020). Applications of sesquiterpene lactones: A review of some potential success cases. *Applied Sciences*, *10*(9), 3001.

Nadda, R. K., Ali, A., Goyal, R. C., Khosla, P. K., & Goyal, R. (2020). *Aucklandia costus* (syn. *Saussurea costus*): Ethnopharmacology of an endangered medicinal plant of the Himalayan region. *Journal of Ethnopharmacology*, *263*, 113199.

Negi, J. S., Bisht, V. K., Bhandari, A. K., Bhatt, V. P., Sati, M. K., Mohanty, J. P., & Sundriyal, R. C. (2013). Antidiarrheal activity of methanol extract and major essential oil contents of *Saussurea lappa* Clarke. *African Journal of Pharmacy and Pharmacology*, *7*(8), 474–477.

Olennikov, D. N., Tankhaeva, L. M., & Rokhin, A. V. (2011). Glucofructans from *Saussurea lappa* roots. *Chemistry of Natural Compounds*, *47*(3), 339–342.

Onifade, A. A., Jewell, A. P., & Adedeji, W. A. (2013). Nigella sativa concoction induced sustained seroreversion in HIV patient. *African Journal of Traditional, Complementary and Alternative Medicines*, *10*(5), 332–335.

Parekh, J., & Sumitra, C. (2007). Antibacterial and phytochemical studies on twelve species of Indian medicinal plants. *African Journal of Biomedical Research*, *10*(2), 175–181.

Pandey, A., & Tripathi, S. (2014). Concept of standardization, extraction and pre phytochemical screening strategies for herbal drug. *Journal of Pharmacognosy and Phytochemistry*, *2*(5).

Pandey, M. M., Govindarajan, R., Rawat, A. K. S., Pangtey, Y. S., & Mehrotra, S. (2004). High-performance liquid chromatographic method for quantitative estimation of an antioxidant principle chlorogenic acid in *Saussarea costus* and *Arctium lappa*. *Natural Product Sciences*, *10*(1), 40–42.

Pandey, M. M., Govindarajan, R., Rawat, A. K. S., & Pushpangadan, P. (2005). Free radical scavenging potential of *Saussarea costus*. *Acta Pharmaceutica*, *55*(3), 297–304.

Pandey, M. M., Rastogi, S., & Rawat, A. K. S. (2007). *Saussurea costus*: botanical, chemical and pharmacological review of an ayurvedic medicinal plant. *Journal of Ethnopharmacology*, *110*(3), 379–390.

Pandey, R. S. (2012). *Saussurea lappa* extract modulates cell mediated and humoral immune response in mice. *Der Pharmacia Lettre*, *4*(6), 1868–1873.

Peng, Z. X., Wang, Y., Gu, X., Xue, Y., Wu, Q., Zhou, J. Y., & Yan, C. (2015). Metabolic transformation of breast cancer in a MCF-7 xenograft mouse model and inhibitory effect of volatile oil from *Saussurea lappa* Decne treatment. *Metabolomics*, *11*(3), 636–656.

Peng, Z., Wang, Y., Fan, J., Lin, X., Liu, C., Xu, Y., & Su, C. (2017). Costunolide and dehydrocostuslactone combination treatment inhibit breast cancer by inducing cell cycle arrest and apoptosis through c-Myc/p53 and AKT/14-3-3 pathway. *Scientific Reports*, *7*(1), 1–16.

Qureshi, R. A., Ghufran, M. A., Gilani, S. A., Sultana, K., & Ashraf, M. (2007). Ethnobotanical studies of selected medicinal plants of Sudhan Gali and Ganga Chotti hills, district Bagh, Azad Kashmir. *Pakistan Journal of Botany*, *39*(7), 2275–2283.

Raafat, K., El-Darra, N., Saleh, F. A., Rajha, H. N., & Louka, N. (2019). Optimization of infrared-assisted extraction of bioactive lactones from *Saussurea lappa* L. and their effects against gestational diabetes. *Pharmacognosy Magazine*, *15*(61), 208.

Rahmat, W., Rahman, K., Raja, N. I., Qureshi, R., & Bussmann, R. W. (2019). A quantitative medico-botanical expedition of Fairy Meadows National Park, Diamir, Gilgit Baltistan, Pakistan. *Ethnobotany Research and Applications*, *18*, 1–30.

Rathore, S., Debnath, P., & Kumar, R. (2021). Kuth *Saussurea costus* (Falc.) Lipsch: A critically endangered medicinal plant from Himalaya. *Journal of Applied Research on Medicinal and Aromatic Plants*, *20*, 100277.

Rasul, A., Parveen, S., & Ma, T. (2012). Costunolide: A novel anti-cancer sesquiterpene lactone. *Bangladesh Journal of Pharmacology*, *7*(1), 6–13.

Rimando, A. M., Olofsdotter, M., Dayan, F. E., & Duke, S. O. (2001). Searching for rice allelochemicals: An example of bioassay-guided isolation. *Agronomy Journal*, *93*(1), 16–20.

Robinson, A., Kumar, T. V., Sreedhar, E., Naidu, V. G. M., Krishna, S. R., Babu, K. S., & Rao, J. M. (2008). A new sesquiterpene lactone from the roots of *Saussurea lappa*: Structure–anticancer activity study. *Bioorganic & Medicinal Chemistry Letters*, *18*(14), 4015–4017.

Saleem, T. M., Lokanath, N., Prasanthi, A., Madhavi, M., Mallika, G., & Vishnu, M. N. (2013). Aqueous extract of *Saussurea lappa* root ameliorate oxidative myocardial injury induced by isoproterenol in rats. *Journal of Advanced Pharmaceutical Technology & Research*, *4*(2), 94.

Sasidharan, S., Chen, Y., Saravanan, D., Sundram, K. M., & Latha, L. Y. (2011). Extraction, isolation and characterization of bioactive compounds from plants' extracts. *African Journal of Traditional, Complementary and Alternative Medicines*, *8*(1).

Semmler, F. W., & Feldstein, J. (1914). Zur Kenntnis der Bestandteile der ätherischen Öle.(Über das Vorkommen einer Säure C_{15} H_{20} O_2 im Costuswurzelöl). *Berichte der deutschen chemischen Gesellschaft*, *47*(3), 2433–2437.

Semwal, R., Joshi, K., Pandian, A., Badoni, P., & Semwal, D. (2020). Biological applications and secondary metabolites of *Saussurea costus* (Falc.) Lipsch. *Journal of Conventional Knowledge and Holistic Health*, *4*, 201.

Singh, J., & Pandey, A. K. (2016). A review: an impact of *Inula racemosa* (puskarmula) on dyslipidemia and obesity. *World Journal of Scientific Research*, *5*, 528–538.

Singh, R., Chahal, K. K., & Singla, N. (2017). Chemical composition and pharmacological activities of *Saussurea lappa*: A review. *Journal of Pharmacognosy and Phytochemistry*, *6*(4), 1298–1308.

Sun, C. M., Syu, W. J., Don, M. J., Lu, J. J., & Lee, G. H. (2003). Cytotoxic sesquiterpene lactones from the root of *Saussurea lappa*. *Journal of Natural Products*, *66*(9), 1175–1180.

Sutar, N., Garai, R., Sharma, U. S., Singh, N., & Roy, S. D. (2011). Antiulcerogenic activity of *Saussurea lappa* root. *International Journal of Pharmacy and Life Sciences*, *2*(1), 516–520.

Tabata, K., Nishimura, Y., Takeda, T., Kurita, M., Uchiyama, T., & Suzuki, T. (2015). Sesquiterpene lactones derived from *Saussurea lappa* induce apoptosis and inhibit invasion and migration in neuroblastoma cells. *Journal of Pharmacological Sciences*, *127*(4), 397–403.

Tag, H. M., Khaled, H. E., Ismail, H. A., & El-Shenawy, N. S. (2016). Evaluation of anti-inflammatory potential of the ethanolic extract of the *Saussurea lappa* root (costus) on adjuvant-induced monoarthritis in rats. *Journal of Basic & Clinical Physiology & Pharmacology*, *27*(1), 71–78.

Thara, K. M., & Zuhra, K. F. (2012). Comprehensive in-vitro pharmacological activities of different extracts of *Saussurea lappa*. *European Journal of Experimental Biology*, *2*(2), 417–420.

Thurnher, D., Turhani, D., Pelzmann, M., Wannemacher, B., Knerer, B., Formanek, M., & Selzer, E. (2003). Betulinic acid: A new cytotoxic compound against malignant head and neck cancer cells. *Head & Neck: Journal for the Sciences and Specialties of the Head and Neck*, *25*(9), 732–740.

Tian, X., Song, H. S., Cho, Y. M., Park, B., Song, Y. J., Jang, S., & Kang, S. C. (2017). Anticancer effect of *Saussurea lappa* extract via dual control of apoptosis and autophagy in prostate cancer cells. *Medicine*, *96*(30).

Trivedi, S., & Dixit, J. (2015). Clinical effects of irrigation with *Saussurea lappa* extract in the treatment of chronic periodontitis: A pilot randomized trial. *Journal of Orofacial Sciences*, *7*(2), 80.

Vadaparthi, P. R., Kumar, K., Sarma, V. U. M., Hussain, Q. A., & Babu, K. S. (2015). Estimation of costunolide and dehydrocostus lactone in *Saussurea lappa* and its polyherbal formulations followed by their stability studies using HPLC-DAD. *Pharmacognosy Magazine*, *11*(41), 180.

Vashisht, I. (2021). *Saussurea lappa*. In N. Malhotra, & M. Singh (Eds.), *Himalayan Medicinal Plants* (pp. 173–197). Academic Press, Elsevier.

Vijayalakshmi, M., Umamaheswari, A., Agalya, B., Jegasubramaniam, S. N., & Prabu, S. L. (2022). Foresight on phytoconstituents and associated pharmacological activities of traditional medicinal plant: *Saussurea costus* (Falc.) Lipschitz. *Current Pharmacology Reports*, *8*(4), 1–9.

Wang, F., Xie, Z. H., Gao, Y., Xu, Y., Cheng, X. L., & Liu, J. K. (2008). Sulfonated guaianolides from *Saussurea lappa*. *Chemical and Pharmaceutical Bulletin*, *56*(6), 864–865.

Wang, J., Yu, Z., Wang, C., Tian, X., Huo, X., Wang, Y., & Xu, Y. (2021). Erratum: Dehydrocostus lactone, a natural sesquiterpene lactone, suppresses the biological characteristics of glioma, through inhibition of the NF-κB/COX-2 signaling pathway by targeting IKKβ. *American Journal of Cancer Research*, *11*(3), 1009–1011.

Xiong, W., Chen, X., Lv, G., Hu, D., Zhao, J., & Li, S. (2016). Optimization of microwave-assisted extraction of bioactive alkaloids from lotus plumule using response surface methodology. *Journal of Pharmaceutical Analysis*, *6*(6), 382–388.

Xu, L. S., Herrando-Moraira, S., Susanna, A., Galbany-Casals, M., & Chen, Y. S. (2019). Phylogeny, origin and dispersal of *Saussurea* (Asteraceae) based on chloroplast genome data. *Molecular Phylogenetics and Evolution, 141*, 106613.

Yadav, N., Sharma, S., Joys, J. S., & Kumar, S. (2020). Microwave assisted extraction of bioactive compounds: A brief review. *Journal of the Indian Chemical Society, 97*(10), 1751–1756.

Yaeesh, S., Jamal, Q., Shah, A. J., & Gilani, A. H. (2010). Antihepatotoxic activity of *Saussurea lappa* extract on D-galactosamine and lipopolysaccharide-induced hepatitis in mice. *Phytotherapy Research, 24*(2), 229–232.

Yu, H. H., Lee, J. S., Lee, K. H., Kim, K. Y., & You, Y. O. (2007). *Saussurea lappa* inhibits the growth, acid production, adhesion, and water-insoluble glucan synthesis of *Streptococcus mutans*. *Journal of Ethnopharmacology, 111*(2), 413–417.

Zahara, K., Tabassum, S., Sabir, S., Arshad, M., Qureshi, R., Amjad, M. S., & Chaudhari, S. K. (2014). A review of therapeutic potential of *Saussurea lappa*—An endangered plant from Himalaya. *Asian Pacific Journal of Tropical Medicine, 7*, 60–69.

Zhang, T. Y., Wang, P., Wang, Q., Liang, Q. K., Wang, Y. D., Ling, X. C., & Li, H. F. (2016). *Saussurea lappa* modulates gastrointestinal motility, motilin and cholecystokinin expression in ulcer rats. *Journal of Gastroenterology and Hepatology Research, 5*(2), 2021–2027.

Zheng, H., Chen, Y., Zhang, J., Wang, L., Jin, Z., Huang, H., & Gao, W. (2016). Evaluation of protective effects of costunolide and dehydrocostuslactone on ethanol-induced gastric ulcer in mice based on multi-pathway regulation. *Chemico-Biological Interactions, 250*, 68–77.

11 Dandelion (*Taraxacum officinale*)

Santwana Palai and Raees ul Haq

11.1 INTRODUCTION

In recent years, the rediscovery of phytoalimurgic plants for use as food ingredients to benefit human health has increased interest in their use. These edible wild plants are grown in their natural environment to produce food throughout the year, and they also include medicinal plants for their curative properties (Grauso et al., 2019).

Dandelion (*Taraxacum officinale*) is a common medicinal plant in Asia, Europe, South & North America. Around 250 species of the plant are known to exist (Dudman & Richards, 1997) particularly found throughout the warm-temperate zones of the Northern Hemisphere. *Taraxacum officinale*, or the common dandelion, has been used as common medicine in China, India, and Russia since ancient times. It has also been used as a traditional medicine in Ayurveda-, Siddha-, and Unani-style systems. In other cultures, it is popularly used in Turkey and Mexico for the treatment of diabetes (Al-Ahmad, 2018).

Taraxacum officinale, or "common dandelion/Bitterwort/Irish Daisy", is an affiliate of the family Asteraceae. The term "dandelion" is derived from the French term "dent de lion", or "lion's tooth". Taraxis, which means "benefit for inflammation", is the scientific name given to the dandelion family. It has been used in traditional medicine since olden times as a hepatoprotective antioxidant, diuretic, choleretic, anti-inflammatory, and laxative plant. It is also used as a common food source due to its relatively high presence of minerals, fibre, vitamins, and essential fatty acids. Around the world, the use of dandelion roots has been recorded in at least 54 different cuisines. It is commonly consumed as a food (in salads) due to its high concentration of micronutrients such as minerals and vitamins (Abdel-Magied et al., 2019). Dandelion is a perennial herbaceous plant (Figure 11.1) which is constituted of a fleshy taproot and a rosette system of leaves, which are narrow, long, and lance-shaped structures. White fuffballs (Figure 11.2) from the plant represent the matured blossoms, which appear from yellow flowers and are full of winged seeds (Yelverton 2017).

Dandelion has many medicinal properties, including in the treatment of blisters, Type 2 diabetes, liver ailments, diseases of the spleen (Wirngo et al., 2016). It is also a well-known liver tonic and hepatoprotective plant. Dandelion leaves or roots are used to treat liver disorders such as jaundice, gallbladder and liver ailments due to their hepatic and hyperglycemic effects. Its roots cleanse the liver and kidneys of toxins, dissolve gallstones, and stimulate the appetite. When taken orally, they act as cholagogues, increasing bile flow. The roots are also combined with other plants in powder form for the regulation of urine discharge and a burning sensation ans as a sedative and blood purifier in the treatment of hepatitis, jaundice, and fever (Mahboubi & Mahboubi, 2020).

The phytochemical components determine a plant's medicinal value, so they must be identified and studied with analytical techniques (Palai & Shekhawat, 2022). The current chapter examines the biochemical constituents found in several parts of the dandelion. It summarises the most recent pharmacological research demonstrating that dandelion extracts have antibacterial, antioxidant,

DOI: 10.1201/9781003292340-11

FIGURE 11.1 Dandelion in natural habitat.

antiviral, cytotoxic, diuretic, anti-inflammatory, and hepatoprotective properties. These findings imply that more research into this herb is needed to reveal potential therapeutic effects.

11.2 OCCURRENCE

The *Taraxacum officinale* complex was recognised first in Greece, or, more probably, in the northern Himalayas. It is spread all across Europe, Asia and North Africa, spanning from subtropical regions to tundra regions. It is found in North and South America's alpine to subalpine and arctic to subarctic zones, as well as in New Zealand and Australia's montane to alpine zones, wet oceanic islands, and seasonal tropics. It now known in approximately every additional region and country on the planet. It is also seen as an unwanted plant in farming areas and anthropogenetic grasslands (Hawkes et al., 2012).

11.3 SYSTEMATIC CLASSIFICATION

Domain	Eukaryota
Kingdom	Plantae
Phylum	Spermatophyta
Subphylum	Angiospermae
Class	Dicotyledonae
Order	Asterales
Family	Asteraceae
Genus	*Taraxacum*
Species	*officinale*

FIGURE 11.2 Flower and clock of the dandelion.

11.4 NUTRITIONAL ATTRIBUTES OF THE PLANT

Taraxacum officinale contains many nutrients and biologically active substances and its chemical composition has been outlined. Vitamins, inulin, lecithin, inositol, sesquiterpenes, triterpenes, phytosterols, amino acids, phenolic compounds and minerals (particularly potassium) are abundant in the dandelion plant, giving it hypolipidemic, antioxidant, anti-hyperglycemic, antiallergic, anti-inflammatory, and anticoagulant properties (Wirngo et al., 2016).

The plant possesses vitamins (A, B-complex, C, D, E, and K), lecithin, inositol, choline, and oligoelements (calcium, iron, potassium, silicon, magnesium, sodium, copper, manganese, zinc, phosphorus). It also has lecithin, choline, micronutrients, and fibre (Dongare et al., 2021).

Taraxacum officinale roots contain carotenoids such as lutein, carbohydrates like inulin, fatty acids like myristic acid, sugars like glucose, choline, fructose, sucrose, mucilage, vitamins, minerals, and pectin. Inulin is a complex carbohydrate or fructo-oligosaccharide which assists in the elimination of pathogens in the gastrointestinal tract as well as the prevention of obesity, cancer, and osteoporosis. Dandelion is also to be found in a variety of foods and supplements and it is used in drinks as a coffee substitute and flavor enhancer. The extracts from the leaves are effective against obesity and cardiovascular disorders (Pierzak-Sominka, 2020).

11.5 PHYTOCHEMICAL OVERVIEW OF THE PLANT

Dandelion comprises a collection of phytochemicals spread in the plant and gifted with potential healthy possessions. The flavonoids isolated from dandelion include luteolin, caffeic acid, luteolin 7-glucoside, and chlorogenic acid. Its chemical composition is critical to its biological activities. Several *Taraxacum* species were used to isolate sesquiterpene lactones, flavonoids, flavonoid glycosides, triterpenes, and phytosterols. Among the phytochemical constituents isolated from *Taraxacum officinale* were eudesmanolides, guaianolides, and germacranolide ester, taraxasteryl acetate, luteolin, luteolin-7-O-glucoside, esculetin, chrysoeriol, daucosterol, diosmetin and 5-hydroxypyrrolidin-2-one (Figure 11.3) (Lee et al., 2011).

Dandelion's bitterness is caused by the presence of sesquiterpene lactones, primarily eudesmanolide and germacranolide, which are solely found in this plant. Taraxacolides, ixerin, dihydro-lactucin, ainslioside, and taraxinic acids are examples of major sesquiterpene lactones which are present as glycosides. The analysis of dandelion using tandem liquid chromatography and mass spectrometry revealed it as abundant bases of β-carotene, which defends cells from oxidation and cellular impairment (Maggi, 2019).

11.6 FACTORS AFFECTING PHYTOCHEMICAL COMPOSITION OF DANDELION

The phytochemical configuration of dandelion varies, being affected by its various parts, harvest location, the season of harvest, the extraction method, solvent, and temperature. Also, it differs significantly according to where it is found on the plant, varying according to whether it is found on the plant's leaves, flowers, or roots. The biological activity and chemical composition of dandelion are directly related; together, they determine the potency of a dandelion as a protective agent (Thalkari et al., 2020).

The full flavonoid and phenolic content present in the dandelion leaf was not affected by the extraction time or temperature. The leaf had the maximum total phenolic content, followed, in turn, by the stem, flower, and roots. The leaf also had the maximum total flavonoid content, followed by the roots, flower, and stem (Hudec et al., 2007).

Chicoric acid is a significant phenolic compound present in dandelion ethanolic extract. Biochemical investigation has recently discovered chicoric acid as the richest component of dandelion in its leaves and stems. Flowers and leaves have greater polyphenol levels than is found in the stems. Young leaf dandelion extract has a greater phenolic concentration than root extract.

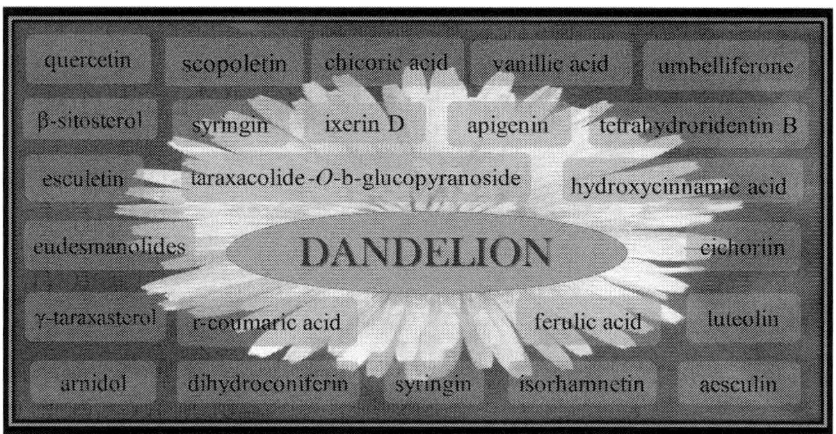

FIGURE 11.3 Phytochemicals of dandelion.

The phenolic and flavonoid content of the 80% hydroalcohol with formic acid was greater than the 80% hydroalcohol without formic acid after 3 hours of extraction at 60 degrees Celsius with or without the presence of formic acid (Lis et al., 2019).

This proof of identity of anthocyanins from the midvein and petiole of dandelion (Garnet Stem) through LC-ESI-HR-QTOF-MS revealed cyanidin-3-(6-malonyl)-glucoside (A-1), cyanidin-3-(6-malonyl)-glucoside (A-2), cyanidin-3-glucoside, and peonidin-3-(malonyl)-glucoside-like phytoconstituents to be at their highest in the leaf, and this varies according to the harvest location (Gomez et al., 2018).

Significant seasonal variation is also observed, with methyl sterols being more prevalent in the winter and sitosterol and cycloartenol esters being more predominant during extended sun exposure time periods. Throughout the year, sitosterol is the richest free sterol in dandelion leaves, followed by stigmasterol and campesterol. Free methyl sterol levels, on the other hand, are highest in the winter, while sitosterol and cycloartenol esters are at their highest in the summer (Alamgir, 2018).

11.7 MEDICINAL USES OF DANDELION

In traditional medicine, dandelion has treated liver complaints, inflammation, and a range of women's illnesses, including breast and uterine cancer. In Chinese traditional medicine, it is also recognised as a nontoxic plant with choleretic, anti-rheumatic, diuretic, and anti-inflammatory effects. Dandelion is a digestive stimulant, immunomodulator, prebiotic, insulin stimulant, antineoplastic, anti-inflammation, antiangiogenic, and demulcent herb (Figure 11.4). Furthermore, dandelion is used to treat dyspepsia and hepatitis B, as well as to improve androgenic hormone metabolism. Because there have, as yet, been few in vivo studies on dandelion, further study is needed to prove its advantages and effectiveness (Qureshi et al., 2017).

Sesquiterpene lactones, taraxerol, chlorogenic acid, taraxasterol, and chicoric acid are all potential bioactive components of dandelion. These components are non-toxic and have the potential to be anti-rheumatic, anti-inflammatory, anti-oxidative, and chloretic. Dandelion has long been utilised in folk medicine. Although sesquiterpene lactones with anti-inflammatory and anticancer effects are the plant's most biologically relevant components, it also contains several phenylpropanoids with inflammation-modulating effects, terpenoids, polysaccharides with immune regulation effects, hepatoprotective effects, and antitumoral activity, and inulin which has immunostimulatory functions (Wirngo et al., 2016).

The plant's different components have a long history of usage as herbal treatments; this herb's roots were originally considered to be useful in aiding digestion. The roots are also popular as a liver tonic, boosting bile production, eliminating toxins, and restoring hydration and electrolyte balance.

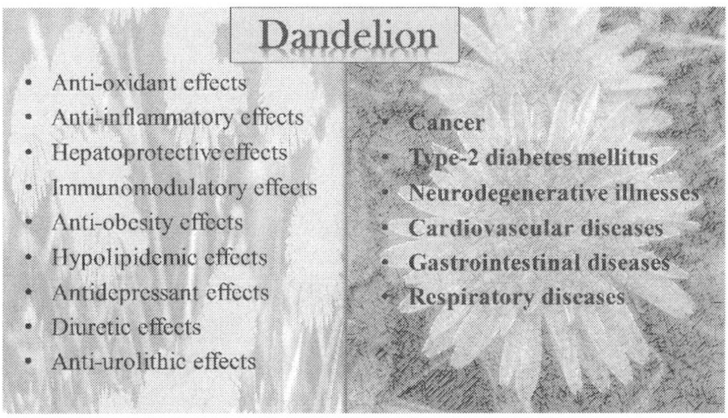

FIGURE 11.4 Pharmacological properties of dandelion.

Dandelion leaves have been used for centuries as a diuretic and digestive stimulant. Dandelion leaves have also been demonstrated to increase production and growth. The villus height/crypt depth ratio, as well as cellular infiltration, have each been found to improve the intestinal mucosa (Martinez et al., 2015).

11.8 APPLICATION OF ACTIVE INGREDIENT

11.8.1 Anti-oxidant Effects

Dandelion contains anti-oxidants. Dandelion leaf and root extracts donate hydrogen, inhibit reactive oxygen species (ROS), and scavenge radicals. Dandelion flower extracts in ethyl acetate scavenge ROS while protecting DNA from ROS-induced damage. The occurrence of active ingredients like luteolin and luteolin 7-O-glucoside prevented oxidative stress. Glucose can cause ROS production in cells, indicating that glucose toxicity is caused by glucose-induced oxidative stress. Excess ROS production through auto-oxidation, glycosylation, oxidative phosphorylation, and glucosamine pathways necessitates anti-oxidant defence, which dandelion extract provides both in vitro and in vivo. Dandelion flowers possess high content of phenolics such as coumaric acid, flavonoids, and ascorbic acid, making them potential anti-oxidant resources. Its leaf extracts are actual hydrogen donors, hydrogen peroxide foragers, and reducing mediators (Wirngo et al., 2016).

Dandelion is an oriental herbal medicine containing phenolic compounds that have antiplatelet, anticoagulant, and antioxidant properties. The extract boosted overall antioxidant capacity and lowered lipid peroxidation, which is a sign of oxidative stress in liver, heart, brain, kidney, etc. The phenolics found in the tested extract influence the quantity and action of nitric oxide synthase via interacting with kinase signalling paths and the intracellular Ca^{2+} involved with NOS phosphorylation and NO generation (NOS). Furthermore, endothelin-1 like vasoconstrictor and endothelial NADPH oxidase may be inhibited by these phenolic compounds, hinting that they may have antioxidant and hypolipidemic effects (Olas, 2022).

Dandelion leaf and root extracts reduce atherosclerosis caused by oxidative stress and also lowers the atherogenic index. Dandelion condensed the degree of atherosclerosis by lowering oxidative stress and serum TC, TG, and LDL-C while increasing serum HDL-C in diet-induced hypercholesterolemic atherosclerosis. Dandelion reduces the hazards for coronary artery disease and helps to prevent hypercholesterolemic atherosclerosis (Choi et al., 2010).

In rats, dandelion fruit extract was found to protect in sodium nitroprusside-made reduced cellular viability and amplified lipid peroxidation in the hippocampus, cortex, and striatum. The slices of cortex, hippocampus, and striatum added with 50 M sodium nitroprusside and 1-20 g/mL of ethanolic extract of dandelion. The extract protected against cellular viability reductions and augmented lipid peroxidation in the hippocampus, striatum, and cortex of the brain. At low doses, the extract displayed DPPH and NO scavenger activity and could protect against H_2O_2- and Fe^{2+}-driven deoxyribose oxidation. Thus, dandelion fruit extract demonstrated an antioxidant effect and rescued brain slices from sodium nitroprusside-induced cellular death (Xie et al., 2018).

11.8.2 Anti-inflammatory Actions

The anti-inflammatory actions of a methanolic extract and its fractions from *Taraxacum officinale* leaves on lipopolysaccharide-induced retorts in the mouse macrophage cell line RAW 264.7 was investigated. The methanol extract and its fractions reduced LPS-induced NO, pro-inflammatory cytokines, and PGE2 production in a dose-dependent way. The chloroform and ethyl acetate fractions strongly reduced NO, PGE2, and two pro-inflammatory cytokines (TNF-alpha and IL-1beta). They also had a dose-dependent effect on LPS-induced iNOS, COX-2, and MAP kinase activation. The methanol extract's chloroform and ethyl acetate fractions displayed maximum anti-inflammatory properties, owing to the down-regulation of NO, PGE2, and pro-inflammatory cytokines, as well as decreased production of iNOS and COX-2 via the MAP kinase deactivation signal path (Koh et al., 2010).

Taraxasterol, a dandelion pentacyclic-triterpene component, was fed to mice before a deadly dose of lipopolysaccharide. Taraxasterol enhanced mouse survival and decreased lung damage at tissue level. TNF, IL-1, IL-6, IFN-α, NO, and PGE2 concentrations were all reduced in endotoxic shock mice sera. Taraxasterol protected mice from endotoxic shock caused by lipopolysaccharide by moderating their inflammatory cytokine and intermediary discharge (Zhang et al., 2014).

11.8.3 Hepatoprotective Effects

Oxidative stress causes irreversible variations in DNA, proteins, and lipids in the liver. Various liver diseases, such as neoplasm, zonal necrosis, granuloma, vascular lesions, hepatitis, cholestasis, and steatosis, all bring about liver disorders. Air pollution, inflammation, triglyceride build-up, microorganisms, obesity, drugs, and insulin resistance all have a significant impact on liver function and disorders. Dandelion has anti-inflammatory, prebiotic, and inhibitory effects against fasting-induced adipose factors, as well as lipogenesis-augmenting effects, which results in a decrease in lipid accumulation and liver inflammation, thereby improving liver functions. Dandelion leaf extract was given to rats orally after sodium dichromate. This sodium dichromate produced acute liver damage, hepatocyte gangrene, and DNA destruction. Dandelion pre-treatment provided hepatoprotection by significantly lowering liver enzymes in the oxidative damage (Hfaiedh et al., 2016).

An ethanolic *Taraxacum officinale* root extract contained sesquiterpenes, triterpenes, phenolic compounds, and phytosterols. Its potential as a preventive hepatoprotective agent is assessed in a rat model of acute or chronic liver failure caused by d-galactosamine and lipopolysaccharide (D-Gal-LPS). *Taraxacum officinale* pre-treatment reduced liver failure or liver damage by reducing oxidative stress and immune liver damage. Only iNOS and eNOS were extremely articulated in liver tissue in acute liver failure, and their expression increased with increasing plasma NO levels. Serum nitrate and nitrite (NOx) levels, as well as 3-nitrotyrosine (3NT) levels, were found to be elevated in D-Gal-LPS-induced acute-on-chronic hepatic failure, confirming the role of excessive NO synthesis in severe liver damage and inflammation. *Taraxacum officinale* treatment condensed serum NOx and 3NT levels, implying significant inhibition of systemic oxidative stress and NO production (Pfingstgraf et al. 2021).

11.8.4 Immunomodulatory Effects

Immune hepatic injury is a disease of liver caused by a disparity in the immune system's T cells and macrophages. Taraxasterol, which is extracted from Taraxacum, has anti-inflammatory and immunomodulatory properties. Taraxasterol treatment significantly reduced the Con A-induced increase in liver index, serum ALT, aspartate aminotransferase (AST), and liver malondialdehyde (MDA) levels while increasing the Con A-persuaded reduction in liver glutathione (GSH) and superoxide dismutase (SOD) creation. Taraxasterol also inhibited the production of pro-inflammatory cytokines such as TNF-, interleukin-6, IL-1b, interferon-c (IFN-c), and IL-4. Taraxasterol reduced Con A-persuaded hepatic histopathological damage and apoptosis. Taraxasterol protects mice from Con A-persuaded acute hepatic injury by hindering the TLRs/NF-jB inflammatory signalling path and endorsing the Bax/Bc1-2 anti-apoptotic signalling path, which is downregulated in autoimmune diseases (Sang et al., 2019).

In vivo, *Taraxacum officinale* extracts improve immune response. In *Trachinotus ovatus*, the properties of dandelion extract as diet supplement on development rate, immune retorts, feed use, body configuration, plasma biochemical indices, hepatic antioxidant capabilities, and confrontation to *Vibrio harveyi* were studied. Dietary supplementation with dandelion extracts may improve specific growth rate, final body weight, weight gain rate, feed efficiency, feed intake, and protein efficiency ratio meaningfully. Dietary dandelion extracts increased plasma total protein, complement C4 content, and alkaline phosphatase, lysozyme, and glutathione reductase (GSR) activity while decreasing triglyceride, low-density lipoprotein, malondialdehyde (MDA), cholesterol, and aspartate aminotransferase activities. In fish-fed dandelion extract supplements, hepatic antioxidant enzymes (SOD, GSR, CAT, GSH-Px) amplified remarkably, but MDA concentration fell dramatically. Fish fed with dandelion extracts supplement had significantly higher post-challenge survival after being challenged with *Vibrio harveyi*. Thus, dandelion extracts could improve its feed utilisation, body protein deposit, growth presentation, immune capability, hepatic and plasma antioxidative enzyme actions and fight against contamination by *Vibrio harveyi* (Tan et al., 2017).

11.8.5 Anti-obesity Effects

Obesity leads to the development of metabolic liver diseases. It can be controlled by inhibiting pancreatic lipase. Dandelion ethanol extract, like orlistat, helped in this regard (Zhang et al., 2008). The extract of dandelion has a minor inhibitory effect on pancreatic lipase. α-Glucosidase inhibitors are used to treat obesity because the inhibiting of glucosidase prevents glucose cleavage from disaccharides and oligosaccharides. Dandelion root extract inhibited α-glucosidase activity and had weak xanthine oxidase and ACE inhibitory effects, confirming anti-pancreatic lipase activity. Adipose tissue expansion indicates adipocyte hypertrophy and hyperplasia. Insulin resistance and lipid buildup in liver are results of systemic effects in the liver. The roots and leaves of dandelion inhibited lipogenesis and adipocyte differentiation in 3T3-L1 pre-adipocytes on lipid metabolism and adipogenesis, thereby reestablishing the role of the liver (Gonzalez-Castejon et al., 2014).

Dandelion root extract reduced adipocyte size and number while increasing lipolysis activity. Both dandelion leaf and unpolished powdered roots extract decreased triglyceride accumulation in mature adipocytes, with the leaf extract having a greater effect than the root extracts. The hydro-alcoholic dandelion root extract inhibited the growth of human primary visceral pre-adipocytes. Dandelion extract induced apoptosis and inhibited adipogenesis in P10 and P20 cell lines, increasing free glycerol freed while decreasing triglyceride accumulation and lipogenesis (Colitti & Stefanon, 2016). As a result, dandelion had anti-obesity effects via multiple mechanisms, including a reduction in liver inflammation.

11.8.6 Hypolipidemic Effects

The antioxidant enzyme and lipid profiles in high-cholesterol-nourished rabbits were measured to assess effects of orally fed dandelion root and leaf on atherosclerosis development. Dandelion shortened the degree of atherosclerosis through lowering oxidative stress and serum TC, TG, and LDL-C while increasing serum HDL-C in diet-induced hypercholesterolemic atherosclerosis. Dandelion reduces the risk of coronary artery disease factors and helps to prevent hypercholesterolemic atherosclerosis (Cho et al., 2002).

The action of dandelion leaf extract on liver steatosis in mice on a high-fat diet showed that hepatic lipid accumulation was significantly reduced. Dandelion supplementation also significantly reduced the high-fat diet-induced insulin, TG, fasting glucose, TC and HOMA-IR. In addition, it suppressed liver lipid accumulation, increased AMPK activation in liver and muscle, and reduced insulin resistance and lipids through the AMPK path. As a result, dandelion leaf extract is beneficial in the prophylaxis and therapy of diseases related to overweightness (Davaatseren et al., 2013).

A diet added with dandelion extract improved the final weight, existence rate, specific growth rate and typical individual weight gain in common carp (*Cyprinus carpio* L.) grown in recirculation system. The addition of dandelion extract to feed reduced both triglyceride and plasma cholesterol levels significantly, promoting a hypolipidemic condition in fish (Sirakov et al., 2019).

11.8.7 Antidepressant Effects

To investigate dandelion's antidepressant effects, mice fed with dandelion leaf and root water extract were assessed for their behavioral variations and adrenal axis hormone variations after tail suspension, swimming, and open field tests. The continued treatment of dandelion significantly reduced immobility time in swimming and tail suspension. Similarly, acute dandelion treatment significantly reduced immobility time in swimming and tail suspension. The open field experiment yielded no results. Chronic dandelion treatment reduced in a rise in corticotropin-releasing hormone (CRH). Long-term treatment slowed the rise in corticosterone. The antidepressant effects of dandelion water extract in behavioral despair models suggested the involvement of the neuroendocrine system (Li et al., 2014).

11.8.8 Appetite-stimulating, Choleretic, and Laxative Effects

Dried or fresh dandelion can be useful as appetite stimulant, and can be used to treat simple stomach disorders. Dandelion root has traditionally been used as a mild digestive bitter to aid digestion, increase bile flow, relieve nausea and vomiting, and improve appetite. The plant's bitter constituents stimulate bile flow and act as an appetite stimulant. The dandelion root act as a mild laxative and can mend digestion. The plant improves the function of liver and gallbladder. Dandelion has both antioxidant and lipid-lowering properties. The root is depurative, choleretic, cholagogue, tonic, digestive stimulant, anti-rheumatic, and alterative. It detoxifies by promoting biliary drainage. It extends hepatoprotective effect in fatty liver (Cheema & Singh, 2021). The success of this plant in the digestive area can be attributed to the stimulation of liver cells and the decongestion of the liver.

11.8.9 Diuretic Effects

The dandelion leaf's ability to be used as a diuretic is supported by ethnobotanical data, demonstrating that numerous dandelion species are extensively used to improve renal fluid elimination in urinary and renal illnesses. The diuretic action of dandelion leaf is better than its root in a murine model and is comparable to furosemide. The diuretic action of dandelion is attributed to numerous compounds with diverse component groups of secondary metabolites, such as phenolics, terpenes, or alkaloids, which act via various diuretic and saluretic pathways (Qadir et al., 2022).

Duke's USDA database states that dandelion contains nearly nine diuretic mixes. Dandelion may provide significant potassium input by substituting for the damage of potassium caused by nearly all diuretics (Clare et al., 2009). Dandelion causes the stimulation of diuresis, which, combined with an increase in liver function, results in an action on general turnover, translating into increased waste elimination from the body.

11.8.10 Anti-urolithic Effect

The effects of taraxasterol, an aqueous extract of aerial part of dandelion, and potassium citrate were studied on calcium oxalate crystallisation in vitro. Taraxasterol is a pentacyclic-triterpene compound of dandelion with anti-urolithic properties. As potassium citrate prevents kidney stone formation, the anti-crystallization activities of taraxasterol, and the aqueous extract of an aerial part of dandelion are compared to those of potassium citrate. The presence of taraxasterol, an aqueous extract of the aerial part of dandelion, and potassium citrate significantly reduced absorbance. Crystal nucleation is inhibited, and the number of CaOx crystals and the number of CaC_2O_4 monohydrates are both reduced. The number of CaC_2O_4 dihydrate crystals increases, while their diameter decreases noticeably. Tarxasterol and the aqueous extract of dandelion aerial part both have anti-crystallization properties. The extract proves more effective than taraxasterol, which could be due to another extract component with synergistic action (Yousefi Ghale-Salimi et al., 2018).

11.9 PHARMACOLOGICAL APPLICATIONS

11.9.1 Cancer

In several cancer cell models/in-vitro systems, aqueous dandelion extracts show anti-cancer potential without causing toxicity to non-cancerous cells. In colon cancer cell models, aqueous dandelion root extract clearly kills cancerous cells. Aqueous dandelion root extract, in particular, induced planned cell death in more than 95% of colon cancer cells. In vivo studies confirmed the extract's anti-cancer efficacy, as the oral feeding of dandelion root extract slowed the development of human colon xenograft models by 90%. The treatment of cancer cells with dandelion root extract activated multiple death pathways as gene expression analyses reveal the expression genes related to programmed cell death. The extract's phytoconstituents analysis revealed a complex composition of bioactive phytochemicals such as taraxasterol, α and β amyrin, lupeol, etc. As a result, this extract can potentially engage and efficiently target the numerous susceptibilities of cancer cells. Thus, dandelion root extract is a non-lethal and actual alternative against cancer which can reduce the numbers of drug-resistant cancerous cells (Ovadje et al., 2016).

Breast cancer stem cells play a significant role in disease progression such as drug resistance metastasis and recurrence. As a result, the targeting of breast cancer stem cells can slow cancer growth. The properties of dandelion extracts on breast cancer stem cell proliferation in vitro in 2D and 3D cell culture platforms were evaluated by evaluating cell death induction, gene expression of death receptor-signalling pathways, and reactive oxygen species production by breast cancer stem cells. These cells formed spheroids as microtumors in vitro and showed tumor-like properties in vivo. Dandelion extracts meaningfully reversed the multiplication of breast cancer stem cells in both 2-D and 3-D models. The IC50 of dandelion extracts in breast cancer stem cells was much higher in the 3D model than in the 2D model. Dandelion extracts amplified tumor necrosis factor-related apoptosis-inducing ligand (TRAIL) and TRAIL receptor 2 (death receptor 5) expression. This increased the expression of DR4. Methanolic dandelion extract increased ROS creation in breast cancer stem cells and it had a better effect than ethanol extract. Dandelion extracts are thus a hopeful alternative in the treatment of breast tumors with high potential for the development of anti-cancer drugs (Trinh et al., 2016).

Methanolic extracts of dandelion root were tested for their antiproliferative activity on HCT116, MCF7, HepG2, and standard Hs27 cell viability. Methanolic extracts of dandelion root significantly inhibited the development of the HepG2 cell line, a lesser effect is seen on the MCF7 and HCT116 cell lines, and no effect at all was seen in the Hs27 cell lines. Methanolic extracts of dandelion root also increased AMPK phosphorylation in HepG2 cells, which is important in the treatment of cancer and other metabolic disorders. The cell viability test was consistent with the percentage of apoptotic cells (HepG2 cells). Methanolic extracts of dandelion root were discovered to have the ability to regulate AMPK in treated cell lines of cancer (Rehman et al., 2017).

11.9.2 Type-2 Diabetes Mellitus

The enormous upsurge in type 2 diabetes prompted a search for less opulent substitutes. Dandelion has a rich profile of bioactive components which are known for their anti-diabetic effects. It is used as a therapeutic in some countries to control and treat type 2 diabetes. Dandelion's anti-diabetic actions are the result of bioactives such as chlorogenic acid, chicoric acid, sesquiterpene lactones, and taraxasterol. The cellular influence of dandelion is inadequate, requiring extra study on clonal cell lines (INS-1E). To explore the physiological, pharmacological, and biochemical mechanisms of dandelion-derived bioactives on type 2 diabetes, extensive in vivo, in vitro, and clinical study is needed (Wirngo et al., 2016).

The root of *Taraxacum officinale* has long been used for the treatment of anorexia, dyspepsia, diabetes, heartburn, and hepatitis. Using antidiabetic and subcutaneous glucose tolerance tests, various extracts of dandelion root were assessed for their antidiabetic action in normoglycemic and alloxan-made diabetic mice. The in vitro glucose uptake assay was done using HepG2 and 2-NDBG, and the phytochemical screening of phytoconstituents was performed using LC-MS/MS. In alloxan-brought diabetic mice, an aqueous root extract of dandelion (400 mg/kg) reduced blood glucose levels significantly, whereas the other extracts had no effect. The aqueous root extract of dandelion also demonstrated significant glucose uptake enhancement with antidiabetic potency comparable to metformin, as well as the activation of insulin-signalling pathways. The aqueous root extract of dandelion had a higher level of phenolic compounds than was to be found in flavonoids with luteolin-7-glucoside, chlorogenic acid and protocatechuic acid as major identifiable phytochemicals. The LC-MS/MS showed chlorogenic acid to be the major constituent of the aqueous extract, with a significant part in glucose metabolism. Other active constituents, such as procatechuic acid and luteolin-7-glucoside, were discovered to play a role in the antihyperglycemic potential of dandelion aqueous root extract. The high antioxidant activity of the detected phytoconstituents may contribute to dandelion root's antihyperglycemic action by decreasing oxidative stress, and thereby affecting glucose homeostasis (Juee & Naqishbandi, 2020).

11.9.3 Neurodegenerative Illnesses

Neuron damage in oxidative stress contributes to pathology and the development of neurodegenerative illnesses. Dandelion has antioxidant properties. The neuroprotective effects of ethanolic extracts of dandelion on glutamate-induced oxidative stress in HT22 cells were studied to determine neuroprotection and the basic molecular pathways in oxidative damage. Cell viability and reactive oxygen species assays revealed that ethanolic extracts dandelion effectually reduced glutamate-induced cytotoxicity and ROS generation. Furthermore, the extracts augmented the expression of heme oxygenase-1 (HO-1) and facilitated the nuclear translocation of nuclear factor erythroid 2-related factor-2 (Nrf2). The HO activity inhibitor, tin protoporphyrin (SnPP), partially reversed the repressive effects of ethanolic extracts of dandelion on glutamate-enthused cell harmfulness and ROS creation. Thus, ethanolic extracts of dandelion can defend HT22 cells from glutamate-brought oxidative damage by making the Nrf2/HO-1 (nuclear factor erythroid 2-related factor and heme Oxygenase-1) pathways, proving it to be an effective agent for the prevention of neurodegenerative diseases (Huang et al., 2018).

11.9.4 Cardiovascular Diseases

Dandelion has been shown to help with cardiovascular illnesses such as atherosclerosis. Excessive oxidative stress is related to the growth of atherosclerosis-related metabolic syndrome. Rabbits were fed a high-cholesterol diet before being treated with the root and leaf of dandelion to evaluate its possible hypolipidemic and antioxidative properties. Positive improvements in plasma antioxidant enzyme activity and lipid contours in cholesterol-fed rabbits revealed hypolipidemic and antioxidant effects. Dandelion root and leaf extracts reduce the atherogenic index and defend against oxidative stress-related atherosclerosis. Dandelion decreased the degree of atherosclerosis by dropping oxidative stress and serum TC, TG, and LDL-C while boosting serum HDL-C in diet-brought hypercholesterolemic atherosclerosis. Dandelion reduces the hazards for coronary artery sickness and helps to avoid hypercholesterolemic atherosclerosis (Choi et al., 2010).

11.9.5 Gastrointestinal Diseases

The liver must be protected because it is a dynamic organ in the body, contradicting oxidative stress and harmful effects. Dandelion is a general plant with hepatoprotective properties because of its antioxidant and anti-inflammatory belongings against chemical agents. Inhibitory effects against the release of lipopolysaccharides, the prebiotic effects of oligofructans, anti-inflammatory effects, fasting induced adipose factor, lowering the rate of lipogenesis induced lipid accretion and improving liver inflammation and functions are some of the documented benefits of dandelion (Mahboubi & Mahboubi, 2020).

Dandelion leaf has the most total phenolics and flavonoids concentration, whereas the root contained the least. HPLC was used to identify eleven phenolic acids and flavonoids, with chicoric acid being the highest prevalent in all parts of the dandelion. The leaf extract possesses the greatest total antioxidant and 2,2-diphenyl-1-picrylhydrazyl (DPPH) scavenging activities. Dandelion leaf extract inhibited ROS formed by hydrogen peroxide in HT-29 cells. Both leaf extract and chicoric acid inhibited lipopolysaccharide-induced NF-B p 65 and cyclooxygenase-2 actions as well as LRR and PYD domains-containing protein 3 (NLRP3) inflammasome initiation and caspase-1 action, as well as NLRP3 inflammasome-mediated IL-1 and IL-18 emission. Chicoric acid, the key component of dandelion, reduced inflammasome activation via ROS scavenging and the anti-inflammatory action primarily responsible for the beneficial effects (Xue et al., 2017).

11.9.6 Respiratory Disease (COVID)

During its continuous transmission among humans, SARS-CoV-2 was constantly mutating. This could ultimately lead to the virus avoiding prevailing therapeutic and preventative measures aimed at the viral spike. Water-based common dandelion extracts effectively inhibit protein-to-protein collaboration between the human virus cell entry receptor ACE2 and SARS-CoV-2 spike with five relevant mutations. This was demonstrated in vitro by the overexpression of ACE2 and ACE2/TMPRSS2 proteins in human kidney (HEK293) and lung (A549) cells, respectively. The extract effectively prevented lung cells infection with SARS-CoV-2 pseudo-typed lentivirus. The ability of dandelion to inhibit protein–protein interactions between spike S1 and the human ACE2 cell surface receptor. The extract effectively prevented lung cells infection with SARS-CoV-2 spike pseudo-typed lentivirus particles, as well as virus-activated pro-inflammatory interleukin 6 secretion. These findings need deeper examination of dandelions' efficacy in SARS-CoV-2 inhibition (Tran et al., 2021).

11.10 NANO FORM OF DANDELION

nTOPL is a novel mannan-lipid nanoparticle which has the possibility to be a functional food and an alternative medicine in the treatment of lung cancer. TOPL is a novel-branched glucomannan with an average molecular weight of 4.29 104 Da, which was isolated from dandelion leaves for the first time. TOPL has a high IC50 against non-small cell lung cancer cell lines A549 and H460. TOPL's drug-like qualities are increased by improving its solubility and counteracting its low bioavailability. The nTOPL nano-functional particle is made up of several nanocarriers, such as liposomes and solid lipid nanoparticles. It has a strong cytotoxic impact on A549 and H460 cells, while having no effect on normal WI-38 cells. The nanoparticle nTOPL enhanced apoptosis and triggered G2/M phase arrest by increasing Beclin-1 and Bax thresholds and inhibiting Bcl-2. expression (El-Emam et al., 2021).

11.11 CLINICAL TRIALS OF DANDELION

Dandelion juice has been used as herbal drug for the past 1000 years. Ingestion of dandelion juice has been used in the treatment of vesicular hand eczema. Development in dyshidrotic hand eczema is assessed by HECSI score in the study provided at National Library of Medicine (2020).

The dandelion root chemical utilised in the Pandey lab is five times more concentrated than the extract available for purchase over the counter, and it has been demonstrated in laboratory animals to eradicate leukemia, melanoma, and pancreatic cancer cells. Caroline Hamm, an oncologist at the Windsor Regional Cancer Centre, petitioned to Health Canada in 2012 for permission to conduct human clinical trials. The research work was also sponsored through the Natural Sciences and Engineering Research Council of Canada (NSERC) and the Canadian Institutes of Health Research (CIHR) (Pandey, 2015).

No clinical trial outcomes had been reported by 2019. The discovery that while dandelion root extract kills breast cancer cells in the test tube, it may accelerate their development in the body because they are hormone-sensitive and impacted by estrogen activity is just one of the numerous reasons why a laboratory positive may become a clinical trial failure. Dandelion tea has been shown to stimulate urination and lower blood sugar levels. None of these are visible in the lab test tube (Peter, 2019).

Extensive in vitro, in vivo, and clinical research is required to further understand the pharmacological, physiological, and biochemical processes that underpin the effects of dandelion-derived compounds on type 2 diabetes (Wirngo et al., 2016).

An aqueous dandelion root extract has anti-cancer potential in multiple cancer cell models using in vitro systems, with little damage to non-cancer cells. An aqueous dandelion root extract's ability has the potential to destroy cancer cells in colon cancer cell cultures. In vivo tests validated the extract's anti-cancer activity, as the oral treatment of dandelion root extract slowed the development of human colon xenograft replicas by 90%. Dandelion root extract engages different signaling pathways in cancer cells which induces the activation of the death-receptor in leukemia and pancreatic cancer cells of humans (Ovadje et al., 2016).

11.12 CONCLUSION

Many phytochemicals are found in dandelion, including eudesmanolides, guaianolides, and germacranolide ester, taraxasteryl acetate, luteolin, luteolin-7-O-glucoside, esculetin, chrysoeriol, daucosterol, and diosmetin. The secondary plant metabolite extends beneficial effects such as antioxidant, anti-inflammatory, hepatoprotective, immunomodulatory, anti-obesity, hypolipidemic, antidepressant, diuretic and anti-urolithic properties. Various lifestyle diseases, including cancer, type-2 diabetes mellitus, neurodegenerative illnesses, cardiovascular diseases, gastrointestinal diseases, and respiratory diseases (especially COVID), are successfully treated through its intervention. In many respects, it offers a safer remedy than modern pharmaceuticals.

REFERENCES

Abdel-Magied, N., Abdel Fattah, S. M., & Elkady, A. A. (2019). Differential effect of *Taraxacum officinale* L. (dandelion) root extract on hepatic and testicular tissues of rats exposed to ionizing radiation. *Molecular Biology Reports*, *46*(5), 4893–4907.

Al-Ahmad, H. (2018). Biotechnology for bioenergy dedicated trees: meeting future energy demands. *Zeitschrift für Naturforschung C*, *73*(1–2), 15–32.

Alamgir, A. N. M. (2018). Phytoconstituents—active and inert constituents, metabolic pathways, chemistry and application of phytoconstituents, primary metabolic products, and bioactive compounds of primary metabolic origin. In A. N. M. Alamgir (Ed.), *Therapeutic Use of Medicinal Plants and their Extracts: Volume 2* (pp. 25–164). Springer.

Cheema, H. S., & Singh, M. P. (2021). The use of medicinal plants in digestive system related disorders—a systematic review. *Journal of Ayurvedic and Herbal Medicine*, *7*, 182–187.

Cho, S. Y., Park, J. Y., Park, E. M., Choi, M. S., Lee, M. K., Jeon, S. M., & Park, Y. B. (2002). Alternation of hepatic antioxidant enzyme activities and lipid profile in streptozotocin-induced diabetic rats by supplementation of dandelion water extract. *Clinica Chimica Acta*, *317*(1–2), 109–117.

Choi, U. K., Lee, O. H., Yim, J. H., Cho, C. W., Rhee, Y. K., Lim, S. I., & Kim, Y. C. (2010). Hypolipidemic and antioxidant effects of dandelion (*Taraxacum officinale*) root and leaf on cholesterol-fed rabbits. *International Journal of Molecular Sciences*, *11*(1), 67–78.

Clare, B. A., Conroy, R. S., & Spelman, K. (2009). The diuretic effect in human subjects of an extract of *Taraxacum officinale* folium over a single day. *The Journal of Alternative and Complementary Medicine*, *15*(8), 929–934.

Colitti, M., & Stefanon, B. (2016). Different anti-adipogenic effects of bio-compounds on primary visceral pre-adipocytes and adipocytes. *EXCLI Journal*, *15*, 362.

Davaatseren, M., Hur, H. J., Yang, H. J., Hwang, J. T., Park, J. H., Kim, H. J., & Sung, M. J. (2013). *Taraxacum officinale* (dandelion) leaf extract alleviates high-fat diet-induced nonalcoholic fatty liver. *Food and Chemical Toxicology*, *58*, 30–36.

Dongare, P. N., Kadu, T. B., Pohane, A. L., Dighade, S. J., & Patil, P. G. (2021). Review on pharmacognosy, phytochemistry and pharmacological activity of *Taraxacum officinale* (Dandelion plant). *World Journal of Advanced Research and Reviews*, *12*(02), 212–218.

Dudman, A. A., & Richards, A. J. (1997). Dandelions of Great Britain and Ireland. *Botanical Society of the British Isles Handbook*, Natural History Museum London.

El-Emam, S. Z., El-Ella, D. M. A., Fayez, S. M., Asker, M., & Nazeam, J. A. (2021). Novel dandelion mannan-lipid nanoparticle: Exploring the molecular mechanism underlying the potent anticancer effect against non-small lung carcinoma. *Journal of Functional Foods*, *87*, 104781.

Gomez, M. K., Singh, J., Acharya, P., Jayaprakasha, G. K., & Patil, B. S. (2018). Identification and quantification of phytochemicals, antioxidant activity, and bile acid-binding capacity of garnet stem dandelion (*Taraxacum officinale*). *Journal of Food Science*, *83*(6), 1569–1578.

González-Castejón, M., García-Carrasco, B., Fernández-Dacosta, R., Dávalos, A., & Rodriguez-Casado, A. (2014). Reduction of adipogenesis and lipid accumulation by *Taraxacum officinale* (Dandelion) extracts in 3T3L1 adipocytes: an in vitro study. *Phytotherapy Research*, *28*(5), 745–752.

Grauso, L., Emrick, S., de Falco, B., Lanzotti, V., & Bonanomi, G. (2019). Common dandelion: a review of its botanical, phytochemical and pharmacological profiles. *Phytochemistry Reviews*, *18*(4), 1115–1132.

Hawkes, J. G., Maxted, N., & Ford-Lloyd, B. V. (2012). *The Ex-Situ Conservation of Plant Genetic Resources*. Springer Science & Business Media.

Hfaiedh, M., Brahmi, D., & Zourgui, L. (2016). Hepatoprotective effect of *Taraxacum officinale* leaf extract on sodium dichromate-induced liver injury in rats. *Environmental Toxicology*, *31*(3), 339–349.

Huang, S., Meng, N., Liu, Z., Guo, L., Dong, L., Li, B., & Ye, Q. (2018). Neuroprotective effects of *Taraxacum officinale* Wigg. extract on glutamate-induced oxidative stress in HT22 cells via HO-1/Nrf2 pathways. *Nutrients*, *10*(7), 926.

Hudec, J., Burdová, M., Kobida, L. U., Komora, L., Macho, V., Kogan, G., & Chlebo, P. (2007). Antioxidant capacity changes and phenolic profile of Echinacea purpurea, nettle (*Urtica dioica* L.), and dandelion (*Taraxacum officinale*) after application of polyamine and phenolic biosynthesis regulators. *Journal of Agricultural and Food Chemistry*, *55*(14), 5689–5696.

Juee, L. Y., & Naqishbandi, A. M. (2020). In vivo and in vitro antidiabetic potential of *Taraxacum officinale* root extracts. *Current Issues in Pharmacy and Medical Sciences*, *33*(3), 168–175.

Koh, Y. J., Cha, D. S., Ko, J. S., Park, H. J., & Choi, H. D. (2010). Anti-inflammatory effect of *Taraxacum officinale* leaves on lipopolysaccharide-induced inflammatory responses in RAW 264.7 cells. *Journal of Medicinal Food*, *13*(4), 870–878.

Lee, S., Han, S., Kim, H. M., Lee, J. M., Mok, S. Y., & Lee, S. (2011). Isolation and identification of phytochemical constituents from *Taraxacum coreanum*. *Journal of the Korean Society for Applied Biological Chemistry*, *54*(1), 73–78.

Li, Y. C., Shen, J. D., Li, Y. Y., & Huang, Q. (2014). Antidepressant effects of the water extract from *Taraxacum officinale* leaves and roots in mice. *Pharmaceutical Biology*, *52*(8), 1028–1032.

Lis, B., Jedrejek, D., Moldoch, J., Stochmal, A., & Olas, B. (2019). The anti-oxidative and hemostasis-related multifunctionality of L-chicoric acid, the main component of dandelion: An in vitro study of its cellular safety, antioxidant and anti-platelet properties, and effect on coagulation. *Journal of Functional Foods*, *62*, 103524.

Maggi, F. (2019). Dandelion. In S. M. Nabvi & A. Silva (Eds.), *Nonvitamin and Nonmineral Nutritional Supplements* (pp. 203–204). Academic Press, Elsevier.

Mahboubi, M., & Mahboubi, M. (2020). Hepatoprotection by dandelion (*Taraxacum officinale*) and mechanisms. *Asian Pacific Journal of Tropical Biomedicine*, *10*(1), 1.

Martinez, M., Poirrier, P., Chamy, R., Prüfer, D., Schulze-Gronover, C., Jorquera, L., & Ruiz, G. (2015). *Taraxacum officinale* and related species—An ethnopharmacological review and its potential as a commercial medicinal plant. *Journal of Ethnopharmacology*, *169*, 244–262.

National Library of Medicine. (2020). *Dandelion Juice in the Treatment of Dyshidrotic Hand Eczema*. ClinicalTrails.gov. https://clinicaltrials.gov/ct2/show/NCT00442091

Olas, B. (2022). New Perspectives on the Effect of Dandelion, Its Food Products and Other Preparations on the Cardiovascular System and Its Diseases. *Nutrients*, *14*(7), 1350.

Ovadje, P., Ammar, S., Guerrero, J. A., Arnason, J. T., & Pandey, S. (2016). Dandelion root extract affects colorectal cancer proliferation and survival through the activation of multiple death signalling pathways. *Oncotarget*, *7*(45), 73080.

Palai S. and Shekhawat N. (2022) Novel extraction techniques for phytochemicals: A comprehensive review. *International Journal of Pharmaceutical Research*. *14*(3), 1.

Pandey, S. (2015). *Human Clinical Trials on for Cancer Killing Dandelion Extract*. University of Windsor. https://www.uwindsor.ca/dailynews/2015-02-18/human-clinical-trials-cancer-killing-dandelion-extract

Peter, K. (2019). *Dandelion Tea Kills Cancer Cells But It Is Not a Cancer Cure*. World Tea News. https://www.worldteanews.com/Insights/dandelion-tea-kills-cancer-cells-it-not-cancer-cure

Pfingstgraf, I. O., Taulescu, M., Pop, R. M., Orăsan, R., Vlase, L., Uifalean, A., & Pârvu, A. E. (2021). Protective effects of *Taraxacum officinale* L. (dandelion) root extract in experimental acute on chronic liver failure. *Antioxidants*, *10*(4), 504.

Pierzak-Sominka, J. (2020) Selected biologically active substances and healing properties of dandelion (*Taraxacum officinale*). In M. Chwil & M. S. Skoczylas (Eds.), *Biologically Active Compounds of Plant Origin in Medicine*, (pp. 26–33). WUP Lublinie.

Qadir, I., Nazir, S., Sheikh, M. A., Naaz, F., Bashir, S., Ovais, S., & Masoodi, M. H. (2022). *Taraxacum officinale*: The Esculent Dandelion as Herbal Medicine. In M. H. Masoodi & M. U. Rehman (Eds.), *Edible Plants in Health and Diseases* (pp. 299–326). Springer.

Qureshi, S., Adil, S., Abd El-Hack, M. E., Alagawany, M., & Farag, M. R. (2017). Beneficial uses of dandelion herb (*Taraxacum officinale*) in poultry nutrition. *World's Poultry Science Journal*, *73*(3), 591–602.

Rehman, G., Hamayun, M., Iqbal, A., Khan, S. A., Khan, H., Shehzad, A., & Lee, I. J. (2017). Effect of methanolic extract of dandelion roots on cancer cell lines and AMP-activated protein kinase pathway. *Frontiers in Pharmacology*, *8*, 875.

Sang, R., Yu, Y., Ge, B., Xu, L., Wang, Z., & Zhang, X. (2019). Taraxasterol from *Taraxacum* prevents concanavalin A-induced acute hepatic injury in mice via modulating TLRs/NF-κB and Bax/Bc1-2 signalling pathways. *Artificial Cells, Nanomedicine, and Biotechnology*, *47*(1), 3929–3937.

Sirakov, I., Velichkova, K., Stoyanova, S., Zhelyazkov, G., & Staykov, Y. (2019). The effect of diet supplemented with dandelion's (*Taraxacum officinale*) extract on the productive and blood parameters of common carp (*Cyprinus carpio* L.), cultivated in the recirculation system. *Macedonian Veterinary Review*, *42*(2), 131–139.

Tan, X., Sun, Z., Chen, S., Chen, S., Huang, Z., Zhou, C., & Wang, A. (2017). Effects of dietary dandelion extracts on growth performance, body composition, plasma biochemical parameters, immune responses and disease resistance of juvenile golden pompano *Trachinotus ovatus*. *Fish & Shellfish Immunology*, *66*, 198–206.

Thalkari, A. B., Karwa, P. N., Zambare, K. K., Thorat, V. M., & Shinde, P. S. (2020). Dandelion: A Natural Booster. *Research Journal of Pharmacognosy and Phytochemistry*, *12*(1), 24–26.

Tran, H. T. T., Gigl, M., Le, N. P. K., Dawid, C., & Lamy, E. (2021). In vitro effect of *Taraxacum officinale* leaf aqueous extract on the interaction between ACE2 cell surface receptor and SARS-CoV-2 spike protein D614 and four mutants. *Pharmaceuticals*, *14*(10), 1055.

Trinh, N. V., Doan-Phuong Dang, N., Hong Tran, D., & Van Pham, P. (2016). *Taraxacum officinale* dandelion extracts efficiently inhibited the breast cancer stem cell proliferation. *Biomedical Research and Therapy*, *3*(7), 1–13.

Wirngo, F. E., Lambert, M. N., & Jeppesen, P. B. (2016). The physiological effects of dandelion (*Taraxacum officinale*) in type 2 diabetes. *The Review of Diabetic Studies: RDS*, *13*(2–3), 113.

Xie, P. J., Huang, L. X., Zhang, C. H., Ding, S. S., Deng, Y. J., & Wang, X. J. (2018). Skin-care effects of dandelion leaf extract and stem extract: Antioxidant properties, tyrosinase inhibitory and molecular docking simulations. *Industrial Crops and Products*, *111*, 238–246.

Xue, Y., Zhang, S., Du, M., & Zhu, M. J. (2017). Dandelion extract suppresses reactive oxidative species and inflammasome in intestinal epithelial cells. *Journal of Functional Foods*, *29*, 10–18.

Yelverton, F. (2017). *Common Dandelion*. NC State Extension Publications. https://content.ces.ncsu.edu/common-dandelion

Yousefi Ghale-Salimi, M., Eidi, M., Ghaemi, N., & Khavari-Nejad, R. A. (2018). Inhibitory effects of taraxasterol and aqueous extract of *Taraxacum officinale* on calcium oxalate crystallization: in vitro study. *Renal Failure*, *40*(1), 298–305.

Zhang, J., Kang, M. J., Kim, M. J., Kim, M. E., Song, J. H., Lee, Y. M., & Kim, J. I. (2008). Pancreatic lipase inhibitory activity of *Taraxacum officinale* in vitro and in vivo. *Nutrition Research and Practice*, *2*(4), 200–203.

Zhang, X., Xiong, H., Li, H., & Cheng, Y. (2014). Protective effect of taraxasterol against LPS-induced endotoxic shock by modulating inflammatory responses in mice. *Immunopharmacology and Immunotoxicology*, *36*(1), 11–16.

12 Sweet Violet (*Viola odorata*)

Kanchan Sharma and Raees ul Haq

12.1 INTRODUCTION

Sweet violet (*Viola odorata*), an angiosperm, has been used for centuries due to its medicinal properties as well as for its unique fragrance in cosmetics. It is found growing widely under the shade, in gardens, lawns, or at the edges of forests. In India, it is commonly known by the name "banafsha," However, it has numerous other names by which it is commonly known across the globe, including English violet, garden violet, luktviol, etc. (USDA-ARS, 2023). The plant is herbaceous and is a common sight in the temperate regions of the world. The plant's leaves are heart-shaped with dark green color, and the petals of flowers are violet in color (Figures 12.1 & 12.2). The blossoms, which appear in early spring, are delicate and scented. It belongs to the genus *Viola*, which includes more than five hundred species of perennial and annual plants. Furthermore, the very existence of hundreds of hybrids under the same genus makes it difficult for the botanist to differentiate in general, Accordingly, violas with the large flowers are generally referred to as pansies whereas those with smaller blooms are often called violets. Both of these have flowers in a variety of colors, but the former ones have large flowers with multicolored and/or distinct markings whereas the latter produce small mounds of foliage topped with abundant blooms, and tend to grow in lower/cool temperatures. "Johnny jump-up" is one of the names given to the third category of violas, which is characterized by the presence of shades of three colors, namely, purple, yellow, and white, and also for its sweet and whimsical flowers. However, the common theme among these three categories of violas is that the flowers of pansies, violets, and "Johnny jump-ups", are edible and they can be added to salads and drinks, and used to garnish cakes, pastries, and ice cubes (Sudhanshu et al., 2012; Hassan & Naeem, 2014; Gautam et al., 2012; Akhbari et al., 2012).

Viola odorata L., commonly known as sweet violet, is a grassy, perennial herbaceous plant and an important member belonging to family. In Kashmir and western Himalayan regions, the plant is distributed at an altitude of around 1500 to 1800 m above sea level. *Viola odorata* has been used by traditional healers (Hakeems) for the treatment of different ailments around the world and also in the Kashmir Valley, but the active ingredients of the plant (specifically from Kashmir region) are still under the cover of darkness as the phytochemical screening of *Viola odorata* from Kashmir region has not yet been scientifically evaluated. The identification and isolation of different plant compounds through the use of advance instrumentation has led to the advancement in different fields and the discovery of potential new compounds for therapeutic uses. The secondary products synthesized by plants, consisting usually of phenolics, serve as defensive moieties by preventing the oxidative damage via scavenging the generated reactive oxygen species (ROS).

 DOI: 10.1201/9781003292340-12

FIGURE 12.1 *Viola odorata* in natural habitat.

FIGURE 12.2 Flower of *Viola odorata*.

12.2 SYSTEMATIC CLASSIFICATION

Kingdom	Plantae
Division	Tracheophytes
Subdivision	Angiospermae
Class	Dicotyledones
Order	Malpighiales
Family	Violaceae
Genus	*Viola*
Species	*odorata*

12.3 ORIGIN

Sweet violet is native to Europe and it also has strong mythological connections. It is cultivated in Europe, North and West Asia, Africa and in some parts of India, proliferating in Kashmir. It has been grown in gardens across many parts of Asia. A cold and moist climate favors the growth of this plant. It is used/consumed in both forms, the whole plant as well as its roots and flowers. Literature revealed that the plants of sweet violet were also grown in the mountains of Java with the aim of controlling soil erosion. Furthermore, its use in Unani medicine for the treatment of some respiratory problems and other ailments has been reported in numerous studies (Vohora, 1986; Vishal et al., 2009; Ziad et al., 2011). Table 12.1 shows some other common names of sweet violet across the world (USDA, 2023).

TABLE 12.1
Vernacular Names of *V. odorata* in Different Languages Across the World

Language	Vernaculars
French	Violette odorantle
English	Sweet violet
	Common violet
	English violet
	Florist's violet
	Garden violet
Hindi (India)	Banafsha
Sanskrit (India)	Banaphsha
Swedish (Sweden)	Luktviol
Russian (Russia)	Fialka
Rōmaji (Japan)	Nioisumire
Persian	Gul-e-banafsha
Dutch	Tamme viol
Unani	Fareer
Japanese	Nioi-Sumire
Spanish	Violeta
Thai	Waiolet
Italian	Mammola

12.4 TRADITIONAL USES

Sweet violet has been extensively used in folk Persian medicine to treat coughs, headaches, epilepsy, common colds, palpitation, fever, dyspnea, insomnia, dysuria, constipation, and skin conditions (Feyzabadi et al., 2017). Additionally, due to the beautiful morphology of the flowers of sweet violet, it has also long been used for ornamental purposes. Owing to its distinct sweet fragrance, the flowers of *Viola odorata* are popular for their use in cosmetics for the production of fragrances and perfumes (Gautam et al., 2012). It has also long been used in the preparation of tea in India as remedial measure against coughs and colds. For this purpose, the stalks of the plant, along with its flowers, are usually dried and then stored for use when required (Kandpal et al., 2017; Vohora, 1986; Vishal et al., 2009).

12.5 COMPOSITION

Chemical analysis of *V. odorata* has confirmed that many compounds, including flavonoids, anthocyanins, saponins, steroids, terpenes, etc., are present in different parts of the plant i.e., leaves, flower, root. The principal component present in its essential oils is monoterpenes and the overall composition of this plant is summarized in Table 12.2.

12.6 IDENTIFICATION AND CHARACTERIZATION

Viola odorata can be identified, morphologically, by several characteristics, including the scent of its flowers. These are either dark violet or white in color, with heart-shaped simple toothed leaves present in a basal arrangement. It has generally short stem that is unnoticeable in the overall morphology of the plant. It propagates with the help of stolons, which are horizontal stems growing above ground that propagates into new plants (Figure 12.3).

The aqua-alcoholic extract of *V. odorata* showed cholinesterase & tyrosinase inhibition, nitric oxide & 2,2- diphenyl-1-picrylhydrazyl, radical scavenging activity, metal chelating activity, and

TABLE 12.2
General Composition of *V. odorata*

S. No.	Major Component	Its Constituents	Reference
1	Flavonoids	Isovitexin, flavanones, kaempferol-6-glucoside, nikotiflorin, apigenin, pulegone, butanone, glycosides,	Ebrahimzadeh et al., 2010; Khatibi et al., 1989; Orhan et al., 2015; Karioti et al., 2011; Akhbari et al., 2012; Vishal et al., 2009; Gautam et al., 2012 Gautam et al., 2017.
2	Polyphenols	Rutin, violarvensin, violanthin, isoschaftoside, schaftoside, neoschaftoside, rhamnopyrosyl, quercetin, isorhamnetin, hyperoside, ellagic acid, coumarin, catechol, arbutin, luteolin, coumaic acid, vanillic acid, salicylic acid,	Ebrahimzadeh et al., 2010; Lamaison et al., 1991; Gautam et al., 2012
3	Pigments	Anthocyanins, chlorophylls, xanthins	Lamaison et al., 1991; Vishal et al., 2009; Orhan et al., 2015
4	Other compounds	Mucilage, terpenes, viridiflorol, saponins, tannins, alkaloids, steroids, cyclotides	Lamaison et al., 1991; Akhbari et al., 2012; Vishal et al., 2009; Gautam et al., 2012; Gautam et al., 2017

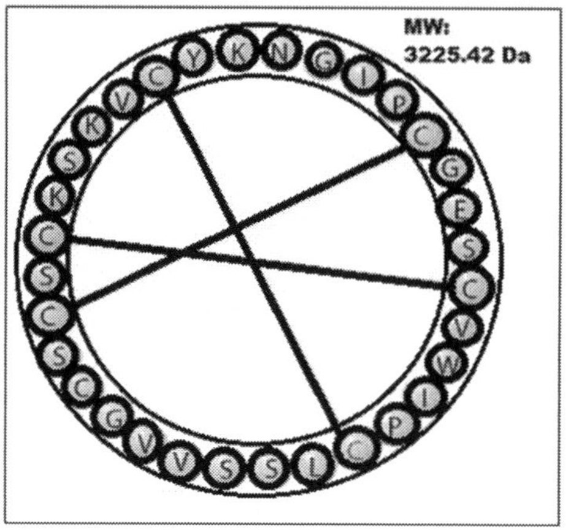

FIGURE 12.3 Cycloviolacin O8 (cyO8) sequence, showing disulfides (Parsley et al., 2018).

antioxidant activity when it undergoes enzyme inhibition assays & HPLC analysis. Additionally, an aqua-ethanol extract of *V. odorata* when it undergoes in vitro analysis outshines significant anti-tyrosinase activity, a result which might be due to the presence of flavonoid derivatives and kaempferol. The HPLC analysis of *V. odorata* shows the presence of vitexin and isovitexin (Orhan et al., 2015).

Drozdova and Bubenchikov (2005) characterized the anti-inflammatory properties of *V. odorata* from water-soluble polysaccharides extracted from the plant. Many studies also confirmed the anti-hypertensive properties of the plant using its leaf extract. It was observed in the studies that it showed hypertensive activity in anaesthetized rats by lowering their mean arterial blood pressure, and inhibiting the force and rate of spontaneous atrial contractions in isolated guinea pigs (Siddiqi et al., 2012). Various cyclotides, cycloviolacins (O2, O13, O14, O15), exhibit hemolytic activity, the most hemolytic being O2 to the least hemolytic O14. Several studies confirmed the diuretic, pulmonary protective, and laxative activity of the aqueous extracts of the aerial parts of the plant of *V. odorata* in animals. Moreover, both its aqueous as well as it methanol extract confirmed its analgesic properties by exhibiting significant effects in the central and peripheral models of pain (Rebuelta et al., 1983; Vishal et al., 2009; Koocheck et al., 2003; Antil et al., 2011). Furthermore, the antimicrobial activity of *V. odorata* might be attributed mainly to the presence of cyclotides (cyO2), which showed growth inhibition of fungi, *Salmonella enterica*, *Staphylococcus aureus*, *Escherchia coli*, and plant-pathogenic bacteria (Zarrabi et al., 2013; Pranting et al., 2010; Hammami et al., 2011). The detailed summary of characterization of various biochemical properties of sweet violet is summarized in Table 12.3. .

12.7 PHARMACOLOGICAL BENEFITS

A bird's-eye view of the pharmacological properties of the sweet violet plant could be obtained from the above discussion. However, it must be noted that the plant has been used for treating health ailments, extensively in the Unani system. Different parts of the sweet violet plant possess health benefits due to which it has been used in different kind of preparations for that purpose. Likewise, the whole plant, generally in dried form, has been used in the preparation of a special kind of *kawah* (a kind of tea) for the treatment of fever, cough, sore throat, bronchitis, etc. Another preparation of sweet violet (*Raughan-e-Banafsha*), particularly its flowers only, is used in the Unani system as an effective remedy against the treatment of cephalagia and insomnia. This preparation, is actually the extract of the sweet violet flowers prepared with sesame oil. Moreover, it has been also used in the form of powder, pills, decoction, poultice, douche, and massage oil. Anti-inflammatory, emollient,

TABLE 12.3
Identification and Characterization of Various Chemical and Pharmacological Activities of Sweet Violet

Study	Assay/Model	Extract/Fraction	Activity Identified/ Characterized	Reference
In vitro	Growth inhibition (24 hr)	Extract of leaves, flower and root of *Viola odorata* with diethyl either	Potent growth inhibition against *Trichomonas vaginalis*	Salehi et al., 2014
	DPPH (N, N-dimethyl-p-phenylendiamine), β-carotene	Essential oil of *Viola odorata* with Methanol, chloroform	Antioxidant and radical scavenging properties	Akhbari et al., 2012
	Radial diffusion and MIC	Small cyclic proteins obtained from *V. odorata*	Potent antimicrobial activity, especially against Gram-negative bacteria	Pranting et al., 2010
	Cell proliferation assay, Breast cancer line (MCF-7, MCF-/ADR)	Small cystic proteins	Potent cytotoxicity	Gerlach et al., 2010
	Radial diffusion, MIC and MBC	Cyclotides	Antimicrobial activity with more potency against Gram-negative and plant pathogenic bacteria	Zarrabi et al., 2013
	DPPH (2,2- diphenyl-1-picrylhydrazyl)	Lyophilized aqueous extract	Radical scavenging action	Stojkovic et al., 2011
	Anti-tyrosine activity	Ethanol extract of *V. odorata*	Tyrosine inhibition at 100 μg/ml	Orhan et al., 2015
	Serum Glucose Deprivation (SGD)-induced neurotoxicity	Hydroalcoholic extract of *V. odorata*	Protective action against SGD-induced cell damage	Mousavi et al., 2016
	Cell viability, tyrosinase (B16F10 cell line)	Ethyl acetate fraction	Anti-tyrosinase activity within the cells; anti-melanogenic activity	Rahimi et al., 2017
In vivo	Rat-induced tongue dysplasia	Syrup of *V. odorata*	Inhibition and reduction in the development of dysplasia	Helli et al., 2016
	Tail immersion and hot plate	Aqueous and methanol extract of *V. odorata*	Analgesic effect	Antil et al., 2011
	Measuring sleep induction and sleep time	Methanol and chloroform extract	Pre-anesthetic and sedative action with increased sleep time	Monadi & Rezaie, 2013
	Paracetamol induced hepatic damage	Aqueous-methanol extract	Reduction in the levels of total bilirubin and serum liver enzymes	Qadir et al., 2014
	Anesthetized rats, tyloxapol-induced dyslipidimia	Aqueous-methanol extract of leaves of *V. odorata*	Reduced mean arterial bold pressure and total cholesterol & triglyceride levels	Siddiqi et al., 2012

(*Continued*)

TABLE 12.3 (Continued)

Study	Assay/Model	Extract/Fraction	Activity Identified/ Characterized	Reference
Human	Clinical trial; benign prostate hyperplasia	Hydroalcoholic extract of *V. odorata*	Effective against benign prostate hyperplasia with reduced intermittency, straining, nocturia and prostate volume	Beiraghdar et al., 2017
	Clinical trial (nasal drop; approx. 66 mg in one month)	Oil of *V. odorata*	Hypnotic activity; increased Insomnia Severity Index scores	Feyzabadi et al., 2014
	Clinical trial of children with intermittent asthma	Syrup of *V. odorata*	Effective against cough suppression and duration	Qasemzadeh et al., 2015

laxative, diaphoretic, expectorant, and diuretic properties are present in the whole plant of *V. odorata* and is taken orally to treat chest problems such as bronchitis, cough, asthma, and respiratory catarrh. Moreover, it is also an effective remedy against breast, lung and gastrointestinal (GI) cancer. For the treatment of infections related to mouth and throat in particular, it is used externally rather than its consumption in any kind. Furthermore, a specific kind of preparation prepared from the whole fresh plant is used to treat spasmodic coughs and rheumatism of the wrist in homeopathy. The oil obtained from the sweet violet plant has shown promising antimicrobial activity upon synergic use with *Santalum austrocaledonicum* extracts (Orchard et al., 2023).

As mentioned earlier, each part of the sweet violet plant has its own medical uses and is, therefore, used in different ways for effective results against the target health problem. The syrup prepared from the petals of its flower, which is demulcent and emollient, is used against infantile disorders. Its flowers are also used to treat biliousness and pulmonary troubles. Moreover, an essential oil obtained from its flowers is used in aromatherapy to treat exhaustion, as well as bronchial and skin problems. In Unani medicine, the poultice of pulverized petals of sweet violet has been used as an emollient and for the treatment of anal swelling and fissures.

The use of the seeds of the plant of *V. odorata* for urinary problems can be attributed to its purgative and diuretic nature. Its roots, which pose strong expectorant properties, are generally gathered and used in dried form. The leaves of the sweet violet plant have been effectively used against dysuria, flank pain, urinary retention and the management of rectal prolapse. However, the expectorant properties of its roots, which make it suitable for the treatment of problem coughs, are much stronger and effective than the other parts of the plant of sweet violet. In addition to these health benefits, it has significant properties to treat problems related to the central nervous system owing mainly to its rich polyphenolic composition. It is, therefore, used to treat sleep disorders, epilepsy, headaches, fever convulsions in children, catarrh, etc. (Hamid et al., 2013; Feyzabadi et al., 2014, 2017).

The recent revolution in successful drug development can be attributed mainly to the therapeutic potential of peptides which specifically and selectively targets imperfect protein–protein interactions that are known for initiating different human diseases. Proteolytic degradation greatly affects bioavailability and their expensive production costs has led to the investigation of different peptide scaffolds as different and novel sources for protein-based therapeutics. The extreme stability of natural peptides has received special attention for the development of ligands with considerably higher affinity in term of capturing proteins. Cyclotides are amongst a new and promising family of plant-based circular peptides isolated mainly from the Violaceae and Rubiaceae families, with the former to date having proved to be the richest source of cyclotides. Antifungal, anti-tumor, antimicrobial, and HIV inhibition are among the different pharmaceutical properties associated with cyclotides. The 3D structure of cyclotides, along with variations in intercysteine residues, has a considerable impact on the cytotoxicity of the peptide on different cancerous cell lines.

12.8 CONCLUSION

Traditionally, Banafsha (*Viola odorata* L.) has been used in treating various health ailments such as cancer, asthma, cough, insomnia, eczema, etc., in traditional medicine systems such as Unani and homeopathy. Both animal and human studies have backed the medicinal potential of the herb; however, a lot remains to be done to reveal the full medicinal potential of the plant, including the potency of cyclotides towards different human pathogens. The phytochemical screening of the plant will help to identify different bioactive compounds, including its phenolic profile that can serve as a basis for future research.

REFERENCES

Akhbari, M., Batooli, H., & Kashi, F. J. (2012). Composition of essential oil and biological activity of extracts of *Viola odorata* L. from central Iran. *Natural Product Research*, *26*(9), 802–809.

Antil, V., Kumar, P., Kannappan, N., Diwan, A., Saini, P., & Singh, S. (2011). Evaluation of the analgesic activity of *Viola odorata* aerial parts in rats. *Journal of Natural Pharmaceuticals*, *2*, 24–27.

Beiraghdar, F., Einollahi, B., & Ghadyani, A. (2017). A two-week, double-blind, placebo-controlled trial of *Viola odorata*, *Echium amoenum* and *Physalis alkekengi* mixture in symptomatic benign prostate hyperplasia (BPH) men. *Pharmaceutical Biology*, *55*(1), 1800–1805.

Drozdova, I., & Bubenchikov, R. (2005). Composition and anti-inflammatory activity of polysaccharide complexes extracted from sweet violet and low mallow. *Pharmaceutical Chemistry Journal*, *39*(4), 197–200.

Ebrahimzadeh, M. A., Nabavi, S. M., Nabavi, S. F., Bahramian, F., & Bekhradnia, A. R. (2010). Antioxidant and free radical scavenging activity of *H. officinalis* L. var. angustifolius, *V. odorata, B. hyrcana* and *C. speciosum*. *Pakistan Journal of Pharmaceutical Sciences*, *23*, 29–34.

Feyzabadi, Z., Ghorbani, F., Vazani, Y., & Zarshenas, M. M. (2017). A critical review on phytochemistry, pharmacology of *Viola odorata* L. and related multi-potential products in traditional Persian medicine. *Phytotherapy Research*, *2*, 1–8.

Feyzabadi, Z., Jafari, F., Feizabadi, P. S., Ashayeri, H., Esfahani, M. M., & Badiee, A. S. (2014). Insomnia in Iranian traditional medicine. *Iranian Red Crescent Medical Journal*, *16*(3), e15981.

Gautam, S.S., Navneet, & Kumar, S. (2012). The antibacterial and phytochemical aspects of *Viola odorata* Linn. extracts against respiratory tract pathogens. *Proceedings of the National Academy of Sciences, India Section B: Biological Sciences*, *82*(4), 567–572.

Gautam, S.S., Navneet, & Kumar, S. (2017). Current aspects on phytochemistry and bioactive constituents of *Viola odorata* Linn. *Indian Journal of Biotechnology and Pharmaceutical Research*, *5*(2), 2347–3266.

Gerlach, S. L., Rathinakumar, R., & Chakravarty, G. (2010). Anticancer and chemosensitizing abilities of Cycloviolacin O2 from *Viola odorata* and psyle cyclotides from *Psychotria leptothyrsa*. *Peptide Science*, *94*(5), 617–625.

Hammami, I., Kamoun, N., & Rebai, A. (2011). Biocontrol of *Botrytis cinerea* with essential oil and methanol extract of *Viola odorata* L. flowers. *Archives of Applied Science Research*, *3*(5), 44–51.

Hassan, F., & Naeem, I. (2014). Biological activity of *Viola odorata* Linn. against *Mycobacterium tuberculosis*. *International Journal of Pharma and Bio Sciences*, *5*(3), 61–69.

Hamid, K. S., Ranjbar, S. H., Esfehani, M., Mohammad, K., & Larijani, B. (2013). A systematic review of the antioxidant, anti-diabetic, and anti-obesity effects and safety of triphala herbal formulation. *Journal of Medicinal Plants Research*, *7*(14), 831–844.

Helli, S., Damghani, H., Mohajeri, D., Mesgari-Abbasi, M., Attaran, R., & Zahed, M. (2016). Evaluation of the effect of two different systemic doses of *Viola odorata* on prevention of induced tongue dysplasia in rats. *Journal of Dentistry*, *17*(3), 185–192

Kandpal, A., Chaubey, S., & Pandey, M. (2017). A brief knowledge of banafsha (*Viola odorata* Linn.) & other *Viola* species. *International Journal of Ayurveda and Pharma Research*, *5*(4), 73–78.

Karioti, A., Furlan, C., Vincieri, F. F., & Bilia, A. R. (2011). Analysis of the constituents and quality control of *Viola odorata* aqueous preparations by HPLC-DAD and HPLC-ESi-MS. *Analytical and Bioanalytical Chemistry*, *399*, 1715–1723.

Khatibi, A., Shah, A. H., Ageel, A. M., Ahmad, M. S., Al-Yahya, M. A., & Tariq, M. (1989). Saudi folk medicine: Phytochemical and antimicrobial screening. *Pakistan Journal of Pharmaceutical Sciences*, *2*(1), 29–34.

Koochek, M., Pipelzadeh, M., & Mardani, H. (2003). The effectiveness of *Viola odorata* in the prevention and treatment of formalin-induced lung damage in the rat. *Journal of Herbs, Spices & Medicinal Plants*, *10*(2), 95–103.

Lamaison, J. L., Petitjean, F. C., Carnat, A. (1991). The violet flower: a comparison between *Viola lutea*, *Viola calerata* and *Viola odorata*. *Plantes Médicinales et Phytothérapie*, *25*(2–3), 79–88.

Monadi, A., & Rezaie, A. (2013). Evaluation of sedative and pre-anesthetic effects of *Viola odorata* Linn. extract compared with diazepam in rats. *Bulletin of Environment, Pharmacology and Life Sciences*, *2*(7), 125–131.

Mousavi, S. H., Naghizade, B., Pourgonabadi, S., & Ghorbani, A. (2016). Protective effect of *Viola tricolor* and *Viola odorata* extracts on serum/glucose deprivation-induced neurotoxicity: role of reactive oxygen species. *Avicenna Journal of Phytomedicine*, *6*(4), 434–447.

Orchard, A., Moosa, T., Motala, N., Kamatou, G., Viljoen, A., & Vuuren, S. V. (2023). Commercially available *Viola odorata* oil, chemical variability and antimicrobial activity. *Molecules*, *28*(4), 1676.

Orhan, E. I., Senol, F. S., Erdem, A. S., Tatli, I., Kartal, M., & Alp, S. (2015). Tyrosinase and cholinesterase inhibitory potential and flavonoid characterization of *Viola odorata* L. (sweet violet). *Phytotherapy Research*, *29*(9), 1304–1310.

Parsley, N. C., Kirkpatrick, C. L., Crittenden, C. M., Rad, J. G., Hoskin, D. W., Brodbelt, J. S., & Hicks, L. M. (2018). PepSAVI-MS reveals anticancer and antifungal cycloviolacins in *Viola odorata*. *Phytochemistry*, *152*, 61–70.

Pranting, M., Loov, C., Burman, R., Goransson, U., & Andersson, D. I. (2010). The cyclotide Cycloviolacin O2 from *Viola odorata* has potent bactericidal activity against Gram-negative bacteria. *Journal of Antimicrobial Chemotherapy*, *65*(9), 1964–1971.

Qadir, M. I., Ali, M., Saleem, M., & Hanif, M. (2014). Hepatoprotective activity of aqueous methanolic extract of *Viola odorata* against paracetamol-induced liver injury in mice. *Bangladesh Journal of Pharmacology*, *9*(2), 198–202.

Qasemzadeh, M.J., Sharifi, H., & Hamedanian, M. (2015). The effect of *Viola odorata* flower syrup on the cough of children with asthma a double-blind, randomized controlled trial. *Evidence-Based Complementary and Alternative Medicine*, *20*(4), 287–291.

Rahimi, V. B., Askari, V. R., Emami, S. A., & Tayarani-Najaran, Z. (2017). Antimelanogenic activity of *Viola odorata* different extracts on B16F10 murine melanoma cells. *Iranian Journal of Basic Medical Sciences*, *20*(3), 242–249.

Rebuelta, M., Vivas, J. M., San-Roman, L., & Serranillos-Fdez, M. (1983). Study of the diuretic effect of different preparations of *Viola odorata*. *Plant Med Phytother*, *17*, 215–221.

Salehi, L., Asghari, G., Yousofi, H., & Yousofi-Darani, H. (2014). The effects of different extracts of *Viola odorata* on *Trichomonas vaginalis* in culture medium. *Journal of Isfahan Medical School*, *31*(266), 2140–2148.

Siddiqi, H. S., Mehmood, M. H., Rehman, N. U., & Gilani, A. H. (2012). Studies on the antihypertensive and antidyslipidemic activities of *Viola odorata* leaves extract. *Lipids in Health and Disease*, *11*, 6.

Stojković, D., Glamočlija, J., Ćirić, A., Šiljegović, J., Nikolić, M., & Soković, M. (2011). Free radical scavenging activity of *Viola odorata* water extracts. *Journal of Herbs, Spices and Medicinal Plants*, *17*(3), 285–290.

Sudhanshu, Mittal, S., Rao, N., & Menghani, E. (2012). Antimicrobial potentials of *Viola odorata*. *World Journal of Pharmaceutical Research*, *1*(3), 635–641.

USDA-ARS. (2023). *Viola odorata* L, *Germplasm Resources Information Network (GRIN Taxonomy)*. National Germplasm Resources Laboratory. https://npgsweb.ars-grin.gov/gringlobal/taxon/taxonomydetail?id=41733

Vishal, A., Parveen, K., Pooja, S., Kannappan, N., & Kumar, S. (2009). Diuretic, laxative and toxicity studies of *Viola odorata* aerial parts. *Pharmacology Online*, *1*, 739–748.

Vohora, S. B. (1986). Unani joshandah drugs for common cold, catarrh, cough, and associated fevers. *Journal of Ethnopharmacology*, *16*(3), 201–211.

Zarrabi, M., Dalirfardouei, R., Sepehrizade, Z., & Kermanshahi, R. K. (2013). Comparison of the antimicrobial effects of semipurified cyclotides from Iranian *Viola odorata* against some of plant and human pathogenic bacteria. *Journal of Applied Microbiology*, *115*, 367–375.

Ziad, D., Elias, A., & Roula, A. M. (2011). Antibacterial activity of *Rheum rhaponticum*, *Olea europaea* and *Viola odorata* on ESBL producing clinical isolates of *Escherichia coli* and *Klebsiella pneumoniae*. *International Journal of Pharmaceutical Sciences and Research*, *2*(7), 1669–1678.

Index

Pages in *italics* refer to figures, pages in **bold** refer to tables

A

Absinthin, **12**, 13, 26
Acetogenins, 34, 38–39, 45
Adenoviruses, 102
Alkaloids, 3, 12, 37–40, *41*, 43, 45, 50–53, 55–56, 60, 64, 80, **81**, 97, 103, **105**, 117, 123, 130, 151, 153–154, 159, 183, **193**
Allelopathic, 64–65, 70
Aloe-emodin, 4, 131, **133**, 136–138, 140
Amylase, 21, 23, **105**, 144
Analgesic, 2–3, 21–22, **26**, 46, 50, 84, 103, 117, 194, **195**
Anorexia, 184
Anthocyanins, 4, 38, 43, 96, 98–100, 104, **105**, 157, 179, 193
Anthraquinones, 4, **81**, 130–132, **133**, 135–136, 138–141, 144, 151, 154
Anthroquinones, 4
anti-urolithic, 4, 183, 187
antibiotic, 24, 115, 122, 137–138
anticancer, 2, 9, 18, 20, 27, 43, 45, 63, 69–70, **105**, 130, 132, 154, 160–161, 167–168, 179
anticonvulsant, 3, 165, 168
antidepressant, 3–4, 17, 20, **26**, 27, 102, 183, 187
antihypertensive, 4, 63, 96, 117, 194
anti-obesity, 4, 182, 187
antioxidant, 2–9, 17–23, **26–27**, 34, 37–38, 43–44, 46, 55, 60–61, 63, 65, **66**, 68–78, 83–85, 96–100, 102, 104, **105**, 106–110, 120–121, 130, 132–137, 145, 159, 163, 168, 175, 177, 180–183, 185–187, 194, **195**
antipyretic, 2–3, 22, **26**, 72, 100, 115
anti-ulcer, 3, 23, 46, 72, 87, 103
antiviral, 4–5, 9, 24, **27**, 38, 63, 83, 85, 96, 99, 115, 117, 122–123, 132, 160, 176
Apigenin, 64–65, **66**, 80, **81**, 82, 98, 100, 153–154, **193**
apoptosis, 18–19, 21, 26, 45, 61, **68**, 69–70, 102, 124, 140, 145, 159–161, 167, 181–182, 184, 186
Arginine, 63
Artemisetin, 20
Artemisinin, **12**, **14**, 17–18, 20, 25, **27**, 44
Ascaris suum, 25
Ascorbic acid, **12**, 37, 43, 69, 72, **81**, 97, **105**, 107, 180
Asia, 3, 5, 9, 18, 50, 78–79, 93–94, 97, 100, 151, 175–176, 192
Atropine, 3, 50–53, 56–57
Auraptene, 64–66
Ayurveda, 1, 3, 100, 103, 128, 151, 165, 175

B

Bacillus subtilis, 24–25, 45, 70, **86**, 102, 137
Betulinic acid, 117, *118*, 153–154, 160, 165

C

Cadinene, 12, **15**, 23, **27**, 38, *42*
Caffeic acid, **12**, 23, *42*, 64, 68, 80, **81**, 82, 107–108, 117, *119*, 121, 124, 178
Caffeoylquinic acids, 5
Caftaric acids, 99
Candida, 18, 24–25, 65, 70–71, 82, **83**, 137, 160
carminative, 79, 84, 94, 97
Carotenoids, 12, 39, 42, 64, **81**, 83, 96, 98, 177
Carvacrol, 65, 70
Caryophyllene, 38, 97, 153–154
Catechin, **81**, **105**, 135
Chicoric acid, 178–179, 184, 186
Chloroform, 11, 22–23, 45, 56–57, **66**, 67, 70–71, 136, 158, 163, 180, **195**
Cholecystitis, 18
Cholinesterase, 20, **26**, 193
Chrysophanol, 4, 131–132, **133**, 136–138, 140, 142
Costunolide, 151, 153–154, 159–164, 167–168
Coumarins, 12, 64–65, 96
Creatinine, 23, 44
Cycloartenol, 179
Cyclopeptides, 39
Cycloviolacin, 194
Cytokines, 20, 103, 140, 142, 144, 160, 180–181
cytotoxic, 5, 20, **66**, 85–86, 117, 137, 151, 154, 160–163, 167–168, 176, 185–186, **195**

D

decongestant, 79, 86
Diabetes mellitus, 4, 44, 184, 187
diacylglycerol, **81**
Dietary fiber, 63, 130, **131**
Dihydrocostunolide, 153–154, 168
Diuretic, 4–5, 98, 115, 128, 132, 142, 175–176, 179–180, 183, 187, 194, 196
Dropsy, 53
Duguevalline, 43
Dyspepsia, 2, 19, 143, 163, 179, 184

E

Emodin, 4, 131–132, **133**, *134*, 136–142
endangered species, 4, 27, 168
Epicatechin, 135

F

Fatty acids, 3, 12, 34, 38–39, 43, 60, 64–65, **68**, 69, 73, 80, **81**, 82, 97, 143, 159, 175, 177
flavonoids, 2–5, 11–12, 17, 21, 23–24, **26–27**, 37–38, 42, 44–45, 63–64, 68, 73, 80, **81**, 82–83, 98–99, 103–104, **105**, 107, 109, 115, 120–121, 130, 135, 151, 153–154, 157, 178, 180, 185–186, 193
Flavonol, 3, 5, 97, 99
Folic acid, 37, 43
Fourier Transform Infrared Spectroscopy, 122

G

Gallic acid, 26, *42*, 68, 71, **81**, 83, 135
Generally Recognized as Safe, 97
Glucosidase, 23, **105**, 123, 134, 182
Glycosides, 12, 37–38, 43, 45, 51, 80, 97, 100, 130–131, 134–135, 178, 193
Guaiazulene, 12, **16**

H

Heartburn, 184
Hemagglutinins, 60
Hepatitis, 19, 21, 24, 102, 130, 160, 162, 175, 179, 181, 184
Herpes, 102, 117, 123
Hesperidin, 117, *119*, 124
Hexane, 18, 22–23, 43, 64–65, **66**, 67, 104, **105**, 136, 153, 158, 161
High Performance Liquid Chromatography, 17, 66, 104, 124, 136, 154, 159
Histidine, 63
HIV, 24, **27**, 99, 122–123, 196
Hydroxycinnamic acid, 18, 65, 117, *120*
Hyoscine, 3, 50
Hyoscyamine, 3, 50–54, 56–57
hypnotic, 4, 18, **196**
hypoglycemic, 2, 4, 9, 23, **27**, 61, 115, 124, 165–166
hypolipidemic, 4, 22, 104, 115, 177, 180, 182, 185, 187

I

immunomodulatory, 2, 4, 18, 21, **26**, 109, 117, 181, 187
immunoregulatory, 4, 122
infectious diseases, 2, 24, **27**, 104
Inositol, 177
insecticidal, 3, 9, 17, 26, **27**, 44, 46, 69, 99
Interleukin, 43, 140, 144, 181, 186
Inulin, 177, 179
Irritable bowel syndrome, 57

K

Kashmir, 3, 9, 18, 60, 115–116, 128, **129**, 151–152, 190, 192
Ketone, 39, 78, 153–154
Klebsiella pneumonia, 24, 70, 102

L

Lecithin, 177
Leukotrienes, 69
Linalool, 12, **16**, 38, *42*, 79, 81, 94, 96–97, 99, 102, 108, 153
Linoleic acid, 34, 60, 64, **68**, 69, 81, 97, 154
lipopolysaccharide, 20, 140, 162, 180–181, 185–186
lipoprotein, 22, 130, 142, 182
lipoxygenase, 103, 139
Listeria monocytogenes, 24
Lysin, 60, 63

M

Malondialdehyde, 102, 181–182
Maximal inhibitory rate, 43
Metal chelators, 97
Methylamine, 56
Microwave assisted extraction, 108, 136, 145, 157
Myocardial infraction, 104
Myrcene, 12, **15**, 153

N

Nanoparticles, 25, 70, 186
Naringenin, 23, 80, **81**, 82
neurodegenerative, 4, 21, 38, 68, 185, 187
neuroprotective, 2, 20, 26, 68, 100, 185
Nuclear Magnetic Resonance, 17, *55*, 122, 154

O

Oleic acid, 34, **68**, 69, 97
Ophthalmology, 53
Organic acid, 4, 12, 64, 80, 115, 131
Ornithine, 51–52, 56

P

Pancreatitis, 57, 140, 143–144
Pepsin, 23, 44, 57, 103
Peptides, 52, 60–69, 71, 196
pharmacopoeia, 18, 104
Phenyllactic acid, 54–55
Phenylpropanoids, 4, 13, 64, 78, 96–97, 115, 179
Physcion, 4, 131–132, **133**, 134, 136–137
Pinene, 12–13, **14**, **16**, 21, 23, **26–27**, 38, *42*, 99, 107, 153
polysaccharides, 4, 17, 21, 24, **26**, 36, 100, 102–103, 115, 121, 123, 179, 185, 194
pressurized liquid extraction, 16, 55, 107, 136–137, 145
Proline, 52, 95
Protocatechuic acid, 64, 68, 185
Prunellin, 117, *118*
Pseudomonas putida, 24
Pulegone, 81–82, **193**

R

Rhein, 131–132, **133**, 134, 136–140, 142
rhizome, 4, 10, 78, 86, 128–132, 136–137, 164
Roemerolidine, 43

S

Saccharomyces cerevisiae, 18, 25
Salmonella typhimurium, 24, 82, **83**, **86**
saponins, 17, 37–38, 45, 60, 63–64, **81**, 97, **105**, 117, 193
Sepsis, 143, 145
Sesquiterpenes, **15**, 38, 40, 64, 84, 151, 153–154, 177, 181
Sesquiterpenoids, 5, 13, 38, 78
Sitosterol, 154, 179

Staphylococcus aureus, 24, 45, 70, 82, **83**, 102, 131, 137, 194
Starch, 3, 23, 63
Steroids, 4, 38, 43, 64, **81**, 98, **105**, 153, 159, 193
Stilbene, 4, 40, 130–131, 134–135
sub-tropical, 4, 128
Succinaldehyde, 56
Supercritical fluid extraction, 16, 67, 108, 137, 145

T

Tannins, 2, 11–12, 38, 40, 63, 68, 97–98, 103, **105**, 117, 139, 141, 152–153, 157, **193**
Taraxasterol, 179, 181, 183–184
Thiobarbituric acid, **26**, 72, 164
Thujone, 9, 11–13, **14**, 18–19, 21, 23, 25, **26**, 27
Tinospora cordifolia, 1
traditional medicine, 1, 3–4, 34, 79, 100, 115, 128, 130–131, 145, 166, 175, 179
triacylglycerol, **81**
Triterpenes, 123, 151, 177–178, 181

U

Ultrasound assisted extraction, 65, 67, 106, 154, 157
Umbelliferone, 64–66, **105**, 117, 120
Umbelliprenin, 64–65, **66**
Uric acid, 44

V

Vasoconstrictor, 86, 180
vomiting, 27, 94, 151, 183

W

Wogonin, 117, *120*

Made in the USA
Las Vegas, NV
19 September 2026

f2580f79-da2d-4b62-a1d4-cf68de512d10R01